Fishes

Fishes

A Guide to Their Diversity

PHILIP A. HASTINGS, H. J. WALKER, and GRANTLY R. GALLAND

UNIVERSITY OF CALIFORNIA PRESS

University of California Press, one of the most distinguished university presses in the United States, enriches lives around the world by advancing scholarship in the humanities, social sciences, and natural sciences. Its activities are supported by the UC Press Foundation and by philanthropic contributions from individuals and institutions. For more information, visit www.ucpress.edu.

University of California Press
Oakland, California

Library of Congress Cataloging-in-Publication Data
Hastings, Philip A., author.
  Fishes : a guide to their diversity / Philip Alan Hastings, H.J. Walker, and Grantly R. Galland.
    pages cm
  Includes bibliographical references and index.
  ISBN 978-0-520-27872-1 (Cloth) — ISBN 978-0-520-28353-4 (Paper) - – ISBN 978-0-520-95933-0 (e-book)
  1. Fishes—Anatomy. 2. Fishes—Classification. 3. Fishes—Identification. I. Walker, H. J. (Harold Jack), 1950– author. II. Galland, Grantly R., 1982– author. III. Title.
  QL615.H37 2014
  597—dc23   2014011258

Manufactured in China

23 22 21 20 19
10 9 8 7 6 5 4 3 2

The paper used in this publication meets the minimum requirements of ANSI/NISO Z39.48–1992 (R 2002) (*Permanence of Paper*). ♾

Cover photographs (top to bottom, left to right): Logperch, *Percina caprodes;* Roosterfish, *Nematistius pectoralis;* Fantail Filefish, *Pervagor spilosoma;* Scribbled Pipefish, *Corythoichthys intestinalis;* Sea Lamprey, *Petromyzon marinus;* Nightlight Lanternfish, *Myctophum lychnobium;* Rockmover Wrasse, *Novaculichthys taeniourus;* Coelacanth, *Latimeria chalumnae;* Humpback Anglerfish, *Melanocetus johnsonii;* Horn Shark, *Heterodontus francisci.* All specimens and photographs are from the Marine Vertebrate Collection, Scripps Institution of Oceanography, University of California, San Diego.

Dedicated to the many students of fishes who helped build and maintain the Scripps Institution of Oceanography's Marine Vertebrate Collection and to all who have built and maintained natural history collections throughout the world

CONTENTS

Phylogenetic hypothesis including all orders covered in this book (inside back cover).

COMPLETE CONTENTS

Phylogenetic hypothesis including all orders covered in this book (inside back cover).

INTRODUCTION

In nearly every body of water around the world, the most abundant vertebrate is a fish. From the deepest parts of the ocean to high alpine streams, fishes live and reproduce, sometimes in places where no other vertebrates can survive. Whether peering out from a submarine while conducting deep-sea research, or stopping for a drink of water during a hike in the mountains, explorers, scientists, and naturalists find fishes.

With well over 30,000 species, fishes account for more than half of the total extant vertebrate diversity on Earth—in other words, there are more living species of fishes than of amphibians, turtles, lizards, birds, and mammals combined. Not only are fishes diverse in number of their species, but they are diverse in the habitats in which they live, the foods that they eat, the ways in which they reproduce, communicate, and interact with their environment, and the behaviors that they exhibit. Fishes can also be extremely abundant: the most abundant vertebrates on the planet are the small bristlemouth fishes (Gonostomatidae) that are common throughout the vast open ocean. In some cases abundant fishes such as cods, tunas, salmons, herrings, and anchovies support massive fisheries that feed hundreds of millions of people. By supporting coastal communities and societies, these fisheries (and the fishes they target) have helped shape human history, becoming the foundation for coastal economies and an engine for global exploration and expansion.

WHAT IS A FISH?

Humans use the term "fish" to refer to several groups of vertebrates that do not have a clear set of diagnostic characteristics unique to them. "Fishes" is not a monophyletic group (i.e., a group made up of an ancestor and all of its descendants) because the tetrapods, which share a common vertebrate ancestor with fishes, are excluded. Thus "fish" typically refers to any vertebrate that is *not* a tetrapod. Fishes (usually) live in water, (usually) obtain oxygen through gills, are (usually) ectothermic (i.e., cold blooded), and (usually) have limbs in the form of fins. Naturally, there are exceptions to each of these rules. Some fishes spend time out of the water, some breathe air, some are endothermic (i.e., warm blooded), and some have no limbs at all.

While there is no clear set of characteristics that distinguishes all fishes from all other vertebrates, there are four groups that collectively make up the fishes. The extant fishes include the jawless fishes (Agnatha), the cartilaginous fishes (Chondrichthyes), the ray-finned fishes (Actinopterygii), and a small portion of the lobe-finned fishes (Sarcopterygii). Of the extant fishes, the ray-finned fishes are by far the most speciose, accounting for more than 30,000 species, the cartilaginous fishes include about 1,200 species, and the jawless fishes include fewer than 100 species. Only eight species of lobe-finned fishes, two species of coelacanths, and six species of lungfishes are considered by most to be "fishes," while the remaining 28,000 or more sarcopterygian species are tetrapods.

WHY THIS BOOK?

This book is intended to be a reference text for students and lovers of fishes to assist them in learning the morphology, diagnostic characters, and basic ecology of fishes. It started as a guide to the systematics of fishes, compiled by the senior author for use in ichthyology courses at Scripps Institution of Oceanography and the University of Arizona. It will serve that purpose, but will also provide an entry into the world of fishes for anyone interested in exploring their diversity. To our knowledge, no comparable volume exists. While numerous excellent regional guides to fishes are available (e.g., Eschmeyer and Herald, 1983; Hart, 1973; McEachran and Fechhelm, 1998, 2005; Page and Burr, 2011; Quéro et al., 1990; Robertson and Allen, 2008; Robins and Ray, 1986; Scott and Crossman, 1973; Scott and Scott, 1988; TeeVan et al., 1948–1989; Whitehead et al., 1986), these lack a global perspective. *Fishes of the World* (Nelson, 2006) covers the entire diversity of fishes, including all of the 515 families, but the scope of that impressive work prohibits the illustration of specimens and key characteristics of various groups. Our goal is to give an overview of the global diversity of fishes, together with more detailed accounts and illustrations of the common groups of fishes, as well as those important to humans and those widely discussed in the ichthyological literature.

The general anatomy of fishes is briefly covered, focusing on external features that help to distinguish major groups. These include external body regions, fin types and positions, body shapes, mouth positions, and selected skeletal features. We then provide accounts of approximately 180 groups of fishes, including all currently recognized orders of fishes and a variety of common and diverse families. We start with the jawless fishes (Agnatha) and progress through the cartilaginous fishes (Chondrichthyes), the lobe-finned fishes (Sarcopterygii), and the ray-finned fishes (Actinopterygii).

SYSTEMATICS OF FISHES

Ichthyologists have been interested in the evolutionary history of fishes for hundreds of years, and classification systems have attempted to capture that history in a hierarchical (Linnaean) system of names. It remains difficult to implement a truly monophyletic classification, one that recognizes only monophyletic groups, for any large group such as fishes, given both the complexity of the tree of life and our continuing uncertainty as to its form. Traditional classifications recognize several hierarchical levels, but students should keep in mind that a particular level in a classification, such as a family, has little meaning other than that it ideally includes all descendants of a common ancestor (i.e., it recognizes a monophyletic group) that are included in a higher level of the classification. For example, although ichthyologists have designated the two species of fangtooths and the 1,700 species of gobies as the families Anoplogastridae and Gobiidae, respectively, these groups clearly differ greatly in diversity, age, and ecological breadth.

In organizing this guide, we have had to face a host of perplexing and often conflicting hypotheses of fish relationships. For chondrichthyan fishes we have elected to follow a somewhat traditional classification of their diversity based primarily on Nelson (2006). Our organization of the ray-finned fishes largely follows the classification provided in Helfman and Collette (2011), which is, in turn, based largely on Nelson (2006), as modified by Wiley and Johnson (2010). Within the Percomorpha, a large group of ray-finned fishes whose relationships remain poorly understood, we have followed the taxonomic levels of Wiley and Johnson (2010) rather than those of Helfman and Collette (2011). In some cases we have modified these classification schemes based on well corroborated studies. However, we have not implemented some recent and radically different classification schemes (e.g., Betancur et al., 2013; Near et al., 2013). We find it difficult and in fact unnecessary to implement certain changes in percomorph classification at this time, and instead treat its hypothesized members in a more or less traditional manner.

Until very recently, our understanding of fish relationships was based almost exclusively on morphological features. With the advent of modern molecular methods, the study of the evolutionary relationships of fishes has grown exponentially, with new studies of various groups appearing at a nearly overwhelming pace. In many cases, the hypotheses generated by these studies conflict with long-held concepts of fish relationships, some to small degrees, others to very great degrees. Too often, these molecular-based phylogenetic hypotheses are not supported by morphology, as the number of molecular-based hypotheses have far outpaced the ability of morphologists to fully explore them (Hastings, 2011). Students of fishes should remember that these published phylogenies are merely hypotheses of relationships, and are subject to testing and refuting. As a consequence of this burgeoning of new ideas about fish relationships, the time is ripe for a morphological renaissance in ichthyology. Emerging molecular hypotheses provide a wealth of testable hypotheses for students with knowledge and expertise in morphology as we continue to refine our understanding of the fish tree of life.

ABOUT THIS BOOK

While ichthyology students often learn regional fish faunas through a series of local field trips, appreciation of the true diversity of fishes is more readily gained by a survey of a wide diversity of preserved specimens from a variety of habitats and from different geographic regions. Consequently our approach in this guide has been to include images of representative preserved specimens, labeled with the most important and easily visible diagnostic characters for the group to which they belong. For several groups, we provide images of more than one species, and in some cases, additional anatomical details to document variation within the group. Each photograph in this book is of a specimen archived in a natural history collection. Because our illustrations are of museum specimens, some are damaged, with broken fins or twisted bodies. This is especially true of many fishes of the deep-sea groups, as they are fragile and frequently damaged by nets during collection. In addition, the preservation methods used by fish collections (fixation in 10% formalin and transfer to alcohol for long-term storage) do not retain the bright colors typical of many living fishes. However, a vast number of images of living and freshly caught fishes are available on the internet, and students are encouraged to use one of the common search engines to locate additional images of groups of fishes of particular interest.

Almost all of the images in this guide are of specimens archived at the Scripps Institution of Oceanography Marine Vertebrate Collection (SIO). Details on the collecting locality and other information for each of these specimens are available online at https://scripps. ucsd.edu/collections/mv/. The Marine Vertebrate Collection is an extraordinary resource with over 2,000,000 specimens of fishes from all over the world. This inventory, supplemented by a few specimens from other collections, permitted us to provide coverage of all 78 currently recognized orders of fishes, as well as an additional 92 families of diverse, common, or otherwise interesting groups. While we have a slight bias towards groups found in North American waters, we also illustrate groups from other areas where possible. We are indebted to fish collections at other institutions for a few of the illustrated specimens. These include the Academy of Natural Sciences of Philadelphia (ANSP), California Academy of Sciences (CAS), Cornell University (CU), Tulane University (TU), the University of Arizona (UAZ), and the University of Michigan (UMMZ), as well as our colleague Dave Ebert (DE).

Each primary account also includes an estimate of the group's diversity based on Eschmeyer and Fong (2013), the approximate distribution of the group (the continents or oceans where they are found), the habitats in which they normally occur (freshwater, coastal marine, oceanic zone), and the portion of the water column where they typically reside (pelagic, neritic, demersal, or benthic). The Remarks section includes information such as the phylogenetic relationships of the group, their reproductive strategies and food preferences, their importance to humans, and in some cases the conservation status of the group. Additional details on the biology of most fishes can be found in the online resource Fishbase (Froese and Pauly, 2000; www.fishbase.org/home.htm). Finally, each account includes a list of some of the most important guides for identification, classic references on the systematics and biology of the group, and recent studies of their phylogeny. We owe a deep debt of gratitude to the late Joseph S. Nelson and his compendium, *Fishes of the World,* now in its fourth edition (2006). This work proved especially useful in compiling key characters for the groups of fishes represented herein. We also benefitted greatly from several classic references on fishes, too numerous to mention here, as well as a number of online resources, especially Eschmeyer's Catalog of Fishes (Eschmeyer, 2013; http://researcharchive.calacademy.org/

research/Ichthyology/catalog/fishcatmain.asp). Additional details on the biology of most fishes can be found in standard ichthyology texts (e.g., Bond, 1996; Bone and Moore, 2008; Helfman et al., 2009; Moyle and Cech, 2004).

Fishes are fascinating animals and have held our interest for most of our lives. We hope that this general survey of the most speciose group of vertebrates on the planet will provide others a greater appreciation of the amazing diversity of fishes, stimulating interest in them and all things ichthyological.

ACKNOWLEDGMENTS

We would like to thank several University of California, San Diego students who helped photograph fish specimens and edit the images used throughout this book, especially Matt Soave, Megan Matsumoto, and Corey Sheredy. Matt led the way with his extraordinary photographic and editing skills as well as his hard work and dedication. Several others provided photographic assistance including Dan Conley and John Snow. A number of colleagues provided specimens illustrated in the book either as loans or as gifts to the Scripps Institution of Marine Vertebrate Collection. These include: John Lundberg and Mark Sabaj (Academy of Natural Sciences of Philadelphia), John Sparks and Barbara Brown (American Museum of Natural History), Dave Catania (California Academy of Sciences), Amy McCune (Cornell University), Dave Ebert (Moss Landing Marine Lab), Hsuan-Ching Ho (National Museum of Marine Biology and Aquarium, Taiwan), Hank Bart and Nelson Rios (Tulane University), Peter Reinthal (University of Arizona), and Douglas Nelson (University of Michigan). We thank Cindy Klepadlo for curatorial assistance and her support in many ways, Tom Near and Leo Smith for providing information on the phylogeny of fishes, Larry Frank and Rachel Berquist for providing images from the Digital Fish Library project, Leo Smith for providing the excellent osteological image, and the National Science Foundation (DBI-1054085) for funds to purchase the MVC digital radiography system. Bruce Collette, Linn Montgomery, and Jackie Webb thoroughly reviewed an early draft of the book and provided many helpful comments. We would also like to thank the staff of the University of California Press, including Kate Hoffman, Merrik Bush-Pirkle, and Blake Edgar for their professional support and expertise in numerous ways; David Peattie of BookMatters for his patience and skill in formatting the book; and Chuck Crumly for his encouragement to pursue this project. We would also like to thank freelance copyeditor Caroline Knapp. Philip A. Hastings would like to thank Marty L. Eberhardt for her support, encouragement and companionship. H. J. Walker thanks Sonja, Tara, and Jeffrey Walker for their love and support, and for their love of snorkeling which led to some of the best fish-times of our lives. Grantly R. Galland would like to thank Gale and Bud Galland for showing him his first fishes and teaching him their names. Finally, we would all like to thank our numerous mentors who over the years have schooled us in our unwavering appreciation of fishes.

ANATOMY OF FISHES

While their anatomy varies greatly, all fishes have several features in common. In this section, we briefly review and illustrate the major features of fish anatomy, focusing on those that are most important for distinguishing among lineages and groups.

External Anatomy

Several external regions of fishes have specific names.

SNOUT   The area of the head between the tip of the upper jaw and the anterior margin of the orbit.

CHEEK   The area of the head below and posterior to the eye, anterior to the posterior margin of the preopercle.

NAPE   Dorsal area just posterior to the head.

OPERCULUM   Plate-like structure covering the branchial chamber and consisting of four bones: the opercle, preopercle, subopercle, and interopercle.

BRANCHIOSTEGALS   Slender, bony elements in the gill membrane, slightly ventral and posterior to the operculum.

ISTHMUS   Area of the throat ventral to the gill openings.

LATERAL LINE   Sensory system consisting of pores and canals along the head and body for the detection of vibrations and water movement, often associated with perforated scales along the body.

CAUDAL PEDUNCLE   Area of the body between the insertions of the dorsal and anal fins and the base of the caudal fin.

ANUS (VENT)   Terminal opening of the alimentary canal.

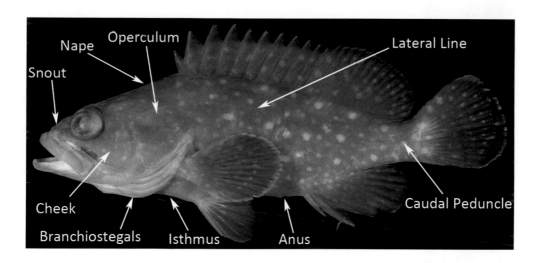

Body Shapes

Many fishes are somewhat elongate, laterally compressed, and oval in cross section. Several specialized shapes are recognized, including the following primary examples:

COMPRESSED   Flattened laterally, sometimes strongly so, and often deep-bodied.
DEPRESSED   Flattened dorsoventrally.
GLOBIFORM   Rounded, often spherical.
ANGUILLIFORM   Greatly elongate and usually tubular.
FUSIFORM   Roughly bullet-shaped, often tapering both anteriorly and posteriorly.

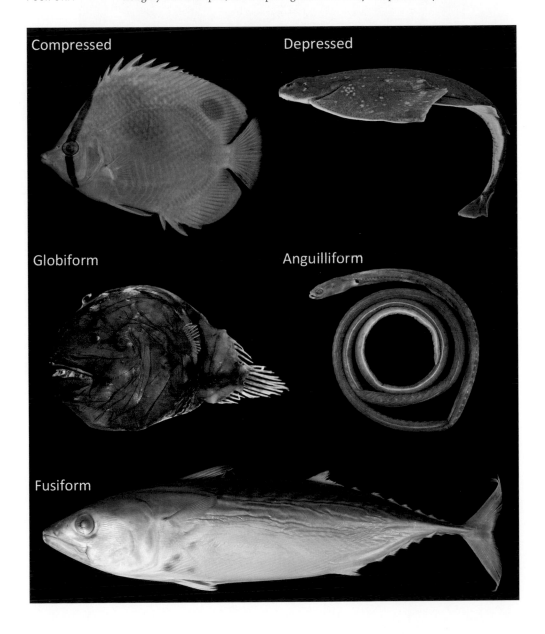

Compressed   Depressed

Globiform   Anguilliform

Fusiform

Fins

The fins of fishes are either unpaired or paired. The unpaired fins, also called median fins, include the dorsal, anal, and caudal fins, as well as the adipose fin in some fishes. The paired fins include the pectoral and pelvic fins.

The fins of actinopterygian fishes are composed of two types of rays: **soft rays**, which have evident segments, are bilaterally divided, are often branched, are typically flexible, and are usually connected by a fleshy membrane; and **spines**, which lack segments, are not bilaterally divided, are never branched, and are usually stiff and sometimes pungent. These fin-ray elements are derived from dermal tissues and are collectively called **lepidotrichia.** The dorsal fin of actinopterygians may be composed of soft rays only or of both spines and soft rays. In the latter case, the two parts of the fin may be continuous, separated by a notch, or completely separate. The fin rays of chondrichthyan fishes are flexible, unsegmented, and derived from epidermal tissues; they are called **ceratotrichia.**

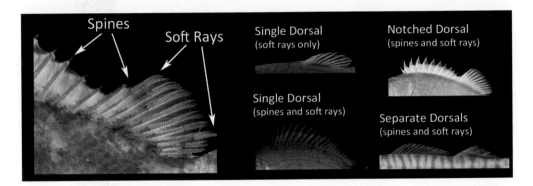

Pelvic-fin Positions

The pelvic fins of fishes vary considerably in their position on the body, a feature useful in distinguishing many groups.

ABDOMINAL — Inserted well posterior to the pectoral fins.
THORACIC — Inserted slightly posterior to or directly under the pectoral fins.
JUGULAR — Inserted slightly anterior to the pectoral fins.
MENTAL — Inserted far forward, often near the symphysis of the lower jaw.

Caudal-fin Shapes

The caudal fins of fishes come in a variety of shapes that are roughly related to a species' swimming behavior. Slow moving fishes often have rounded caudal fins, while fast swimming fishes have deeply forked fins with stiff upper and lower lobes. Most sharks and the early lineages of ray-finned fishes have a **heterocercal** caudal fin in which the vertebral column is deflected dorsally and extends along the upper, larger, caudal-fin lobe. Most ray-finned fishes have a **homocercal** caudal fin, which is externally symmetrical and supported by a series of laterally flattened bones. A few specialized groups such as the flyingfishes have a **hypocercal** caudal fin in which the lower lobe is larger than the upper lobe. Shapes of caudal fins include the following examples:

ROUNDED	No sharp or straight edges, convex posteriorly.
TRUNCATE	Posterior profile vertical.
EMARGINATE	Upper and lower rays slightly longer than central rays.
FORKED	Separate upper and lower lobes that join at a sharp angle.
LUNATE	Crescent-shaped posteriorly, with extremely large upper and lower lobes.
HETEROCERCAL	Vertebral column is deflected dorsally and extends along the upper, larger caudal-fin lobe.

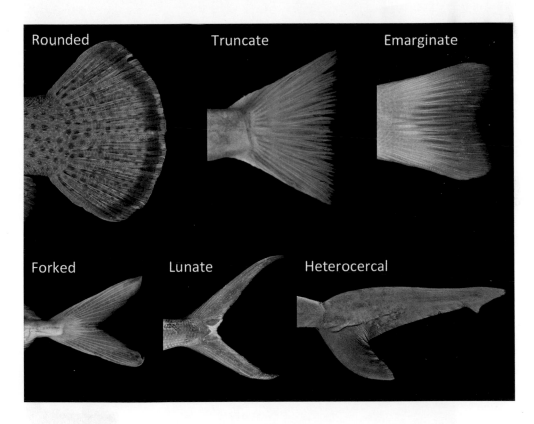

Mouth Positions

In addition to the size of the gape and the size and type of teeth, the position of a fish's mouth provides clues to its feeding habits. These include the following:

TERMINAL   Mouth located at the tip of the snout.
SUBTERMINAL   Mouth located below the tip of the snout.
INFERIOR   Mouth opens ventrally, well posterior to the snout.
SUPERIOR   Mouth opens dorsally.

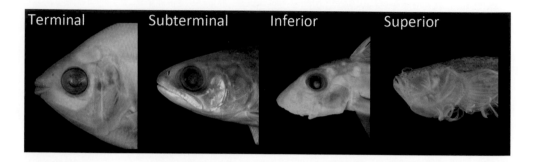

Oral and Pharyngeal Jaw Diversity

In the chondrichthyan fishes, the upper jaw is formed by the palatoquadrate cartilage, while in the ray-finned fishes, it is formed by two bones, the maxilla and the premaxilla. In early lineages, both of these bones bear teeth and are included in the gape, while in more derived ray-finned fishes, only the premaxilla bears teeth and the toothless maxilla is excluded from the gape. In addition to these "**oral jaws**," ray-finned fishes have a second set of jaws, the "**pharyngeal jaws,**" located anterior to the esophagus, comprising bones associated with the upper and lower gill arches. In many of these fishes, the oral jaws function to grasp and/or ingest prey, while the pharyngeal jaws are often specialized for processing prey.

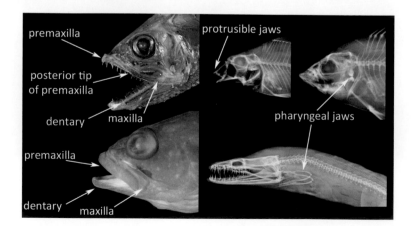

Standard Meaurements

Several standard measurements are used to document the size and shape of fishes (Hubbs and Lagler, 1958; Strauss and Bond, 1990). These include the following:

TOTAL LENGTH (TL)
Horizontal distance from the most anterior point on the head to the tip of the longest lobe of the caudal fin. The most anterior point is often the tip of the snout, but may be the tip of the lower jaw in some species.

FORK LENGTH (FL)
Horizontal distance from the most anterior point on the head to the end of the central caudal-fin rays.

STANDARD LENGTH (SL)
Horizontal distance from the tip of the snout to the central base of the caudal fin (i.e., the end of the hypural plate). The latter can often be located as a crease formed when the caudal fin is slightly bent.

HEAD LENGTH
Horizontal distance from the tip of the snout to the posterior margin of the operculum.

SNOUT LENGTH
Horizontal distance from the tip of the snout to the anterior margin of the orbit.

BODY DEPTH
Maximum vertical distance between the dorsal and ventral outlines of the body.

SNOUT-VENT LENGTH
Distance from the tip of the snout to the anterior margin of the vent.

DISK WIDTH
In batoid fishes (rays), the maximum distance between the lateral margins of the left and right pectoral fins.

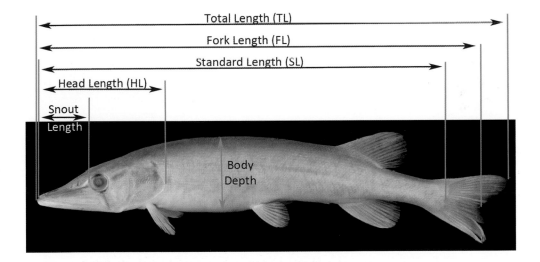

Sensory Systems

Fishes have the full array of sensory systems common to all vertebrates (olfaction, taste, vision, and hearing), as well as some unusual ones such as the lateral line and electroreception. Details of these systems are often useful in diagnosing various lineages of fishes. Numerous reviews of the sensory biology of fishes are available, including several chapters in volume 1 of the *Encyclopedia of Fish Physiology,* edited by Farrell (2011).

OLFACTION
: Fishes have left and right olfactory organs (paired in most fishes, unpaired in agnathans) that are chemoreceptive. Each side includes incurrent and excurrent nostrils (or nares) that may have a divided single opening or paired openings.

TASTE
: Fishes have chemoreceptive taste buds located inside the mouth, and in many groups also on the gill arches, barbels, fin rays, and the skin.

VISION
: The eyes of fishes come in a variety of sizes and forms and frequently are reflective of a species' habitat and habits. Eyes are often large in nocturnal species, upwardly directed in mesopelagic fishes, and small or sometimes absent in fishes from dark habitats including the deep sea and cave environments.

HEARING AND BALANCE
: Fishes have an inner ear with one (hagfishes), two (lampreys), or three (all other fishes) semicircular canals that function in maintaining balance and orientation. The main organs of hearing are the paired otolith organs, each of which consists of a sensory epithelium with an overlying calcium carbonate otolith (bony fishes) or otoconia (cartilaginous fishes). Sound waves are propagated from the water, through the tissues of the head, to the otoliths or otoconia, whose vibrations are detected by the sensory epithelium. A variety of so-called "otophysic connections" between the inner ear and the gas bladder serve to amplify sound reception in some fishes. These include anterior projections of the gas bladder that extend close to or, in some cases, into the otic capsule, and the Weberian apparatus, a mechanical linkage formed from modified anterior vertebrae, stretching between the gas bladder and inner ear of otophysans (Braun and Grande, 2008).

LATERAL LINE
: The mechanosensory lateral-line system of fishes detects water flow and vibrations made by movements of other organisms. Its sensory organs, called neuromasts, are located in pored lateral-line canals on the head (cephalic lateral line) and body (trunk lateral line), as well as on the skin (superficial neuromasts). Their expression in fishes varies greatly, but their configuration provides clues to the habits of many species (Webb, 1989, 2013).

ELECTRORECEPTION
: Receptors that detect weak electrical fields produced by other organisms are present in lampreys, all cartilaginous fishes, and some bony fishes (Kramer, 1996). In the cartilaginous fishes they are called ampullae of Lorenzeni, and involve sensory cells located at the base of canals filled with conductive jelly and open to the surface. They are especially common on the ventral side of the head, where they facilitate detection and capture of prey items. In teleosts the electroreceptive sense detects electrical fields in the environment, including those generated by conspecifics, as well as potential prey.

The skeletal structure of fishes has been studied extensively for clues to both phylogeny and function. The skeletal structure of cartilaginous fishes was recently reviewed by Claeson and Dean (2011). Several excellent guides to the osteology of ray-finned fishes are available, including the classic text by Gregory (1933) and a recent overview by Hilton (2011). For ray-finned fishes, several superficial bones of the head are especially useful in identifying various groups of fishes (illustrated below). The major components of the caudal fin and posterior vertebral column are also illustrated below.

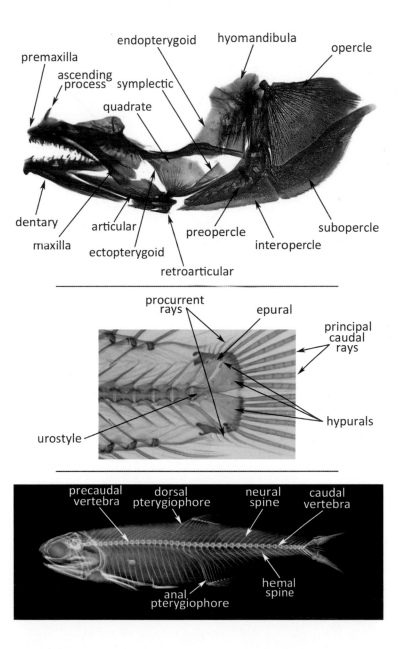

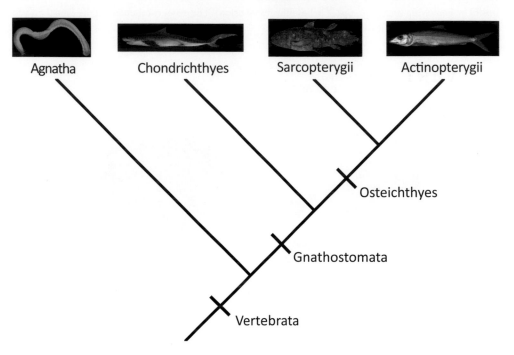

Hypothesized phylogenetic relationships of the major lineages of vertebrates.

THE FISHES

VERTEBRATA—VERTEBRATES

The Vertebrata is one of the most successful lineages of animals, dominating both aquatic and terrestrial habitats around the globe. This diverse group, with well over 60,000 species, is characterized by the presence of ossifications surrounding and often occluding the notochord (in most living species), a well-developed brain, a notochord that is restricted posterior to the brain, a chambered heart, and a host of other features (Forey, 1995; Nelson, 2006). Aquatic representatives number well over 30,000 species and include the jawless fishes (Agnatha), cartilaginous fishes (Chondrichthyes), and ray-finned fishes (Actinopterygii). Terrestrial habitats are largely the domain of the Tetrapoda, the dominant clade of the Sarcopterygii, which also includes a handful of aquatic lung fishes and the coelacanths. Relationships among these major lineages of vertebrates have been discussed for decades and a consensus has been reached (Meyer and Zardoya, 2003). The lobe-finned and ray-finned fishes form a monophyletic group (called the Osteichthyes or "bony fishes"); they together with the cartilaginous fishes make up the "jawed vertebrates" or the Gnathostomata. The jawless fishes are the sister group of all other extant vertebrates. This book covers all major lineages of the Vertebrata with the exception of the Tetrapoda.

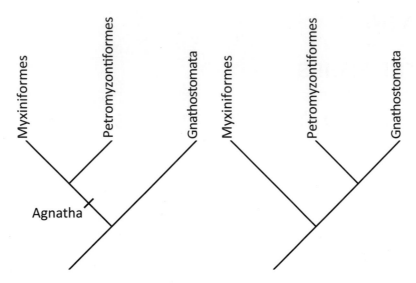

Phylogenetic relationships depicting a monophyletic Agnatha (left), and lampreys as the sister-group to the Gnathostomata (right) (after Heimberg et al., 2010).

AGNATHA (CYCLOSTOMATA)

Jawless Fishes

As their name implies (*a* = without; *gnathos* = jaw), agnathans lack jaws, and instead possess a rounded mouth, a fact reflected in the older term for the group, the Cyclostomata (*cyclo* = round; *stoma* = mouth). Extant members lack pelvic fins, have pore-like rather than slit-like gill openings and an elongate, eel-like body. Agnathans have a well-developed notochord; a rudimentary vertebral column is present only in the lampreys. The group has a rich fossil record, and many of the extinct members had a bony external skeleton that is lacking in living representatives whose entire skeleton is cartilaginous. Extant agnathans include two major lineages, the hagfishes (Myxiniformes) and the lampreys (Petromyzontiformes). Analyses of morphological features imply that the lampreys, though not the hagfishes, are the sister group of the jawed vertebrates (e.g., Forey, 1995; Janvier, 1996). However, extensive molecular data (e.g., Heimberg et al. 2010; Kuraku and Kuratani, 2006) overwhelmingly support the sister-group relationship of hagfishes and lampreys and thus the monophyly of the Agnatha. This finding implies that the extant representatives, especially the Myxiniformes, are reductive in a number of features, confounding efforts to reconstruct their phylogenetic relationships based solely on morphology. Their biology was summarized by Hardisty (1979).

MYXINIFORMES : MYXINIDAE—Hagfishes

DIVERSITY: 1 family, 6 genera, 76 species

REPRESENTATIVE GENERA: *Eptatretus, Myxine, Nemamyxine*

DISTRIBUTION: Atlantic, Indian, and Pacific oceans

HABITAT: Marine; tropical to temperate; inshore to deep sea, benthic, in or on soft substrates

REMARKS: The single family of hagfishes is one of two groups of living jawless or agnathan fishes. In addition to the features listed above, they are characterized by a single nostril, and two features, one semicircular canal, and body fluid isosmotic with seawater, unique among the Vertebrata. Their eyes are degenerate, lacking a lens and extrinsic eye muscles. Their conspicuous slime glands contain both mucous and thread cells and serve to thwart predators. Hagfishes are known to prey on benthic organisms but generally are considered scavengers. They are able to remove chunks of flesh from carcasses using the paired tooth plates on the tongue, gaining leverage by tying their body in a knot. Hagfishes have a few very large eggs and, lacking a larval phase, the hatchlings resemble small adults (Jorgensen et al., 1998). Hagfishes are utilized by the fish leather industry (Grey et al., 2006).

REFERENCES: Fernholm, 1998; Fernholm, in Carpenter, 2003; Fernholm and Paxton, in Carpenter and Niem, 1998; Grey et al., 2006; Jorgensen et al., 1998; Kuo et al., 2003; Kuraku and Kurtani, 2006; Wisner and McMillan, 1995.

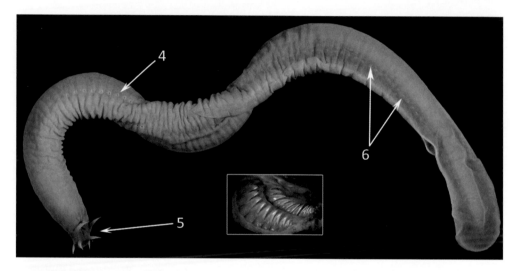

MYXINIFORM CHARACTERISTICS:
1) body eel-like, naked
2) paired fins and dorsal fin absent
3) lateral line absent in adults
4) one to sixteen pairs of external pore-like gill openings
5) oral barbels distinctive
6) numerous mucous pores on body

ILLUSTRATED SPECIMEN:
Eptatretus stoutii, SIO 87–125, 145 mm TL
Inset: Tooth plates of *Myxine capensis,* showing keratinous cusps, SIO 92–107

PETROMYZONTIFORMES—Lampreys

The 46 species of living lampreys are found in temperate areas of both hemispheres. The monotypic Geotriidae and the three species of Mordaciidae are found in the Southern

Hemisphere, while the more diverse and well-known Petromyzontidae is restricted to the Northern Hemisphere (Renaud, 2011).

REFERENCES: Gill et al., 2003; Renaud, 2011.

PETROMYZONTIFORMES : PETROMYZONTIDAE—Northern Lampreys

DIVERSITY: 8 genera, 42 species

REPRESENTATIVE GENERA: *Ichthyomyzon, Lampetra, Petromyzon*

DISTRIBUTION: North America, Europe, and Asia

HABITAT: Freshwater lakes, rivers, and streams or anadromous; temperate; demersal, or benthic on soft substrates

REMARKS: Lampreys are characterized by two semicircular canals, an otic capsule anterior to the first branchial opening, and body fluid hyposmotic to seawater. Unlike hagfishes, lampreys lay numerous small eggs; their larva, called an ammocoete, filter-feeds on detritus. Lampreys include both parasitic and free-living species. In general, the 22 free-living species (called brook lampreys) remain in small streams and rivers throughout life, though they cease feeding after metamorphosis. Parasitic species have a similar lifestyle in their young stages, but as adults they migrate to the ocean or large lakes where they use their round mouth to attach to other fishes and their rows of teeth to rasp away flesh. Transitions between these life history patterns have occurred repeatedly, as several pairs of closely related species include a parasitic and a free-living form (Potter, 1980). Gill et al. (2003) explored the phylogeny of lampreys based on morphological features and, more recently, Renaud (2011) reviewed their systematics and biology.

REFERENCES: Gill et al., 2003; Hardisty and Potter, 1971; Potter, 1980; Renaud, 1997, 2011.

PETROMYZONTID CHARACTERISTICS:
1) body eel-like, naked
2) no paired fins, one or two dorsal fins
3) lateral line absent
4) seven pairs of external pore-like gill openings
5) oral barbels absent
6) oral disk and tongue bearing rows of teeth
7) single nostril located between eyes, anterior to pineal eye
8) cloaca located under anterior half of second dorsal fin or posterior lobe of single dorsal fin

ILLUSTRATED SPECIMEN:
Petromyzon marinus, SIO 74–134, 124 mm TL

GNATHOSTOMATA

Jawed Vertebrates

The Gnathostomata is an extraordinarily successful lineage of over 60,000 species characterized by upper and lower jaws that are derived from modified gill arches. Most gnathostomes also possess paired pectoral and pelvic limbs, vertebral centra, slit-like gill openings at some stage of development, and three semicircular canals (Forey, 1995; Nelson, 2006). These and a host of other features mark them as active, mobile predators with more powerful sensory abilities than agnathans (Gans, 1987; Gans and Northcutt, 1983; Shimeld and Holland, 2000). Living gnathostomes include the Chondrichthyes (sharks and rays), Sarcopterygii (lobe-finned fishes and tetrapods), and Actinopterygii (ray-finned fishes). Phylogenetic relationships of gnathostomes were summarized by Stiassny et al. (2004).

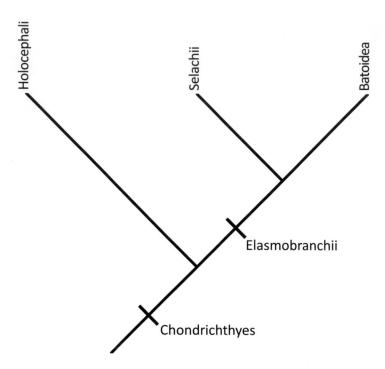

Hypothesized phylogenetic relationships of the major lineages of the Chondrichthyes (after Aschliman et al., 2012).

Chondrichthyes

Cartilaginous Fishes

The cartilaginous (or chondrichthyan) fishes are a major group that includes 14 orders and more than 1,200 living species. These fishes are distinguished by a skeleton made entirely of cartilage, and a neurocranium with no sutures. Their teeth are derived from placoid scales and are replaced serially. Their fin rays, termed ceratotrichia, are usually soft and somewhat flexible, are always unsegmented, and are developmentally epidermal. They have a well-developed electroreceptive sense, with numerous pores of the ampullae of Lorenzeni often evident, especially surrounding the mouth. Males have pelvic-fin claspers for use in mating, as all species of chondrichthyan fishes have internal fertilization. Some groups are oviparous, releasing protective keratinized egg cases in which embryos develop. Others retain developing embryos within the body of the female, where they develop solely from nutrition supplied in the egg (yolk-sac viviparity). Still others supplement the nutrition of the embyos in an astounding variety of ways (Musick, 2011), including mucous and lipid secretions from the uterine lining (mucoid and lipid histotrophy), additional eggs released by the mother that are ingested by the embryos (oophagy), and transfer of nutrients from the mother to the embryo via a placenta (placental viviparity; Hamlett et al., 2005). Interestingly, Musick and Ellis (2005) concluded that the primitive condition in chondrichthyans is yolk-sac viviparity, from which all other forms, including oviparity, evolved. Most chondrichthyan species are marine, although some can enter freshwater and a very small number are restricted to freshwater. Cartilaginous fishes usually have a high concentration of urea in their blood relative to bony fishes, in order to maintain osmotic balance with seawater. The biology of sharks and rays has been summarized by several researchers including Carrier et al. (2012), Hamlett (2005), and Klimley (2013).

This monophyletic group is sister to all other living jawed vertebrates, the Osteichthyes, a group that includes the ray-finned fishes and lobe-finned fishes. While comprising only approximately 3% of fish species diversity, this group includes more than 15% of fish orders, implying a high level of fundamental differences in morphology among relatively few species. There are two distinct evolutionary lines of chondrichthyan fishes, the Elasmobranchii (sharks, skates, and rays) and the Holocephali (chimaeras or ratfishes). Among these fishes, the elasmobranchs include 96% of the diversity, while the Holocephali comprises only 4%. The Batoidea (skates and rays) account for 54% of the total chondrichthyan diversity, leaving 42% to the Selachii (shark-like species).

HOLOCEPHALI—Chimaeras

This chondrichthyan lineage includes one extant order, described below, that is the sister group to the Elasmobranchii (Lund and Grogan, 1997).

CHIMAERIFORMES—Chimaeras

The Chimaeriformes includes three families, six genera, and approximately 50 species of generally deep-sea predators characterized by a single gill opening and an upper jaw fused to the neurocranium (holostylic jaw suspension). The plownose chimaeras (Callorhinchidae) are restricted to the Southern Hemisphere, while the longnose chimaeras (Rhinochimaeridae) and the shortnose chimaeras (Chimaeridae) are more widespread. As their common names imply, the shortnose chimaeres have a blunt snout, the longnose chimaeras have long, pointed snouts, and the plownose chimaeras have long, hook-shaped snouts. One member of the latter family (*Callorhinchus milli*) has become a model organism for comparative genomics because of its relatively compact genome (Venkatesh et al., 2007; Tan et al., 2012). The shortnose chimaeras, detailed below, are the most speciose lineage in this group. Many of the characteristics described for them apply to the other families as well.

REFERENCES: Grogan and Lund, 2004; Grogan et al., 1999; Lund and Grogan, 1997; Patterson, 1965; Tan et al., 2012; Venkatesh et al., 2007.

CHIMAERIFORMES : CHIMAERIDAE—Ratfishes, Shortnose Chimaeras

DIVERSITY: 2 genera, 38 species

REPRESENTATIVE GENERA: *Chimaera, Hydrolagus*

DISTRIBUTION: Atlantic, Indian, and Pacific oceans

HABITAT: Marine; tropical to temperate; lower continental shelf to bathyal, demersal usually over soft substrates

REMARKS: Ratfishes, one of two major groups of chondrichthyan fishes, are characterized by a lack of a stomach and the presence of separate anal and urogenital openings. They are deep-sea predators with tooth plates for crushing hard-bodied prey such as benthic mollusks and crustaceans. Ratfishes are oviparous and produce keratinoid egg cases with a long, pointed end and small hooks that anchor them to the substrate.

REFERENCES: Compagno, in Carpenter and Niem, 1998; Didier, in Carpenter, 2003; Didier, 2004; Lund and Grogan, 1997; Patterson, 1965.

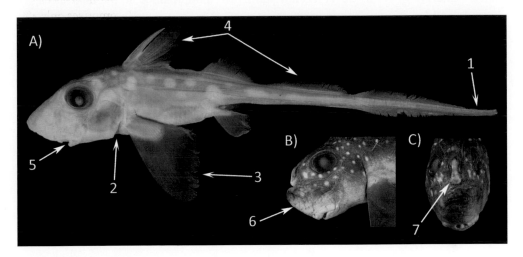

CHIMAERID CHARACTERISTICS:
1) body elongate with a whip-like tail, body usually naked
2) one external gill opening, anterior to pectoral fin
3) pectoral fins broad and wing-like
4) two dorsal fins: the first high with an erectile spine, the second low with a long base
5) mouth inferior
6) conspicuous lateral-line canals on snout
7) males with a club-like clasper on top of head
8) pelvic claspers bi-lobed
9) spiracles absent

ILLUSTRATED SPECIMENS:
A) *Hydrolagus colliei,* SIO 49–121,134 mm TL (tip of tail missing)
B) *Hydrolagus colliei,* SIO 85–73, 448 mm TL (lateral view of head)
C) *Hydrolagus colliei,* SIO 85–73, 448 mm TL (frontal view of head)

ELASMOBRANCHII—Sharks and Rays

The Elasmobranchii, with 12 extant orders of sharks and rays, is the sister group to the Holocephali. In contrast to that group, elasmobranchs have five to seven separate gill openings, and the upper jaw is not fused to the neurocranium (amphistylic or hyostylic jaw suspension). Additionally, males of this group lack a cephalic clasper organ. The phylogenetic relationships of elasmobranchs, although intensively studied by numerous researchers using a variety of data sets including morphology, molecular data, and the fossil record, have been controversial (e.g., Maisey, 2012; Naylor et al., 2005; Shirai, 1996). Among the main contentious issues have been questions about the monophyly of the sharks (the Selachii) as a group, and the monophyly of the rays (the Batoidea) as a group. Naylor et al. (2005) and more recently Aschliman et al. (2012) concluded that these groups are reciprocally monophyletic, that is, each is monophyletic and they are sister groups, sharing a unique common ancestor. Naylor et al. (2012) recently summarized information on the valid species of elasmobranchs.

SELACHII—Sharks

The Selachii includes all species of sharks and is characterized by lateral gill openings and pectoral fins separate from the head (Nelson, 2006). This group includes 518 species, classified in eight orders, 35 families, and over 100 genera (Compagno, 1984a, 1984b, 2005; Compagno et al., 2005; Eschmeyer and Fong, 2013; Naylor et al., 2012). Phylogenetic relationships of sharks have been hypothesized by several researchers (e.g., de Carvalho, 1996; Naylor et al., 2005; Shirai, 1996; Vélez-Zuazo and Agnarsson, 2011), including Maisey et al. (2004), who recognized two major lineages, the Squalomorphii and the Galeomorphii.

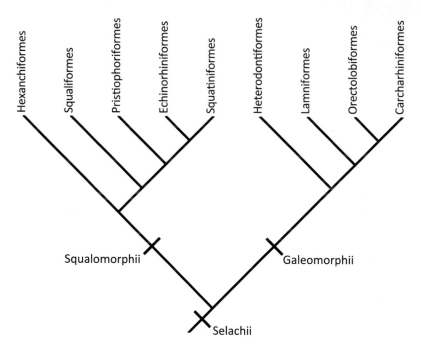

Hypothesized phylogenetic relationships of the Selachii (sharks) after Maisey et al. (2004) and Musick and Ellis (2005).

HETERODONTIFORMES : HETERODONTIDAE—Bullhead Sharks

DIVERSITY: 1 genus, 9 species

REPRESENTATIVE GENUS: *Heterodontus*

DISTRIBUTION: Indian and Pacific oceans

HABITAT: Marine; tropical to warm temperate; continental shelf (one species on continental slope), benthic to demersal including shallow rocky or coral reefs

REMARKS: Members of the family of bullhead sharks are characterized by their distinctive dorsal fins and blunt snouts. They are usually nocturnal and generally feed on benthic invertebrates and occasionally on small fishes. Bullhead sharks are oviparous (Musick, 2011; Musick and Ellis, 2005) and produce distinctive screw-shaped, keratinoid egg cases.

REFERENCES: Compagno, 2001, 2005; Compagno, in Carpenter, 2003; Compagno and Niem, in Carpenter and Niem, 1998; Compagno et al., in Fischer et al., 1995; Compagno et al., 2005; Musick, 2011; Musick and Ellis, 2005.

HETERODONTIFORM CHARACTERISTICS:
1) five external gill slits, two to three behind pectoral-fin origin
2) two dorsal fins, each with a broad-based spine
3) crests above eyes
4) eyes without a nictitating membrane
5) spiracles small
6) nostrils and mouth connected by a groove

ILLUSTRATED SPECIMEN:
Heterodontus francisci, SIO 64–33, 765 mm TL

INSET: Lower jaw of *Heterodontus francisci* (SIO 60–23, 865 mm TL) showing anterior rows of pointed teeth and posterior rows of pavement-like teeth.

ORECTOLOBIFORMES—Carpet Sharks

The carpet sharks comprise seven families, 14 genera, and 42 species of mostly benthic sharks, noted for sitting perfectly still on the bottom of the ocean. Their nostrils have barbels and are connected to the relatively small mouth by a groove. The wobbegongs (Orectolobidae) are sit-and-wait predators and have a head covered in skin flaps, cryptic coloration, large spiracles, and large fang-like teeth. Their phylogenetic relationships were studied by Goto (2001) and Corrigan and Beheregaray (2009). Two families (Ginglymostomatidae and Rhincodontidae) are described in more detail below. The remaining families (Parascyliidae, Brachaeluridae, Hemiscyliidae, and Stegostomatidae) include relatively few species and occur in the tropical Indo-West Pacific.

REFERENCES: Corrigan and Beheregaray, 2009; Goto, 2001.

ORECTOLOBIFORMES : GINGLYMOSTOMATIDAE—Nurse Sharks

DIVERSITY: 1 family, 3 genera, 3 species

REPRESENTATIVE GENERA: *Ginglymostoma, Nebrius, Pseudoginglymostoma*

DISTRIBUTION: Atlantic, Indian, and Pacific oceans

HABITAT: Marine; tropical to subtropical; continental shelf, benthic on shallow reefs and adjacent sandy areas

REMARKS: Nurse sharks are characterized by their brownish coloration and by dorsal fins located far back on the body. They are generally nocturnal and can be observed resting in small groups on the reef or sandy surfaces during the day. The small mouth with a large oral cavity is capable of suction-feeding benthic invertebrates and small fishes. Nurse sharks are yolk-sac viviparous (Musick, 2011; Musick and Ellis, 2005).

REFERENCES: Compagno, 2001, 2005; Compagno, in Carpenter, 2003; Compagno, in Carpenter and Niem, 1998; Compagno et al., in Fischer et al., 1995; Compagno et al., 2005; Musick, 2011; Musick and Ellis, 2005.

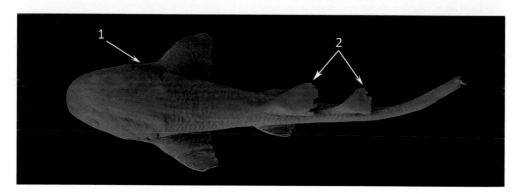

GINGLYMOSTOMATID CHARACTERISTICS:
1) five small gill openings (slits), fifth slit nearly overlapping fourth
2) dorsal fins without spines, positioned posteriorly on body
3) mouth short, subterminal, not extending to level of eyes
4) eyes without nictitating membrane
5) spiracles small, located just behind eyes
6) nostrils with barbels

ILLUSTRATED SPECIMEN:
Ginglymostoma cirratum, SIO 64–229, 521 mm TL (dorsal view)

ORECTOLOBIFORMES : RHINCODONTIDAE—Whale Sharks

DIVERSITY: 1 genus, 1 species

REPRESENTATIVE GENUS: *Rhincodon*

DISTRIBUTION: Atlantic (absent from Mediterranean), Indian, and Pacific oceans

HABITAT: Marine; tropical to warm temperate; usually epipelagic but occasionally meso-pelagic or neritic

REMARKS: While the Whale Shark is distinguished by its huge size (the largest fish on Earth), it feeds on planktonic organisms and fish eggs, using its long, thin gill rakers for filter feeding. Its brain anatomy was studied by Yopak and Frank (2009). The Whale Shark is yolk-sac viviparous (Musick, 2011; Musick and Ellis, 2005), but, curiously, produces kera-tized egg capsules that hatch within the female. These enormous fishes are highly fecund compared to other members of the Chondrichthyes; one female caught by fishermen held

RHINCODONTID CHARACTERISTICS:
1) body covered in yellow or white spots
2) head broad and flattened
3) snout short, mouth nearly terminal, anterior to small eyes
4) mouth and gill openings especially large
5) spiracles small
6) longitudinal ridges on body of adults

ILLUSTRATED SPECIMEN:
Rhincodon typus, SIO 85–20, 601 mm TL (dorsal and lateral views)

300 pups. In many areas, the Whale Shark has been overfished; it is considered "vulnerable" by the IUCN (2013) and it has been given protected status by many nations.

REFERENCES: Compagno, 2001, 2005; Compagno, in Carpenter, 2003; Compagno, Carpenter and Niem, 1998; Compagno et al., in Fischer et al., 1995; Compagno et al., 2005; Musick, 2011; Musick and Ellis, 2005; Yopak and Frank, 2009.

LAMNIFORMES—Mackerel Sharks

The mackerel sharks have two dorsal fins without spines, an anal fin, five gill slits (the last two often above the pectoral fin), eyes without a nictitating membrane, a large mouth that extends well behind the eyes, and small spiracles usually present behind the eyes. This distinctive lineage includes only 15 species, classified in ten genera and seven families, indicating a high level of morphological diversity among a small number of closely related species. For example, the Basking Shark (Cetorhinidae) and the Megamouth Shark (Megachasmidae) are large-bodied filter feeders, while the White Shark (Lamnidae) is one of the ocean's top predators. The Goblin Shark (Mitsukurinidae), characterized by an elongate snout, and the Crocodile Shark (Pseudocarchariidae), which has extremely large eyes, are generally found in deeper water. The sand tiger sharks (Odontaspididae) are some of the few sharks that have exposed teeth when the mouth is closed, giving them a ferocious appearance. In contrast with the otherwise similar Carcharhiniformes, members of the Lamniformes do not have a nictitating membrane protecting the eyes. The phylogenetic relationships of lamniforms have been studied by a number of workers (e.g., Compagno, 1990; Naylor et al., 1997; Shimada, 2005; Shimada et al., 2009), while variation in their caudal-fin anatomy was documented by Kim et al. (2013). Lamniforms have an unusual reproductive mode in which developing embryos eat eggs (oophagy) and sometimes other embryos (Musick, 2011; Musick and Ellis, 2005) produced by the mother. The thresher sharks (Alopiidae) and the mackerel sharks *sensu stricto* (Lamnidae) are further described below.

REFERENCES: Compagno, 1990, 2001; Kim et al., 2013; Musick, 2011; Musick and Ellis, 2005; Shimada, 2005; Shimada et al., 2009.

LAMNIFORMES : ALOPIIDAE—Thresher Sharks

DIVERSITY: 1 genus, 3 species

REPRESENTATIVE GENUS: *Alopias*

DISTRIBUTION: Atlantic, Indian, and Pacific oceans

HABITAT: Marine; tropical to temperate; neritic to epipelagic

REMARKS: Thresher sharks use their long, whip-like caudal fin to disable small pelagic fishes and squids during feeding (Kim et al., 2013). Like mackerel sharks, at least one species (*Alopias vulpinus*) has the ability to maintain a higher body temperature than ambient seawater (Sepulveda et al., 2005). Thresher sharks are oophagous (Musick, 2011; Musick and Ellis, 2005), giving birth to small numbers of relatively large pups. These sharks support limited commercial fisheries.

REFERENCES: Compagno, 2001, 2005; Compagno, in Carpenter, 2003; Compagno, in Carpenter and Niem, 1998; Compagno et al., in Fischer et al., 1995; Compagno et al., 2005; Kim et al., 2013; Musick, 2011; Musick and Ellis, 2005; Sepulveda et al., 2005.

ALOPIID CHARACTERISTICS:
1) upper lobe of caudal fin long and whip-like, equaling length of body
2) mouth subterminal, relatively small
3) second dorsal fin and anal fin small, pectoral fins large
4) caudal peduncle with a precaudal pit
5) gill slits small, fourth and fifth above pectoral-fin base

ILLUSTRATED SPECIMEN:
Alopias vulpinus, SIO 64–804, 1,448 mm TL

LAMNIFORMES : LAMNIDAE—Mackerel Sharks

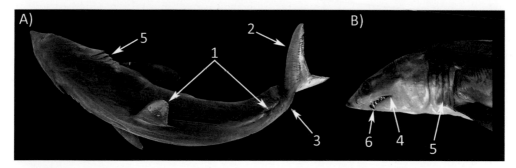

LAMNID CHARACTERISTICS:
1) second dorsal fin much smaller than first dorsal fin
2) caudal fin nearly symmetrical (approaching lunate)
3) caudal peduncle strongly depressed with a lateral keel and precaudal pit
4) mouth large, extending past level of eyes
5) gill openings large
6) teeth large
7) gill rakers absent

ILLUSTRATED SPECIMEN:
A) *Isurus oxyrinchus,* SIO 55–85, 875 mm TL (dorsal view)
B) head of *Isurus oxyrinchus,* SIO 55–85 (lateral view)

(account continued)

DIVERSITY: 3 genera, 5 species

REPRESENTATIVE GENERA: *Carcharodon, Isurus, Lamna*

DISTRIBUTION: Atlantic, Indian, and Pacific oceans

HABITAT: Marine; tropical to temperate; neritic to epipelagic

REMARKS: Mackerel sharks are strong-swimming, large-bodied predators that prey mainly on fishes, birds, marine mammals, and cephalopods. These sharks are one of a few groups of fishes known to have body temperatures higher than their surroundings, allowing them to remain active in very cold water (Bernal et al., 2001). Mackerel sharks are oophagous (Musick, 2011; Musick and Ellis, 2005) and give birth to large, well-developed pups. For example, newly born White Shark pups can be up to 1.3 m long (Domeier, 2012). The White Shark is responsible for a number of attacks on humans each year (Domeier, 2012). Some laminid species support commericial fisheries.

REFERENCES: Bernal et al., 2001; Compagno, 2001; Compagno, in Carpenter, 2003; Compagno, in Carpenter and Niem, 1998; Compagno et al., in Fischer et al., 1995; Compagno et al., 2005; Domeier, 2012; Musick, 2011; Musick and Ellis, 2005.

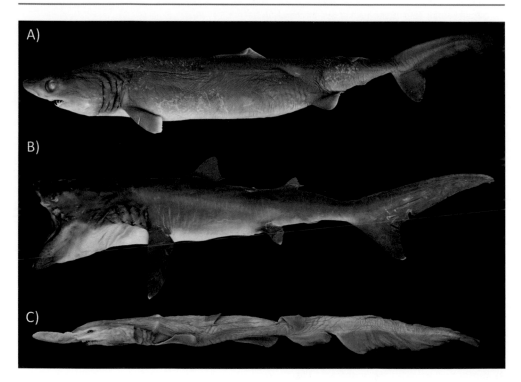

LAMNIFORM DIVERSITY:

A) PSEUDOCARCHARIIDAE—crocodile sharks: *Pseudocarcharias kamoharai,* SIO 97–221, 952 mm TL
B) MEGACHASMIDAE—megamouth sharks: *Megachasma pelagios,* SIO 07–53, 2,150 mm TL
C) MITSUKURINIDAE—goblin sharks: *Mitsukurina owstoni*, SIO 07–46, 1,150 mm TL

CARCHARHINIFORMES—Ground Sharks

The ground sharks comprise the most speciose order of sharks, with eight families, approximately 50 genera, and at least 287 species. More than half of the species are small-bodied cat sharks (Scyliorhinidae and Proscylliidae), named for their horizontally elongate eyes. The false cat sharks (Pseudotriakidae) are characterized by a deep groove anterior to their elongate eyes, while the monotypic Barbled Houndshark (Leptochariidae) has notably long labial furrows and nostrils with barbels. The weasel sharks (Hemigaleidae) are characterized by their wavy, upper caudal-fin lobe. The Carcharhiniformes as a group are difficult to characterize, but all included species have two dorsal fins without spines, an anal fin, five gill slits (last one to three positioned over the pectoral fin), and a large mouth extending behind the eyes (Compagno, 1988). In contrast with the similar Lamniformes, the eyes of most carcharhiniforms are protected by a nictitating membrane and most species lack spiracles. Three families (Carcharhinidae, Sphyrnidae, and Triakidae) are described in more detail below.

REFERENCES: Compagno, 1988, 2001.

CARCHARHINIFORMES : TRIAKIDAE—Hound Sharks

DIVERSITY: 9 genera, 47 species
REPRESENTATIVE GENERA: *Galeorhinus, Mustelus, Triakis*

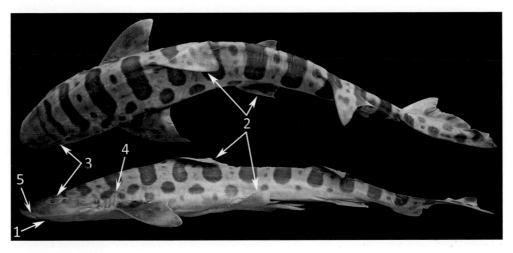

TRIAKID CHARACTERISTICS:
1) mouth small, subterminal, snout long
2) dorsal-fin base well anterior to pelvic fins
3) eyes oval with nictitating membranes
4) fourth and fifth gill slits over pectoral-fin base
5) nostrils with flaps, often broad, not barbel-like

ILLUSTRATED SPECIMEN:
Triakis semifasciata, SIO 62–213, 1,035 mm TL (dorsal and lateral views)

DISTRIBUTION: Atlantic, Indian, and Pacific oceans

HABITAT: Marine, occasionally in river mouths; tropical to temperate; coastal to continental slope, demersal over rocky reefs and soft substrates

REMARKS: Hound sharks are moderately sized (to 2.4 m) and feed on benthic and midwater invertebrates and fishes. Some species specialize on crustaceans and others on cephalopods. Hound sharks are either yolk-sac viviparous or placental viviparous (Musick, 2011; Musick and Ellis, 2005) and produce litters with as many as 52 pups. Their phylogenetic relationships were studied by Lopez et al. (2006).

REFERENCES: Compagno, 1988, 2001, 2005; Compagno, in Carpenter, 2003; Compagno and Niem, in Carpenter and Niem, 1998; Compagno et al., in Fischer et al., 1995; Compagno et al., 2005; Lopez et al., 2006; Musick, 2011; Musick and Ellis, 2005.

CARCHARHINIFORMES : CARCHARHINIDAE—Requiem Sharks

DIVERSITY: 12 genera, 60 species

REPRESENTATIVE GENERA: *Carcharhinus, Galeocerdo, Prionace, Rhizoprionodon*

DISTRIBUTION: Atlantic, Indian, and Pacific oceans

HABITAT: Marine and occasionally freshwater; tropical to temperate; neritic to epipelagic to demersal over reefs and adjacent soft substrates

REMARKS: Requiem sharks are medium to large (up to 7.4 m) predators and are well known for their occasional migrations into freshwater (sometimes exceeding 1,000 km). Species in this family are among those most commonly encountered by divers and sport fishers, and several have been known to attack swimmers. The Bull Shark, Tiger Shark, and Oceanic Whitetip Shark are responsible for most human fatalities. Requiem sharks are generalist predators, taking a wide variety of prey. While some species are yolk-sac viviparous, most species are placental (Musick, 2011; Musick and Ellis, 2005). Some species are highly fecund: the Blue Shark (*Prionace glauca*) is known to have litters of up to 100 pups.

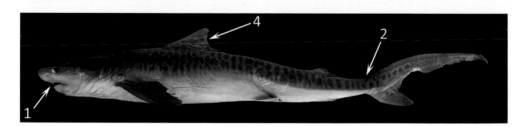

CARCHARHINID CHARACTERISTICS:
1) mouth large, subterminal
2) caudal peduncle with a precaudal pit
3) eyes with nictitating membrane
4) dorsal fin with lateral undulations along posterior margin
5) spiracles usually absent
6) nasal grooves and barbels absent

ILLUSTRATED SPECIMEN:
Galeocerdo cuvier, SIO 66–44, 1,310 mm TL

REFERENCES: Compagno, 1988, 2001, 2005; Compagno et al., 2005; Compagno et al., in Fischer et al., 1995; Compagno and Niem, in Carpenter and Niem, 1998; Garrick, 1982; Musick, 2011; Musick and Ellis, 2005; Naylor, 1992.

CARCHARHINIFORMES : SPHYRNIDAE—Hammerhead Sharks

DIVERSITY: 2 genera, 9 species

REPRESENTATIVE GENERA: *Eusphyra, Sphyrna*

DISTRIBUTION: Atlantic, Indian, and Pacific oceans

HABITAT: Marine; tropical to warm temperate; continental shelf and seamounts, neritic or demersal over reefs and soft bottoms

REMARKS: Their characteristic hammer-shaped heads distinguish these sharks from all other fishes. This unusual feature increases capabilities of both vision and the electromagnetic sense and also improves maneuverability. The phylogenetic relationships of hammerheads were studied by Naylor (1992) and Lim et al. (2010). Hammerheads are top predators that feed on bony fishes, sharks and rays, and squids and other invertebrates. Unlike most sharks, hammerheads often form large schools near seamounts during the day, likely dispersing to hunt individually at night. They are viviparous, with yolk-sac placentas (Musick, 2011; Musick and Ellis, 2005). Hammerheads are particularly vulnerable to overfishing and are often captured as bycatch in longline and net fisheries.

REFERENCES: Compagno, 1988, 2005; Compagno, in Carpenter, 2003; Compagno, in Carpenter and Niem, 1998; Compagno et al., in Fischer et al., 1995; Compagno et al., 2005; Gilbert, 1967; Lim et al., 2010; Musick, 2011; Musick and Ellis, 2005; Naylor, 1992.

SPHYRNID CHARACTERISTICS:
1) head flattened and broad, hammer-shaped
2) eyes and nostrils near ends of hammer-like extensions
3) mouth relatively small, subterminal
4) usually one or two gill slits above pectoral-fin base
5) spiracles absent

ILLUSTRATED SPECIMEN:
A) *Sphyrna zygaena*, SIO 64–528, 1,035 mm TL (dorsal view)
B) head of *Sphyrna zygaena*, SIO 64–528 (lateral view).

HEXANCHIFORMES—Six-gill Sharks

The hexanchiforms were once thought to be the most primitive extant shark group, but recent research (e.g., Naylor et al., 2005; Vélez-Zuazo and Agnarsson, 2011) has shown them to be allied with the squaliform and related sharks. This group is characterized by six or seven gill slits, a single, spineless dorsal fin originating posterior to the origin of the pelvic fins, a large mouth, and small spiracles, which are located above and well posterior of the eyes. The Hexanchiformes comprises two families, four genera, and six species. The two species of frill sharks (Chlamydoselachidae) have a terminal mouth and the first pair of gill slits meeting across the throat. The cow sharks (Hexanchidae) are described in more detail below.

REFERENCES: de Carvalho, 1996; Naylor et al., 2005; Shirai, 1996; Vélez-Zuazo and Agnarsson, 2011.

HEXANCHIFORMES : HEXANCHIDAE—Cow Sharks

DIVERSITY: 3 genera, 4 species

REPRESENTATIVE GENERA: *Heptranchias, Hexanchus, Notorynchus*

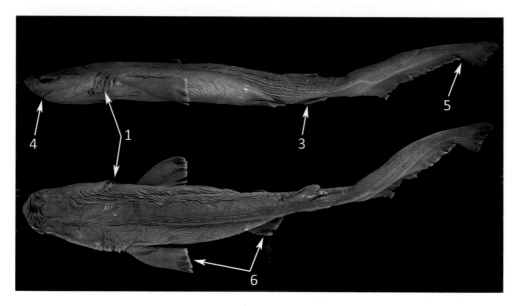

HEXANCHID CHARACTERISTICS:
1) six to seven long gill slits, all anterior to the pectoral fins
2) teeth of lower jaw compressed, wide, and serrated
3) anal fin smaller than dorsal fin, originating posterior to origin of dorsal fin
4) mouth inferior
5) distinct, subterminal notch in caudal fin
6) pectoral fins larger than pelvic fins

ILLUSTRATED SPECIMEN:
Hexanchus griseus, SIO 74–176, 838 mm TL (dorsal and lateral views)

DISTRIBUTION: Atlantic, Indian, and Pacific oceans

HABITAT: Marine; tropical to temperate; continental shelf to slope, occasionally deeper, usually demersal over soft or rocky bottoms

REMARKS: The cow sharks are known for their numerous gill slits (six or seven), and each of the four species is found over a broad geographic range. These sharks are yolk-sac viviparous (Musick, 2011; Musick and Ellis, 2005) with two species known to have litters of over 100 pups. Hexanchids feed on a variety of prey, from squids, crustaceans, and small bony fishes to elasmobranchs, seals, and small cetaceans. Some species are utilized both for their meat and liver oil, and they often are displayed in public aquariums.

REFERENCES: Compagno, in Carpenter, 2003; Compagno and Niem, in Carpenter and Niem, 1998; Compagno et al., in Fischer et al., 1995; Compagno et al., 2005; Musick, 2011; Musick and Ellis, 2005; Shirai, 1992a, 1996.

ECHINORHINIFORMES : ECHINORHINIDAE—Bramble Sharks

DIVERSITY: 1 family, 1 genus, 2 species

REPRESENTATIVE GENERA: *Echinorhinus*

DISTRIBUTION: Atlantic, Indian, and Pacific oceans

HABITAT: Marine; tropical to temperate; continental shelf to slope, benthic or demersal over soft bottoms

ECHINORHINIFORM CHARACTERISTICS:
1) many skin denticles large and thorn-like, in various places on body
2) two relatively small dorsal fins, positioned posteriorly, spines absent
3) all five gill slits anterior to pectoral fin
4) anal fin absent
5) head broad and depressed
6) spiracles small, well posterior to eyes

ILLUSTRATED SPECIMEN:
Echinorhinus cookei, SIO 60–378, 1,700 mm TL.

INSET: Close-up of skin showing thorn-like denticles

REMARKS: The bramble sharks, so-called because of their large denticles, were formerly placed in the Squaliformes, but recent authors have them independent of that group, and possibly closely related to the sawsharks (Vélez-Zuazo and Agnarsson, 2011). Both species have a broad geographic range, but they are rarely seen. These sharks are yolk-sac viviparous (Musick, 2011; Musick and Ellis, 2005), with one female *E. cookei* recorded with a litter of 114 pups; *E. brucus* is known to have up to 26 pups. Bramble sharks feed on bony fishes, small chondrichthyans, crustaceans, octopods, and squids, and *E. brucus* occasionally is utilized in fisheries.

REFERENCES: Compagno, in Carpenter, 2003; Compagno and Niem, in Carpenter and Niem, 1998; Compagno et al., in Fischer et al., 1995; Compagno et al., 2005; de Carvalho, 1996; Musick, 2011; Musick and Ellis, 2005; Vélez-Zuazo and Agnarsson, 2011.

SQUALIFORMES—Dogfish Sharks

The dogfish sharks include six families, at least 24 genera, and 130 species. Some authors (e.g., Compagno, 2005; Ebert, 2003; Eschmeyer and Fong, 2013) also include the two species of bramble sharks (Echinorhinidae) among the squaliforms, but others (e.g., de Carvalho, 1996; Vélez-Zuazo and Agnarsson, 2011) place the bramble sharks in their own order. Both groups are characterized by the lack of an anal fin, presence of spiracles, and the location of all five gill slits anterior to the origin of the pectoral fin. Among the squaliforms, the gulper sharks (Centrophoridae) and the lantern sharks (Etmopteridae) are characterized by two dorsal fins, each with a single grooved spine, and by very large eyes. The lantern sharks usually have light organs along the ventral aspect of the body. Some species of sleeper sharks (Somniosidae) have dorsal-fin spines while others do not, and all species are characterized by abdominal ridges between the pectoral and pelvic fins. The roughsharks (Oxynotidae) are easily distinguished by their triangular body (in cross section) and their large, sail-like dorsal fins. The dogfish sharks *sensu stricto* (Squalidae) and the kitefin sharks (Dalatiidae) are described below.

REFERENCES: de Carvalho, 1996; Vélez-Zuazo and Agnarsson, 2011.

SQUALIFORMES : SQUALIDAE—Dogfish Sharks

DIVERSITY: 2 genera, 30 species

REPRESENTATIVE GENERA: *Cirrhigaleus, Squalus*

DISTRIBUTION: Atlantic, Indian, and Pacific oceans

HABITAT: Marine; tropical to temperate; continental shelf to continental slope and sea-mounts, coastal, demersal over soft bottoms

REMARKS: Dogfish sharks have strong dorsal-fin spines, some of which are venomous. Some species commonly form large, social schools that are known to attack and dismember large prey. All dogfishes are yolk-sac viviparous (Musick, 2011; Musick and Ellis, 2005), and one species, *Squalus acanthias,* has extremely large eggs and a gestation period of up to two years. This low reproductive potential, together with their propensity for forming large schools, makes dogfishes quite vulnerable to overfishing, particularly given their importance as a commercially exploited group of sharks (IUCN, 2013).

REFERENCES: Compagno, 2001, 2005; Compagno, in Carpenter, 2003; Compagno and Niem, in Carpenter and Niem, 1998; Compagno et al., in Fischer et al., 1995; Compagno et al., 2005; Musick, 2011; Musick and Ellis, 2005; Shirai, 1992a.

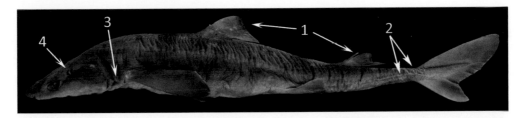

SQUALID CHARACTERISTICS:
1) two dorsal fins, each with a strong, smooth spine
2) caudal peduncle with lateral keels and a precaudal pit
3) all five gill slits usually anterior to pectoral-fin base
4) spiracles large
5) anal fin absent
6) nictitating membrane absent

ILLUSTRATED SPECIMEN:
Squalus suckleyi, SIO 08–138, 740 mm TL

SQUALIFORMES : DALATIIDAE—Kitefin Sharks

DIVERSITY: 6 genera, 10 species

REPRESENTATIVE GENERA: *Dalatius, Isistius, Squaliolus*

DISTRIBUTION: Atlantic, Indian, and Pacific oceans

HABITAT: Marine; tropical to temperate; continental shelf to continental slope, coastal to oceanic, epipelagic to bathypelagic

REMARKS: Kitefin sharks are some of the smallest of all sharks; species in the genus *Squaliolus* reach maximum sizes of only 22–28 cm TL. These cigar-shaped sharks generally eat small fishes, squids, and crustaceans, but the highly specialized cookiecutter sharks (*Isistius* spp.) attack much larger prey, removing bite-sized pieces from live marine mammals and large-bodied fishes. Kitefin sharks are yolk-sac viviparous (Musick, 2011; Musick and Ellis, 2005) and probably use their ventral light organs to achieve countershading, decreasing their visual profile from below.

REFERENCES: Compagno, 2005; Compagno, in Carpenter, 2003; Compagno and Niem, in Carpenter and Niem, 1998; Compagno et al., 2005; Musick, 2011; Musick and Ellis, 2005; Shirai, 1992a.

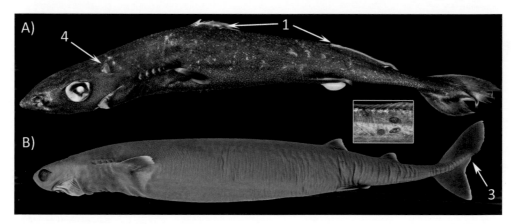

DALATIID CHARACTERISTICS:
1) two dorsal fins, without spines, or with a spine in the first dorsal fin only
2) luminous organs usually present along ventral surface
3) caudal fin with a pronounced notch
4) spiracles large
5) anal fin absent
6) nictitating membrane absent

ILLUSTRATED SPECIMENS:
A) *Squaliolus aliae*, DE 0508, 117 mm TL
B) *Isistius brasiliensis*, SIO 69–345, 470 mm TL

INSET: Flank of Oarfish (*Regalecus russelii*, SIO 13–259) with multiple wounds caused by *Isistius brasiliensis*.

SQUATINIFORMES : SQUATINIDAE—Angel Sharks

DIVERSITY: 1 family, 1 genus, 22 species

REPRESENTATIVE GENUS: *Squatina*

DISTRIBUTION: Atlantic, Indian, and Pacific oceans

HABITAT: Marine; tropical to temperate; continental shelf to continental slope, benthic on soft substrates

REMARKS: These distinctive sharks resemble rays in many respects, but unlike rays, their pectoral fins are not attached to their heads. In keeping with their benthic lifestyle, their well-developed spiracles are used for respiration, and there are barbels associated with their nostrils. Angel sharks are ambush predators that rely on crypsis to attack their unsuspecting prey (Fouts and Nelson, 1999). They are one of only a few groups of sharks that have protrusible jaws and are capable of producing strong negative pressure for suction feeding. Angel sharks are yolk-sac viviparous (Musick, 2011; Musick and Ellis, 2005). Shirai (1992b) studied their phylogenetic relationships.

REFERENCES: Compagno, 2001, 2005; Compagno, in Carpenter, 2003; Compagno and Niem, in Carpenter and Niem, 1998; Compagno et al., 2005; Fouts and Nelson, 1999; Musick, 2011; Musick and Ellis, 2005; Shirai, 1992b.

SQUATINIFORM CHARACTERISTICS:
1) body strongly depressed, ray-like
2) pectoral fins separate from head
3) eyes dorsal
4) two dorsal fins, roughly equal in size, positioned posteriorly
5) spiracles large
6) caudal peduncle with strong, lateral keels
7) anal fin absent

ILLUSTRATED SPECIMEN:
Squatina californica, SIO 65–305, 435 mm TL (dorsal view)

PRISTIOPHORIFORMES : PRISTIOPHORIDAE—Saw Sharks

DIVERSITY: 1 family, 2 genera, 7 species

REPRESENTATIVE GENERA: *Pliotrema, Pristiophorus*

DISTRIBUTION: Atlantic, Indian, and Pacific oceans (excluding eastern Pacific)

HABITAT: Marine; tropical to temperate; continental shelf to slope, benthic on soft substrates

REMARKS: Saw sharks, so named because of their saw-like snouts, are superficially similar to the sawfishes (Pristiformes), but differ in having the pectoral fins separate from the head, lateral gill slits, and long barbels on the ventral side of the rostrum. These sharks are yolk-sac viviparous (Musick, 2011; Musick and Ellis, 2005) and have litters of 7–17 pups. The large rostral teeth lie flat until after birth. The one species of *Pliotrema, P. warreni,* is unusual in having six gill slits. These sharks feed on small fishes, crustaceans, and squids, and occasionally are utilized as food fishes.

REFERENCES: Compagno, 1984a; Compagno, in Carpenter and Niem, 1998; Compagno, in Carpenter, 2003; Compagno, 2005; Compagno et al., 2005; Musick, 2011; Musick and Ellis, 2005.

PRISTIOPHORIFORM CHARACTERISTICS:
1) snout extremely long, depressed, with rows of lateral and ventral teeth
2) barbels long, ventral and anterior to nostrils
3) all five (or six) gill slits lateral and usually anterior to pectoral fin
4) spiracles large
5) lateral ridge on caudal peduncle
6) anal fin absent
7) head depressed, body cylindrical

ILLUSTRATED SPECIMEN:
Pristiophorus japonicus, SIO 92–164, 1,128 mm TL (dorsal view)

BATOIDEA—Skates and Rays

The Batoidea includes over 650 species of skates and rays that are classified in four orders, 17 families, and over 70 genera. They are characterized by a variety of features including a dorso-ventrally flattened head and body, enlarged pectoral fins contiguous with the head, mouth and gill slits opening on the ventral side of the head, and eyes and spiracles placed on the dorsal side of the head. The monophyly of this group is well established (Aschliman et al., 2012; McEachran and Aschliman, 2004; McEachran et al., 1998; Naylor et al., 2005) and the group has long been recognized under a variety of names including the Hypotremata, Batidoidimorpha, and Rajiformes *sensu lato* (Nelson, 2006).

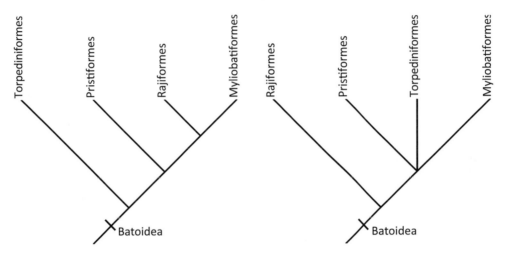

Hypothesized phylogenetic relationships of the Batoidea (rays) after (left) McEachran and Aschliman (2004) and (right) Aschliman et al. (2012).

TORPEDINIFORMES—Electric Rays

There are four families, 11 genera, and 67 species of electric rays, characterized by a nearly circular disc, electric organs located on the pectoral fins, and a completely naked body (without denticles or spines). The Narcinidae is described in more detail below.

REFERENCES: McEachran and Aschliman, 2004

DIVERSITY: 4 genera, 31 species

REPRESENTATIVE GENERA: *Benthobatis, Diplobatis, Discopyge, Narcine*

DISTRIBUTION: Atlantic, Indian and Pacific oceans

HABITAT: Marine; tropical to warm temperate; continental shelf to continental slope, benthic on soft substrates

REMARKS: Numbfishes use their electric organs for both defense and feeding. They can be distinguished from the similar torpedo electric rays (Torpedinidae) by their similar-sized dorsal fins (first fin larger in torpedo rays), relatively smaller caudal fin, and thin (rather than thick) outer margin of the pectoral-fin disc. Numbfishes prey on benthic invertebrates and small fishes, using their protrusible mouths to provide suction for removing organisms from soft sediments. Numbfishes are yolk-sac viviparous (Musick, 2011; Musick and Ellis, 2005).

REFERENCES: Compagno, 2005; de Carvalho et al., in Carpenter and Niem, 1999; McEachran, in Fischer et al., 1995; McEachran and de Carvalho, in Carpenter, 2003; Musick, 2011; Musick and Ellis, 2005.

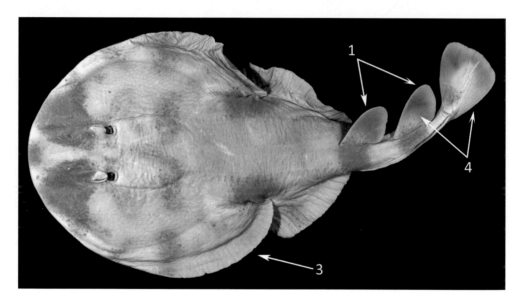

NARCINID CHARACTERISTICS:
1) two dorsal fins of equal size, positioned posteriorly
2) mouth slot-shaped, jaws protrusible
3) pectoral fins thin around outer edges
4) caudal fin and dorsal fins similar in size
5) lobes of caudal fin continuous

ILLUSTRATED SPECIMEN:
Narcine brasiliensis, SIO 67–89, 300 mm TL

PRISTIFORMES : PRISTIDAE—Sawfishes

DIVERSITY: 1 family, 2 genera, 7 species

REPRESENTATIVE GENERA: *Anoxypristis, Pristis*

DISTRIBUTION: Atlantic, Indian, and Pacific oceans

HABITAT: Marine and occasionally in freshwater; tropical; continental shelf and coastal, demersal and benthic on soft substrates

REMARKS: The remarkably large rostral blade with teeth on either side distinguishes sawfishes from all other fishes except sawsharks, from which they can be differentiated by their absence of barbels and by their pectoral fins being connected to the head. These impressive predators use their "saw" for disabling swimming prey or digging for buried prey in soft sediments. Sawfishes are known to enter freshwater, with some individuals captured more than 1,000 km upriver, and are known to reproduce in at least one freshwater lake (Thorson, 1976). Sawfishes are yolk-sac viviparous (Musick, 2011; Musick and Ellis, 2005). All seven species are listed as critically endangered by the IUCN (2013) as a result of overfishing (sawfishes are especially susceptible to gill-nets) and habitat degradation.

REFERENCES: de Carvalho, 2003; Compagno, 2005; Compagno and Last, in Carpenter and Niem, 1999; McEachran and de Carvalho, in Carpenter, 2003; Musick, 2011; Musick and Ellis, 2005; Thorson, 1976.

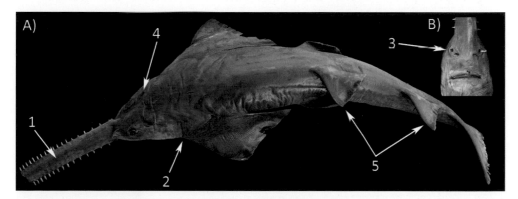

PRISTIFORM CHARACTERISTICS:
1) snout elongate, blade-like, with single row of large teeth on each side
2) body large and shark-like, with anterior margin of pectoral fins attached to head
3) nostrils well anterior and not connected to mouth
4) spiracles large
5) two large dorsal fins of equal size, widely separated
6) barbels absent

ILLUSTRATED SPECIMEN:
A) *Pristis pectinata,* UAZ uncatalogued, 940 mm to end of broken snout (dorsal view)
B) head of *Pristis pectinata,* UAZ uncatalogued (ventral view)

RAJIFORMES—Skates

The skates are the most diverse order of chondrichthyan fishes, with 361 species and over 30 genera in four families. Their disc is diamond- or heart-shaped, and the dorsal surface of the tail has one or more longitudinal rows of thorns but no stinging spine. Skates have large, well-developed spiracles that usually contain visible pseudobranchs.

REFERENCES: Aschliman et al., 2012; McEachran and Aschliman, 2004.

RAJIFORMES : RHINOBATIDAE—Guitarfishes

DIVERSITY: 11 genera, 62 species

REPRESENTATIVE GENERA: *Aptychotrema, Platyrhinoides, Rhina, Rhinobatos, Zapteryx*

DISTRIBUTION: Atlantic, Indian, and Pacific oceans

HABITAT: Marine, rarely entering freshwater; tropical to temperate; continental shelf to continental slope, benthic on soft substrates and rocky reefs

REMARKS: Guitarfishes are elongate batoids with a wedge-shaped head and a wide tail. Their lower caudal-fin lobe is not well defined. In addition to dermal denticles covering the body and fins, they can have enlarged thorn-like spines on the dorsal surface of the snout and the midline of the body and tail. Some authors (e.g., McEachran and Aschliman, 2004; Nelson, 2006) separate them into two or more families, considered here as subfamilies. These include the monotypic Bowmouth Guitarfish (Rhininae), the six species of wedgefishes (Rhynchobatinae), and the six species of thornbacks (Platyrhininae), once

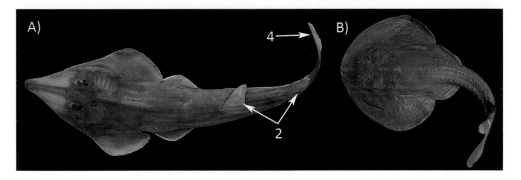

RHINOBATID CHARACTERISTICS:
1) body slightly depressed, more shark-like than ray-like, tail essentially an extension of body in most
2) two dorsal fins of equal size, usually widely separated
3) nostrils well anterior and usually not connected to mouth
4) caudal fin well developed
5) jaws protrusible
6) dorsal surface of body and fins covered with dermal denticles

ILLUSTRATED SPECIMENS:
A) *Rhinobatos productus,* SIO 09–201, 817 mm TL (Rhinobatinae)
B) *Platyrhinoidis triseriata,* SIO 54–188, 660 mm TL (Platyrhininae)

considered members of the Myliobatiformes (Nelson, 2006). Guitarfishes feed on a variety of benthic invertebrates, as well as small bony fishes. They are yolk-sac viviparous (Musick, 2011; Musick and Ellis, 2005).

REFERENCES: Compagno, 2005; Compagno and Last, in Carpenter and Niem, 1999; McEachran, in Fischer et al., 1995; McEachran and Aschliman, 2004; McEachran and de Carvalho, in Carpenter, 2003; Musick, 2011; Musick and Ellis, 2005.

RAJIFORMES : RAJIDAE—Skates

DIVERSITY: 18 genera, 179 species

REPRESENTATIVE GENERA: *Bathyraja, Breviraja, Dipturus, Raja*

DISTRIBUTION: Arctic, Atlantic, Indian, and Pacific oceans

HABITAT: Marine (one species occurs in freshwater); tropical to polar; continental shelf to abyssal plain, benthic on soft substrates

REMARKS: The Rajidae is the largest family of chondrichthyan fishes and constitutes approximately 15% of all chondrichthyan diversity (Ebert and Compagno, 2007). Skates exhibit a wide variety of disc shapes and often have sharp dermal thorn-like spines,

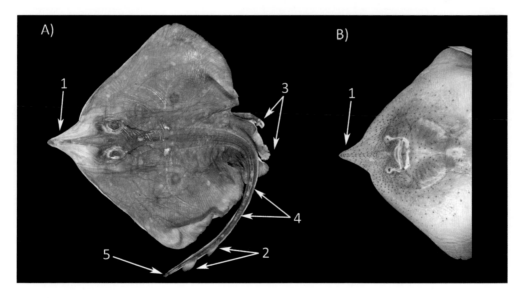

RAJID CHARACTERISTICS:
1) conspicuous, strong rostral cartilage extending from cranium
2) usually two dorsal fins, positioned posteriorly on tail
3) pelvic fins with two lobes or occasionally with a single lateral lobe
4) thorn-like spines usually present on dorsal surface (at least along midline of tail)
5) caudal fin generally reduced

ILLUSTRATED SPECIMEN:
A) *Raja stellulata,* SIO 61–513, 654 mm DW (dorsal view)
B) head of *Raja stellulata,* SIO 61–513 (ventral view)

particularly along the tail. Males can have additional spines near the margins of the pectoral fins. Their phylogenetic relationships have been studied by Ebert and Compagno (2007) and McEachran and Dunn (1998) and their biology was reviewed by Ebert and Sulikowski (2008). Skates are predatory, generally feeding on a wide variety of benthic invertebrates and bony fishes. They are oviparous (Musick, 2011; Musick and Ellis, 2005), laying rectangular, keratinoid egg cases, commonly known as "mermaid's purses."

REFERENCES: Compagno, 2005; Ebert and Compagno, 2007; Ebert and Sulikowski, 2008; Last and Compagno, in Carpenter and Niem, 1999; McEachran, in Fischer et al., 1995; McEachran and de Carvalho, in Carpenter, 2003; McEachran and Dunn, 1998; Musick, 2011; Musick and Ellis, 2005.

MYLIOBATIFORMES—Stingrays

The stingrays include eight families, 27 genera, and 220 species. The distinguishing character of this group is the serrated, often venomous spine located on the tail of most species. The sixgill stingrays (Hexatrygonidae) are characterized by six gill openings and an extremely elongate snout; the river stingrays (Potamotrygonidae) are restricted to South American freshwaters. The round stingrays (Urolophidae) are morphologically similar to the American round stingrays (Urotrygonidae, described below), but occur exclusively in the western Pacific. The Deepwater Stingray (Plesiobatidae) and the river stingrays also resemble the American round stingrays. Along with the urotrygonids, three additional families (Myliobatidae, Gymnuridae, and Dasyatidae) are described below. Their phylogenetic relationships were studied by Dunn et al. (2003) and Aschliman et al. (2012).

REFERENCES: Aschliman et al., 2012; de Carvalho et al., 2004; Dunn et al., 2003.

MYLIOBATIFORMES : UROTRYGONIDAE—American Round Stingrays

DIVERSITY: 2 genera, 17 species

REPRESENTATIVE GENERA: *Urobatis, Urotrygon*

DISTRIBUTION: Western Atlantic and eastern Pacific oceans

HABITAT: Marine, occasionally in freshwater; tropical to warm temperate; continental shelf, benthic on soft bottoms

REMARKS: The well-developed caudal fin, tail of moderate length, and serrated, venomous spines distinguish American round stingrays from nearly all other rays except the Urolophidae. The closely related urolophids (round stingrays) are restricted to the western Pacific Ocean, while the Urotrygonidae are a New World group (McEachran et al., 1996). American round stingrays are predators, feeding primarily on benthic crustaceans and bottom fishes. These rays are lipid histotrophs (Musick, 2011; Musick and Ellis, 2005).

REFERENCES: Compagno, 2005; McEachran, in Fischer et al., 1995; McEachran and Aschliman, 2004; McEachran and de Carvalho, in Carpenter, 2003; McEachran et al., 1996; Musick, 2011; Musick and Ellis, 2005.

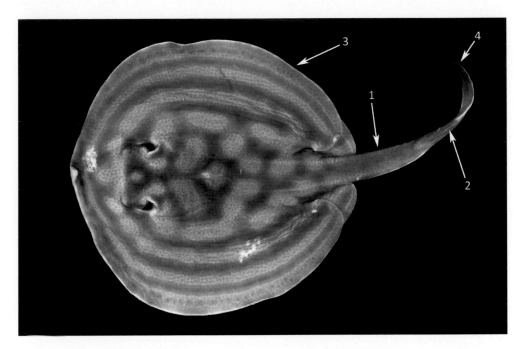

UROTRYGONID CHARACTERISTICS:
1) tail relatively thick at base, its length approximately equal to disc length
2) one or more serrated, venomous spines on tail
3) disc more or less circular
4) caudal fin well developed
5) dorsal fins absent

ILLUSTRATED SPECIMEN:
Urobatis concentricus, SIO 65–297, 180 mm DW

MYLIOBATIFORMES : DASYATIDAE—Whiptail Stingrays

DIVERSITY: 8 genera, 88 species

REPRESENTATIVE GENERA: *Dasyatis, Himantura, Neotrygon, Taeniura, Urogymnus*

DISTRIBUTION: Atlantic, Indian, and Pacific oceans

HABITAT: Marine to freshwater; tropical to temperate; continental shelf to continental slope, normally benthic over soft bottoms but one species oceanic (pelagic)

REMARKS: The long, whip-like tail and serrated, venomous spines distinguish whiptail stingrays from nearly all other rays. Like other chondrichthyan fishes, whiptail stingrays are predatory, feeding primarily on benthic invertebrates. Some marine species are euryhaline, entering freshwater. Whiptail stingrays are lipid histotrophs (Musick, 2011; Musick and Ellis, 2005), with gestation periods up to one year.

REFERENCES: Compagno, 2005; de Carvalho et al., 2004; Last and Compagno, in Carpenter and Niem, 1999; Lovejoy, 1996; McEachran, in Fischer et al., 1995; McEachran and de Carvalho, in Carpenter, 2003; Musick, 2011; Musick and Ellis, 2005; Rosenberger, 2001.

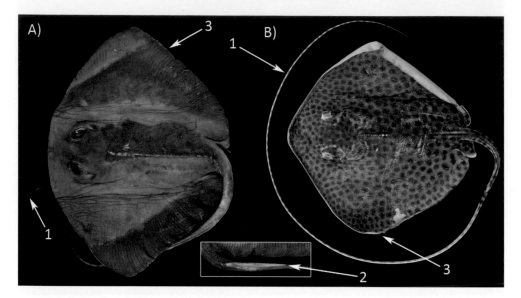

DASYATID CHARACTERISTICS:
1) tail extremely long, usually much longer than disc length
2) one or more serrated, venomous spines on tail
3) pectoral fins thinning toward margins
4) dorsal and caudal fins absent
5) fleshy papillae present in mouth

ILLUSTRATED SPECIMENS:
A) *Pteroplatytrygon violacea,* SIO 72–82, 226 mm DW
B) *Himantura uarnak,* DE 0508, 345 mm DW

INSET: Spine of *Pteroplatytrygon violacea* (SIO 74–79)

MYLIOBATIFORMES : GYMNURIDAE—Butterfly Rays

DIVERSITY: 1 genus, 14 species

REPRESENTATIVE GENUS: *Gymnura*

DISTRIBUTION: Atlantic, Indian, and Pacific oceans

HABITAT: Marine, rarely in freshwater; tropical to warm temperate; continental shelf, benthic on soft bottoms

REMARKS: Butterfly rays are characterized by the unmistakable, wide disc and greatly reduced tail. Jacobson and Bennett (2009) recently reviewed their systematics and synonomized *Aetoplatea* with *Gymnura*, recognizing only a single genus. These rays are predatory and feed primarily on benthic invertebrates including crustaceans and bivalves, as well as small bottom fishes. Like other myliobatiforms, butterfly rays are lipid histotrophs (Musick, 2011; Musick and Ellis, 2005).

REFERENCES: Compagno, 2005; Compagno and Last, in Carpenter and Niem, 1999; Jacobson and Bennett, 2009; McEachran, in Fischer et al., 1995; McEachran and de Carvalho, in Carpenter, 2003; Musick, 2011; Musick and Ellis, 2005.

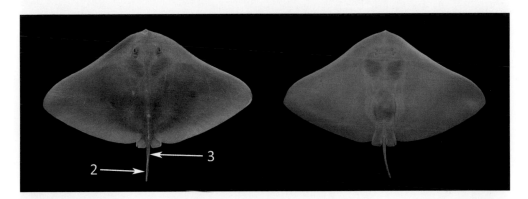

GYMNURID CHARACTERISTICS:
1) disc extremely wide, much wider than long
2) tail greatly reduced
3) venomous spines on tail present or absent
4) dorsal fin reduced or absent
5) caudal fin absent

ILLUSTRATED SPECIMEN:
Gymnura marmorata, SIO 13–237, 215 mm DW (dorsal and ventral views)

MYLIOBATIFORMES : MYLIOBATIDAE—Eagle Rays

DIVERSITY: 7 genera, 44 species

REPRESENTATIVE GENERA: *Aetobatis, Manta, Mobula, Myliobatis, Rhinoptera*

DISTRIBUTION: Atlantic, Indian and Pacific oceans

HABITAT: Marine; tropical to warm temperate; coastal to oceanic, continental shelf to continental slope, pelagic or demersal over soft bottoms and reefs

REMARKS: The eagle rays are divided into three distinctive groups: the cownose rays (Rhinopterinae), the mantas and devil rays (Mobulinae), and the true eagle rays (Myliobatinae), the latter considered by some (e.g., Compagno, 2005; Naylor et al., 2012) to be a separate family. These fishes can be very large bodied, with *Manta birostris,* the largest ray in the world, reaching a disc width of over 7 m. Oceanic species filter feed on large zooplankton and small fishes, while coastal species often specialize on benthic invertebrates, especially bivalves. All species are lipid histotrophs (Musick, 2011; Musick and

Ellis, 2005), with litters of up to six young; the mantas and devil rays have litters of only one.

REFERENCES: Compagno, 2005; Compagno and Last, in Carpenter and Niem, 1999; de Carvalho et al., 2004; McEachran and de Carvalho, in Carpenter, 2003; McEachran et al., 1998; McEachran and Notarbartolo-Di-Sciara, in Fischer et al., 1995; Musick, 2011; Musick and Ellis, 2005; Notarbartolo-Di-Sciara, 1987.

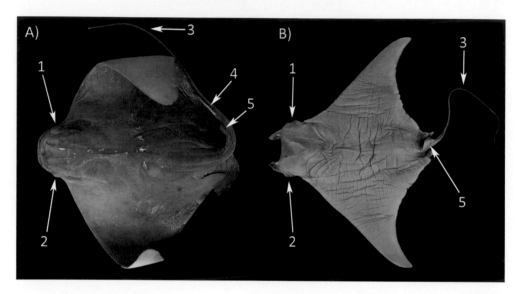

MYLIOBATID CHARACTERISTICS:
1) head raised above surface of disc
2) eyes and spiracles on sides of head
3) tail slender, often whip-like, usually longer than disc
4) serrated, venomous spine on tail (absent in some)
5) small to moderately sized dorsal fin on base of tail
6) caudal fin absent

ILLUSTRATED SPECIMENS:
A) *Myliobatis californica,* SIO 50–26B, 720 mm DW
B) *Mobula tarapacana,* SIO 83-113, 405 mm DW

Osteichthyes

Bony Fishes

The clade Osteichthyes may at first seem misnamed as it includes not only what are readily recognized as "bony fishes" but also the entire lineage of tetrapods. However, Osteichthyes refers to the ancestor and all descendants (i.e., a monophyletic group) of a lineage that is the sister group of the cartilaginous fishes. These two great lineages of fishes differ in several fundamental features, most notably in the composition of their skeleton. Osteichthyans have a bony skeleton while chondricthyans have a skeleton formed entirely of cartilage. In addition, the neurocranium of osteichthyans has evident sutures (sutures are absent in chondrichthyans); their fin rays (if present) are segmented and derived from the dermis (termed "lepidotrichia"), while those of chondrichthyans are unsegmented and epidermal in origin (termed "ceratotrichia"). Finally, osteichthyans typically have a gas bladder or its derivative (lungs in tetrapods), a structure lacking in chondrichthyans. The Osteichthyes comprises two large lineages, the Sarcopterygii, or lobe-finned fishes and tetrapods, and the Actinopterygii, or ray-finned fishes (Stiassney et al., 2004).

SARCOPTERYGII—Lobe-finned Fishes

The Sarcopterygii is a major group of vertebrates that includes the coelacanths, lungfishes, and tetrapods and comprises over 25,000 species. These "lobe-finned fishes" are characterized by the presence of enamel on the teeth, a unique skeletal support for the paired fins (or limbs) that includes a central axis of bone, and autostylic jaw suspension in which the upper jaw is fused with the skull. We treat the two most "fish-like" orders of sarcopterygians, the Ceratodontiformes, or lungfishes, and the Coelacanthiformes, or coelacanths. The tetrapods are not considered here.

COELACANTHIFORMES : LATIMERIIDAE—Coelacanths

DIVERSITY: 1 family, 1 genus, 2 species

REPRESENTATIVE GENUS: *Latimeria*

DISTRIBUTION: Indian and western Pacific oceans, off southern Africa and Indonesia

HABITAT: Marine; tropical to temperate; continental shelf and continental slope, demersal on deep rocky reefs

REMARKS: The living coelacanths represent a group thought to have become extinct 80 million years ago. An extant member of this unique group of fishes was first discovered in

COELACANTHIFORM CHARACTERISTICS:
1) second dorsal, anal, pelvic, and pectoral fins lobe-like
2) first dorsal fin with hollow spines
3) caudal fin diphycercal and in three lobes
4) double gular plate between left and right sides of lower jaw
5) large, bony, cosmoid scales
6) intracranial joint

ILLUSTRATED SPECIMEN:
Latimeria chalumnae, SIO 75–347, 950 mm TL

1938 in the western Indian Ocean, while a second species was first captured by scientists in 1998 in Indonesia (Holder et al., 1999; Pouyaurd et al., 1999). Coelacanths have been and continue to be studied extensively. They are the only vertebrates with an intracranial joint, possibly allowing vertical movement of the head in order to increase the size of the gape. They are piscivorous and utilize an electroreceptive sense to enhance their predation on small fishes. Coelacanths are unusual in having a rectal gland and high levels of urea in the blood. They are internal fertilizers (Smith et al., 1975), may be monogamous (Lampert et al., 2013), and the females give birth to 5–26 well-developed young. Their entire genome was recently sequenced (Amemiya et al., 2013). They are endangered as a result of their low reproductive potential and small geographic range; *Latimeria chalumnae* is listed as critically endangered by the IUCN. The stories of the discovery of both species are full of drama and intrigue (Nelson, 2006; Smith, 1956; Thomson, 1991).

REFERENCES: Amemiya et al., 2013; Bruton, 1995; Cloutier and Ahlberg, 1996; Forey, 1980, 1991, 1998; Holder et al., 1999; Lampert et al., 2013; McCosker and Lagios, 1979; Musick et al. 1991; Pouyaurd et al., 1999; Smith et al., 1975; Smith, 1940, 1956; Thomson, 1991.

CERATODONTIFORMES—Lungfishes

DIVERSITY: 3 families, 3 genera, 6 species

REPRESENTATIVE GENERA: *Lepidosiren, Neoceratodus, Protopterus*

DISTRIBUTION: Sub-Saharan Africa (*Protopterus*), South America (*Lepidosiren*), and Australia (*Neoceratodus*)

HABITAT: Freshwater; tropical to subtropical; benthic to demersal over soft bottoms

REMARKS: The three families of living lungfishes, also called the Dipnoi, have one or two lungs used for either facultative or, in some cases, obligate air breathing (Graham, 1997). The African lungfishes (Protopteridae) are characterized by the presence of slender, elongate pectoral and pelvic fins, small scales, and six gill arches, while the montotypic South American Lungfish (Lepidosirenidae) also has slender pectoral and pelvic fins, small scales, and paired lungs, but only five gill arches. In contrast, the monotypic Australian Lungfish (Ceratodontidae) has paddle-like pectoral and pelvic fins, larger scales, and a single, unpaired lung. The African lungfishes are the sister group of the single species of South American Lungfish, and are placed together in the order Lepidosireniformes by some authors. These two lineages are the sister group of the single extant Australian Lungfish (Ceratodontidae). Nelson (2006) includes all three families in the order Ceratodontiformes. African lungfishes are large, with at least one species reaching lengths up to 1.8 m TL. During the dry season, individuals burrow into the mud and aestivate inside mucous cocoons. The pelvic fins of reproductive males of the South American Lungfish become highly vascularized and feather-like, infusing oxygen into the water where young are developing. Lungfishes are omnivorous, feeding on other fishes, frogs, mollusks, and in some cases, plant material, including seeds.

REFERENCES: Bemis et al., 1987; Cloutier and Ahlberg, 1996; Graham, 1997; Nelson, 2006.

CERATODONTIFORM CHARACTERISTICS:
1) pectoral and pelvic fins slender (A and B) or flattened (C)
2) gular plate absent
3) premaxilla and maxilla absent
4) one or two lungs
5) five or six gill arches

ILLUSTRATED SPECIMENS:
A) *Protopterus aethiopicus,* CAS 46377, 854 mm TL (Protopteridae—African lungfishes)
B) *Lepidosiren paradoxa,* CAS 14001-7, 549 mm TL (Lepidosirenidae—South American Lungfish)
C) *Neoceratodus forsteri,* CAS 18189, 748 mm TL (Ceratodontidae—Australian Lungfish)

ACTINOPTERYGII

Ray-finned Fishes

The Actinopterygii includes approximately 33,000 valid species (Eschmeyer and Fong, 2013) classified by Nelson (2006) in 453 families. They are found in all aquatic habitats occupied by vertebrates and range vastly in body size from a tiny species of the gobiiform genus *Schindleria* that grows no longer than 1 cm (and ca 0.7 mg) and spends its entire short life in the plankton, to the Oarfish (*Regalecus*) that grows to over 8 m in length, to the Ocean Sunfish (*Mola mola*) that weighs up to 2,300 kg. Given this enormous diversity, it is not surprising that the group is difficult to characterize morphologically. Extant actinopterygians have enlarged basal elements in the pectoral fins and fused basal elements in the pelvic fins (Lauder and Liem, 1983; Patterson, 1982). Early lineages (i.e., non-teleosts) have a single dorsal fin (variously lost or divided in many) and ganoid scales (variously lost or modified in most species). Several studies (e.g., Faircloth et al., 2013; Hurley et al., 2007; Inoue et al., 2003; Lauder and Liem, 1983) have examined the relationships among the major lineages of the Actinopterygii.

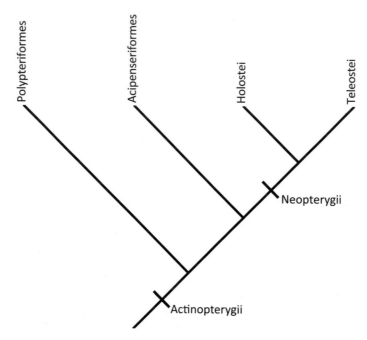

Hypothesized phylogenetic relationships of the early lineages of
actinopterygians (after Faircloth et al., 2013).

Actinopterygii I

Lower Ray-finned Fishes

POLYPTERIFORMES : POLYPTERIDAE—Bichirs

DIVERSITY: 1 family, 2 genera, 12 species

REPRESENTATIVE GENERA: *Erpetoichthys* (= *Calamoichthys*), *Polypterus*

DISTRIBUTION: Africa

HABITAT: Freshwater; tropical; demersal over soft bottoms

REMARKS: Bichirs are thought to be the sister group of all other actinopterygians, exhibiting many unique characters. All bichirs have lungs, an intestinal spiral valve, a skeleton of mostly cartilage, and a uniquely divided dorsal fin. Pelvic fins are present in most species but absent in one (*Erpetoichthys calabaricus*). Bichirs are carnivorous and feed on other

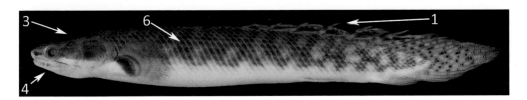

POLYPTERIFORM CHARACTERISTICS:
1) five to eighteen dorsal finlets, each with a single spine and one or more soft rays
2) four gill arches
3) spiracles large
4) two gular plates
5) branchiostegal rays absent
6) ganoid scales
7) maxilla fused to skull

ILLUSTRATED SPECIMEN:
Polypterus palmas, CU 87580, 126 mm TL

fishes, mollusks, and crustaceans. They are restricted to Africa, are known to reach lengths of up to 90 cm, and are often seen in the aquarium trade.

REFERENCES: Britz and Johnson, 2003; Daget et al., 2001; Gayet et al., 2002; Gosse, 1984, 1988.

ACIPENSERIFORMES—Sturgeons and Paddlefishes

The sturgeons and paddlefishes are an ancient lineage of fishes that evolved in the Permian over 250 million years ago. They and the Polypteriformes have a complex mixture of traits not seen in other ray-finned fishes including a heterocercal tail, skeleton of mostly cartilage, and spiral valve in the intestine. In addition, the Acipenseriformes have a well-developed rostrum, lack an opercle, have a reduced or absent preopercle, and lack branchiostegal rays. The group includes two families, six genera, and 28 extant species. Living acipenseriforms are restricted to the Northern Hemisphere where they inhabit coastal areas and large river systems (Bemis et al., 1997).

REFERENCES: Bemis et al., 1997; Chen and Arratia, 1994; Grande and Bemis, 1996

ACIPENSERIFORMES : ACIPENSERIDAE—Sturgeons

DIVERSITY: 4 genera, 26 species

REPRESENTATIVE GENERA: *Acipenser, Huso, Pseudoscaphirhynchus, Scaphirhynchus*

DISTRIBUTION: Northern Hemisphere temperate (except Greenland)

HABITAT: Anadromous or freshwater; temperate; demersal to benthic on soft bottoms

REMARKS: Five rows of bony scutes along the body distinguish sturgeons from all other fishes. These large freshwater/anadromous fishes inhabit lakes, slow-moving rivers, and coastal areas and feed on mollusks, crustaceans, insect larvae, and occasionally plants. The long-lived species of the genus *Huso* are some of the largest freshwater fishes in the world, reaching 8 m TL and over 1,500 kg, and are the source of what is considered the world's finest caviar. As a result of the caviar fishery and habitat destruction, nearly all sturgeons are vulnerable to extinction and at least 16 species are listed as critically endangered by the IUCN (2013).

REFERENCES: Bemis et al., 1997; Birstein et al., 2002; Boschung and Mayden, 2004; Choudhury and Dick, 1998; Grande and Bemis, 1996; Hilton et al., 2011; Hochleithner and Gessner, 2001; IUCN, 2013.

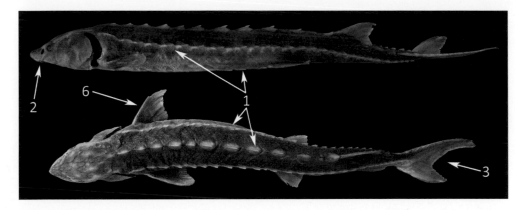

ACIPENSERID CHARACTERISTICS:
1) five rows of bony scutes along body
2) four oral barbels
3) caudal fin heterocercal
4) gular plates absent
5) teeth absent in adults
6) pectoral fin with soft rays fused into an anterior spine-like element
7) skeleton largely composed of cartilage

ILLUSTRATED SPECIMEN:
Acipenser medirostris, SIO 62–155, 960 mm TL (lateral and dorsal views).

ACIPENSERIFORMES : POLYODONTIDAE—Paddlefishes

DIVERSITY: 2 genera, 2 species

REPRESENTATIVE GENERA: *Polyodon, Psephurus*

DISTRIBUTION: Mississippi River and Yangtze River basins

HABITAT: Freshwater but can tolerate some salinity; temperate; pelagic in slow-moving rivers

REMARKS: Paddlefishes are large (up to 3 m in length, with reports of much larger specimens), riverine species characterized by a long spatula-like snout. Although similar in appearance to its Chinese counterpart, the American Paddlefish is planktivorous with a non-protrusible mouth and numerous, long gill rakers, while the Chinese Paddlefish eats other fishes, has a protrusible mouth, and fewer, shorter gill rakers. Both species, however, use their paddles for electroreception in murky river waters (Wilkens et al., 2002). As a result of habitat destruction, pollution, and overfishing, the Chinese Paddlefish is critically endangered and may be extinct (IUCN, 2013; Zhang et al., 2009).

REFERENCES: Boschung and Mayden, 2004; Grande and Bemis, 1991, 1996; Inoue et al., 2003; IUCN, 2013; Wilkens et al., 2002; Zhang et al., 2009.

(account continued)

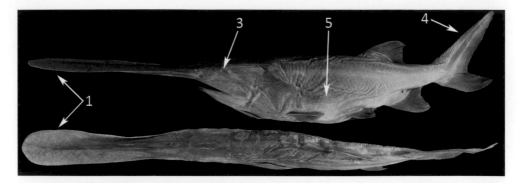

POLYODONTID CHARACTERISTICS:
1) snout long, flat, paddle-like
2) body essentially naked, with few scales
3) spiracle above and behind eye
4) caudal fin heterocercal
5) gill cover extended posteriorly
6) snout and gill cover with conspicuous electrosensory pores

ILLUSTRATED SPECIMEN:
Polyodon spathula, TU 37062-7, 882 mm TL (lateral and dorsal views)

HOLOSTEI—Gars and Bowfins

The Holostei includes the gars, with seven extant species, and the single extant species of bowfin. This group was recognized in early classifications of fishes (e.g., Patterson, 1973), but for a number of years, the Holostei was thought to be a paraphyletic group, with the bowfin more closely related to the teleost fishes. However, a recent exhaustive study of extant and fossil species by Grande (2010) provides morphological support (e.g., presence of a paired vomer) for the monophyly of the Holostei, in agreement with several analyses of molecular data (Inoue et al., 2003; Meyer and Zardoya, 2003). The Holostei is the sister group of the Teleostei; together they are called the Neopterygii.

LEPISOSTEIFORMES : LEPISOSTEIDAE—Gars

DIVERSITY: 1 family, 2 genera, 7 species

REPRESENTATIVE GENERA: *Atractosteus, Lepisosteus*

DISTRIBUTION: North America and Cuba

HABITAT: Freshwater to coastal marine; tropical to temperate; pelagic in rivers, lakes, and estuaries

REMARKS: Gars are elongate, toothy fishes that live predominantly in freshwater or brackish water and feed on other fishes, some benthic invertebrates, and waterfowl. Several small, toothed bones (including the infraorbitals) form the upper jaw. Gars use their lung-like gas bladders to supplement respiration (Graham, 1997) and are known to spend time in a stationary position near the surface. Opisthocoelous vertebrae (anterior end convex, posterior end concave) are unique to this group, one genus of the Blenniidae, and some reptiles.

REFERENCES: Boschung and Mayden, 2004; Graham, 1997; Grande, 2010; Wiley, 1976; Wiley, in Carpenter, 2003.

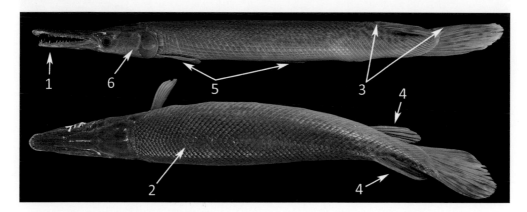

LEPISOSTEIFORM CHARACTERISTICS:
1) elongate jaws with fanglike teeth
2) bony, ganoid scales in oblique rows
3) row of median scales along the first ray of the dorsal, anal, and caudal fins
4) body elongate, with the dorsal and anal fins located posteriorly
5) pectoral fins low on body, pelvic fins abdominal
6) cheek with numerous bony plates
7) three branchiostegal rays

ILLUSTRATED SPECIMEN:
Atractosteus spatula, TU 124963, 666 mm TL (lateral and dorsal views)

AMIIFORMES : AMIIDAE—Bowfins

DIVERSITY: 1 family, 1 genus, 1 species

REPRESENTATIVE GENUS: *Amia*

DISTRIBUTION: Eastern North America

HABITAT: Freshwater in streams, rivers, and swamps; temperate; demersal over soft bottoms

REMARKS: The Bowfin, *Amia calva,* is the only living representative of the Amiidae and Amiiformes. It is restricted to eastern North America, though fossil forms are known world-wide (Grande and Bemis, 1999). It inhabits still or slow-moving freshwaters and can swim by either undulating its long dorsal fin or utilizing its strong tail. The Bowfin is predatory and feeds on other fishes, reptiles, amphibians, birds, snails, and crayfishes. It is among the many groups of air-breathing fishes and utilizes its lung-like gas bladder to supplement respiration (Graham, 1997). Males construct a nest where they defend eggs and young after they hatch.

REFERENCES: Boschung and Mayden, 2004; Graham, 1997; Grande and Bemis, 1998, 1999.

(account continued)

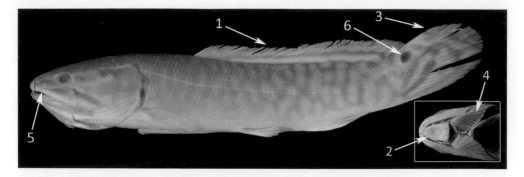

AMIIFORM CHARACTERISTICS:
1) body cylindrical, with long dorsal fin
2) single gular plate
3) caudal fin heterocercal
4) ten to thirteen flattened branchiostegal rays
5) maxilla included in gape
6) males with prominent ocellus near upper base of caudal fin

ILLUSTRATED SPECIMEN:
Amia calva, SIO 69–491, 138 mm TL

INSET: Head of *Amia calva,* SIO uncatalogued (ventral view)

TELEOSTEI—Teleosts

The Teleostei has long been recognized as a monophyletic group. Its composition, distinctive features, and relationships have been discussed at length by a variety of authors (e.g., Arratia, 1997, 1999, 2001; de Pinna, 1996; Fujita, 1990; Gosline, 1971; Greenwood et al., 1966;

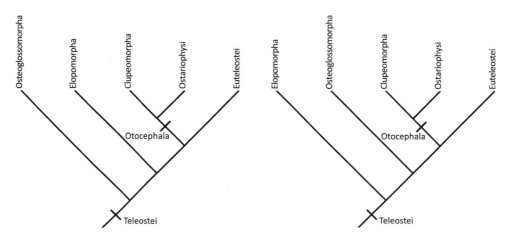

Hypothesized phylogenetic relationships of the early lineages of the Teleostei. The traditional hypothesis (left), based on morphology (Nelson, 2006) and supported by Inoue et al. (2003), differs from the recent molecularly based hypothesis (right) of Faircloth et al. (2013) and Chen et al. (2014) in the placement of the Osteoglossomorpha and Elopomorpha.

McAllister, 1968; Nelson et al., 2010; Patterson and Johnson, 1995; Wiley and Johnson, 2010; Winterbottom, 1974a). Most extant members have a somewhat to highly mobile premaxilla that is free from the skull, unpaired basibranchial tooth plates, a unique caudal-fin skeleton with elongate uroneurals (modified neural spines), and a number of other features. The extraordinary diversification of the teleosts has been attributed, in part, to a genome duplication event that occurred early in their evolution (Hoegg et al., 2004; Hurley et al., 2007; Meyer and Van de Peer, 2005; Santini et al., 2009). The diversity of the Teleostei is immense, including approximately 96% of all living "fish" species (Nelson, 2006). Several recent studies (e.g., Betancur et al., 2013; Faircloth et al., 2013; Inoue et al., 2001; Ishiguro et al., 2003; Near et al., 2012, 2013) have hypothesized the relationships among the major lineages of teleosts based on molecular data.

OSTEOGLOSSOMORPHA—Bonytongues and Mooneyes

The bonytongues and relatives are a broadly distributed early lineage of ray-finned fishes with a substantial fossil record. Their evolution predates the breakup of Gondwana, with extant representatives found in freshwaters of North America, South America, Africa, Asia, and Australia. At least one group, the Notopteridae or Old World knifefishes, also occurs in brackish waters. They are characterized by several internal features (Hilton, 2003; Li and Wilson, 1996; Wiley and Johnson, 2010; Wilson and Murray, 2008) including a unique "shearing bite" between the basihyal and lateral pterygoquadrate teeth (Greenwood et al., 1966). Most species in the group provide parental care to eggs, and in some cases, young (Britz, 2004). Two lineages are recognized within the Osteoglossomorpha: the Osteoglossiformes, with four families (Mormyridae, or elephantfishes, Notopteridae, or Old World knifefishes, Osteoglossidae, or bonytongues, and Gymnarchidae, or African knifefishes), and the Hiodontiformes, with a single family (Hiodontidae, or mooneyes).

OSTEOGLOSSIFORMES—Bonytongues

DIVERSITY: 4 families, 28 genera, 234 species

REPRESENTATIVE GENERA: *Arapaima, Chitala, Gymnarchus, Mormyrus, Pantodon*

DISTRIBUTION: Circumtropical

HABITAT: Freshwater, rarely brackish, rivers, streams and still waters; tropical; near surface and over soft bottoms

REMARKS: The bonytongues exhibit remarkable variety in form among and within the four included families. The Osteoglossidae can have long dorsal and anal fins and attain a length of more than 2.5 m (*Arapaima gigas*) or have short dorsal and anal fins with a maximum length of 10 cm (*Pantodon buchholzi*). The long pectoral fins of *P. buchholzi* (Freshwater Butterflyfish) allow it to glide over the surface up to 2 m. The Notopteridae have an extremely long anal fin, but a short or absent dorsal fin, and the pelvic fins may be present or absent. The Mormyridae have moderately long dorsal and anal fins, considerable variation in the snout and lower jaw, and in some cases, electric organs. The monotypic Gymnarchidae (*Gymnarchus niloticus*) has no anal, pelvic, or caudal fins, but also has electric organs. Air breathing has been documented in all four families (Graham, 1997).

REFERENCES: Graham, 1997; Hilton, 2003; Li and Wilson, 1996; Li et al., 1997; Wilson and Murray, 2008; Zhang, 2006.

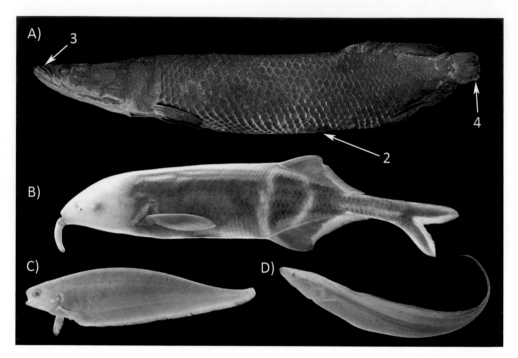

OSTEOGLOSSIFORM CHARACTERISTICS:
1) glossohyal (tongue bone) usually with teeth
2) pelvic fins abdominal or (rarely) absent
3) premaxilla fixed to skull
4) branched caudal-fin rays usually fewer than 16
5) three to seventeen branchiostegal rays
6) six (or fewer) hypurals

ILLUSTRATED SPECIMENS:
A) *Arapaima gigas,* SIO 76-343, 809 mm SL (Osteoglossidae—bonytongues)
B) *Gnathonemus petersii,* SIO 64-228, 118 mm SL (Mormyridae—elephantfishes)
C) *Xenomystus nigri,* SIO 64-228, 132 mm SL (Notopteridae—featherfin knifefishes)
D) *Gymnarchus niloticus,* SIO 64-228, 165 mm SL (Gymnarchidae—Aba or African Knifefish)

HIODONTIFORMES : HIODONTIDAE—Mooneyes

DIVERSITY: 1 family, 1 genus, 2 species

REPRESENTATIVE GENUS: *Hiodon*

DISTRIBUTION: North America

HABITAT: Freshwater; temperate; pelagic in slow-moving rivers and lakes

REMARKS: Mooneyes characteristically have large eyes, their diameter greater than the length of the snout, and silvery or golden bodies. Unlike similar families, in mooneyes the single dorsal fin is located well posterior to the origin of the abdominal pelvic fins. Mooneyes migrate either upstream or to lake shallows in order to spawn. These fishes are visual predators and feed near the surface at night and during low light, primarily on insects, crustaceans, small fishes, frogs, and small mammals. The anal fin is sexually dimorphic, with males having thickened anterior rays.

REFERENCES: Boschung and Mayden, 2004; Britz, 2004; Greenwood, 1970a; Hilton, 2003; Li and Wilson, 1994; Li et al., 1997.

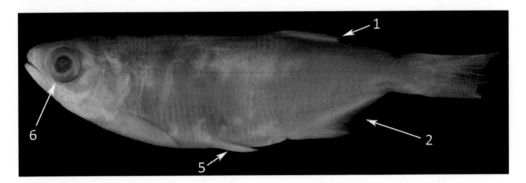

HIODONTIFORM CHARACTERISTICS:

1) single, short-based dorsal fin situated relatively far posteriorly
2) anal-fin base much longer than dorsal-fin base
3) subopercle with a small spine
4) seven to ten branchiostegal rays
5) pelvic fins with seven rays
6) eyes large

ILLUSTRATED SPECIMEN:
Hiodon tergisus, SIO 74–131, 99 mm SL

ELOPOMORPHA

This large group of morphologically diverse fishes is united in sharing a distinctive larval type, the leptocephalus. These ribbon-shaped larvae have small heads and elongate bodies that in some may be as long as 2 m (Böhlke, 1989). At metamorphosis, they shrink in size and take up the general form of their respective lineages. In addition, elopomorphs have numerous branchiostegal rays (15 or more) and teeth on the parasphenoid. The group includes over 850 species classified in five orders, 24 families, and 156 genera. The largest of these orders is the Anguilliformes or true eels, with nearly 800 species. While the monophyly of the Elopomorpha has been questioned by some, recent molecular data and morphological evidence support its monophyly (Chen et al., 2014; Forey, 1973; Forey et al., 1996; Inoue et al., 2004; Obermiller and Pfeiler, 2003; Wiley and Johnson, 2010).

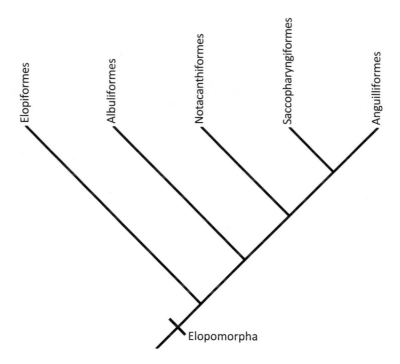

Hypothesized phylogenetic relationships of the Elopomorpha after Chen et al. (2014); that study placed the Saccopharyngiformes within the Anguilliformes.

ELOPIFORMES—Tenpounders and Tarpons

Members of Elopiformes have an elongate body, abdominal pelvic fins, long jaws with toothed premaxillae and maxillae in the gape, wide gill openings, a single gular plate, cycloid scales, and numerous (23–35) branchiostagal rays. Like those of the Albuliformes, their leptocephali have a well-developed, forked caudal fin. This group includes two families, the well-known tarpons (Megalopidae) and the tenpounders (Elopidae).

ELOPIFORMES : ELOPIDAE—Tenpounders and Ladyfishes

DIVERSITY: 1 genus, 7 species

REPRESENTATIVE GENUS: *Elops*

DISTRIBUTION: Atlantic, Indian and Pacific oceans

HABITAT: Primarily marine but occasionally in brackish estuaries and freshwater; tropical to subtropical; coastal pelagic

REMARKS: Tenpounders are elongate, silvery fishes with a large, terminal mouth and a deeply forked caudal fin. These predatory fishes reach lengths of approximately 1 m and feed primarily on small fishes and some crustaceans. Unlike the closely related tarpons, tenpounders do not breathe air but do possess a large pseudobranch. There are currently only seven described species of tenpounders, but there are likely more undescribed, cryptic species in the Indo-Pacific and elsewhere (McBride et al., 2010). Tenpounders are generally not targeted commercially but are considered a good sport fish.

REFERENCES: McBride et al., 2010; Smith, in Carpenter, 2003; Whitehead, 1962.

ELOPID CHARACTERISTICS:
1) body elongate and somewhat rounded in cross section
2) last ray of single dorsal fin not elongate
3) mouth terminal, large, extending past eye
4) pelvic fins abdominal
5) caudal fin deeply forked
6) single gular plate
7) scales small, approximately 100 in lateral line

ILLUSTRATED SPECIMEN:
Elops affinis, SIO 64–326, 172 mm SL

ELOPIFORMES : MEGALOPIDAE—Tarpons

DIVERSITY: 1 genus, 2 species

REPRESENTATIVE GENUS: *Megalops*

DISTRIBUTION: Atlantic and Indo-West Pacific oceans

HABITAT: Primarily marine but occasionally in brackish estuaries and freshwater; tropical to subtropical; coastal pelagic

REMARKS: The tarpons are large, silvery fishes with large scales, a slightly superior mouth, and a deeply forked caudal fin. They grow to well over 2 m in length and are predatory, feeding primarily on fishes and invertebrates. They are able to respire via air gulped at the surface and passed into the physostomous gas bladder (Graham, 1997). There are two known species, one in the Atlantic and a second in the Indo-West Pacific. They are very popular sport fishes.

REFERENCES: Ault, 2008; Forey et al., 1996; Graham, 1997; Greenwood, 1970b; Seymour et al., 2008; Smith, in Carpenter, 2003; Wade, 1962.

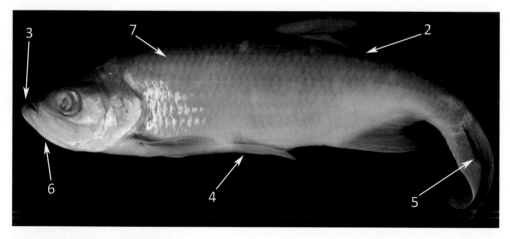

MEGALOPID CHARACTERISTICS:

1) body elongate, laterally compressed
2) last ray of single dorsal fin elongate
3) mouth terminal to superior, large, extending past eye
4) pelvic fins abdominal
5) caudal fin deeply forked
6) single gular plate
7) scales large, less than 50 in lateral line

ILLUSTRATED SPECIMEN:

Megalops atlantica, SIO 78–124, 270 mm SL

ALBULIFORMES—Bonefishes

This group includes two families, the Albulidae (covered below) and the Pterothrissidae, two genera, and about 13 species. They are silvery fishes with a forked caudal fin in adults as well as in their leptocephalus larvae (also found in the Elopiformes) and are distinctive in a few osteological features (Forey et al., 1996; Wiley and Johnson, 2010).

REFERENCES: Forey et al., 1996; Wiley and Johnson, 2010.

ALBULIFORMES : ALBULIDAE—Bonefishes

DIVERSITY: 1 genus, 11+ species

REPRESENTATIVE GENUS: *Albula*

DISTRIBUTION: Atlantic, Indian, and Pacific oceans

HABITAT: Marine, occasionally in freshwater; tropical to warm temperate; coastal, demersal over soft bottoms

REMARKS: Bonefishes are coastal fishes characterized by their single dorsal fin, deeply forked caudal fin, and inferior mouth. As indicated by their mouth position, these fishes feed on or near the bottom, on crustaceans and other small invertebrates, as well as fishes. Reaching a maximum size of over 1 m, bonefishes are highly regarded as sport fishes, but are rarely eaten and generally not targeted commercially. Recent researchers have used molecular and morphological characters to determine the presence of several cryptic species (e.g., Hidaka et al., 2008; Pfeiler et al., 2008).

REFERENCES: Ault, 2008; Hidaka et al., 2008; Pfeiler et al., 2008; Smith, in Carpenter, 2003.

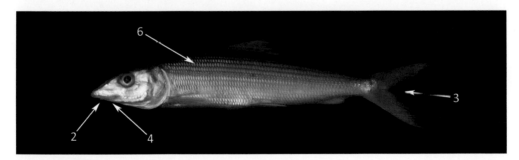

ALBULID CHARACTERISTICS:
1) body elongate
2) mouth inferior, small, not extending past eye
3) tail deeply forked
4) single gular plate
5) six to sixteen branchiostegal rays
6) scales small

ILLUSTRATED SPECIMEN:
Albula sp., SIO 62–213, 159 mm SL

NOTACANTHIFORMES—Spiny Eels and Halosaurs

DIVERSITY: 2 families, 6 genera, 27 species

REPRESENTATIVE GENERA: *Aldrovandia, Halosaurus, Lipogenys, Notacanthus*

DISTRIBUTION: Atlantic, Indian, and Pacific oceans

HABITAT: Marine; tropical to temperate; lower continental shelf to abyssal plain, demersal over soft bottoms

REMARKS: Spiny eels (Notacanthidae) and halosaurs (Halosauridae) are elongate, deep-sea fishes that, in addition to the features below, are characterized by a large connective tissue nodule intercalated between the pterygoid arch and the maxilla. Their long anal fin extends nearly half the body length and includes numerous spines. Spiny eels and halosaurs associate with the benthos and feed on small invertebrates (e.g., crustaceans, echinoderms,

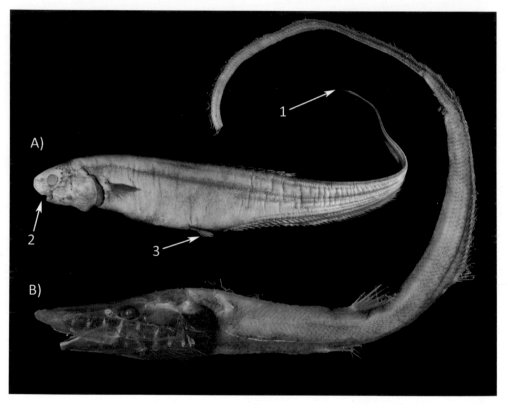

NOTACANTHIFORM CHARACTERISTICS:
1) tail elongate, tapering posteriorly to a point; caudal fin absent
2) mouth small, inferior, not extending past eye
3) pelvic fins abdominal, connected along ventral midline with a membrane
4) maxilla toothed, in gape, with a posteriorly directed spine
5) base of anal fin extremely long

ILLUSTRATED SPECIMENS:
A) *Notacanthus chemnitzii,* SIO 87–84, 385 mm TL (Notacanthidae—spiny eels)
B) *Aldrovandia phalacra,* SIO 68–463, 189.5 mm TL (Halosauridae—halosaurs; tail broken)

polychaetes) and detritus. These fishes can regenerate their long tails when broken or injured. The leptocephalus larvae of some notacanthids are known to reach an amazing length of 2 m (Böhlke, 1989).

REFERENCES: Böhlke, 1989; Crabtree et al., 1985; Forey et al., 1996; Smith, in Carpenter, 2003; Sulak et al., 1984.

ANGUILLIFORMES—Eels

Historically, the true eels have been called the Apodes because they lack pelvic fins and a pelvic girdle. The pectoral fins and caudal fin may be present or absent. Scales are absent in most eels, but if present they are cycloid and imbedded. The gill openings of eels are narrow and gill rakers are absent. Eel leptocephali have a rounded caudal fin that is contiguous with the dorsal and anal fins. The Anguilliformes includes 15 families, 141 genera, and nearly 800 species. They are found in all major aquatic habitats of the world. *Protoanguilla palau*, recently described from deep reefs of Palau, represents a new genus and unique family of eels, the Protoanguillidae, that is hypothesized to be the sister group of all other eels (Johnson et al., 2012).

REFERENCES: Böhlke, 1989; Johnson et al., 2012; Robins, 1989; Santini, Kong et al., 2013.

ANGUILLIFORMES : ANGUILLIDAE—Freshwater Eels

DIVERSITY: 1 genus, 18 species

REPRESENTATIVE GENUS: *Anguilla*

DISTRIBUTION: Atlantic, Indian, and Pacific oceans, and all continents except Antarctica

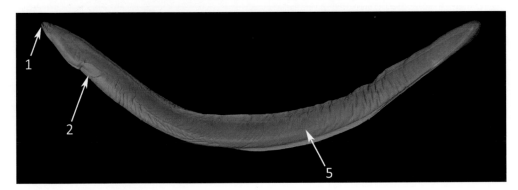

ANGUILLID CHARACTERISTICS:
1) lower jaw projecting beyond upper jaw
2) pectoral fins well developed
3) scales small, embedded in skin
4) teeth small, in bands
5) trunk lateral-line canal complete

ILLUSTRATED SPECIMEN:
Anguilla japonica, SIO 85–138, 356 mm TL

(account continued)

HABITAT: Marine and freshwater; tropical to temperate; catadromous; demersal to benthic in lakes, rivers, and estuaries, benthopelagic in open ocean

REMARKS: Though referred to as freshwater eels, the species in the Anguillidae are generally catadromous. While they live most of their adult lives in freshwater, they spawn in open ocean, far from land (Tsukamoto et al., 2011). Freshwater eels are also semelparous, meaning they die soon after spawning. Recent studies have shown that they are related to a group of deep-sea marine eels, the Nemichthyidae and Serrivomeridae (Inoue et al., 2010). Anguillids are generalized predators and feed on other fishes and benthic invertebrates. Freshwater eels are commercially important food fishes and several species support a large aquaculture enterprise.

REFERENCES: Inoue et al., 2010; Smith, 1989; Tesch, 1977; Tsukamoto et al., 2011.

ANGUILLIFORMES : MURAENIDAE—Moray Eels

DIVERSITY: 15 genera, 198 species

REPRESENTATIVE GENERA: *Echidna, Gymnothorax, Muraena, Uropterygius*

DISTRIBUTION: Atlantic, Indian, and Pacific oceans

HABITAT: Marine, occasionally freshwater; tropical to temperate; continental shelf to upper continental slope, benthic on or in coral and rocky reefs, as well as soft bottoms

REMARKS: Unlike most eels, morays can be quite colorful with distinctive markings. They range in size from ~20 cm to 3.75 m in total length, and are among the world's largest eels. In addition to their laterally compressed head and body, morays have a characteristic raised-head profile behind the eyes. These eels are both predators and scavengers and eat living or recently dead fishes or crustaceans. Piscivorous species are characterized by long, needle-like teeth and highly mobile pharyngeal jaws (Mehta and Wainwright, 2008), while species specializing on crustaceans may have molariform teeth. Care should be taken when eating large morays as they are responsible for many cases of ciguatera poisoning. The phylogenetic relationships of morays were hypothesized by Tang and Fielitz (2013) based on molecular data.

REFERENCES: Böhlke et al., 1989; Böhlke and McCosker, 2001; Böhlke and McCosker, in Carpenter and Niem, 1999; Mehta and Wainwright, 2008; Smith, 2012; Tang and Fielitz, 2013.

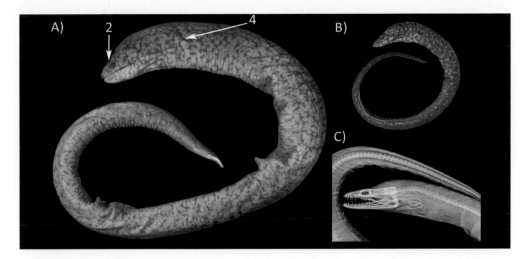

MURAENID CHARACTERISTICS:
1) pectoral fins absent
2) posterior nostril high on head
3) head and body laterally compressed for entire length
4) gill opening reduced to small round hole or slit
5) scales absent
6) trunk lateral-line canal absent

ILLUSTRATED SPECIMENS:
A) *Uropterygius versutus,* SIO 59–7, 289 mm TL
B) *Gymnothorax moringa,* SIO 71–275, 377 mm TL
C) *Enchelycore octaviana,* SIO 65–33, radiograph

ANGUILLIFORMES : OPHICHTHIDAE—Snake Eels and Worm Eels

DIVERSITY: 52 genera, 318 species

REPRESENTATIVE GENERA: *Muraenichthys, Myrichthys, Ophichthus*

DISTRIBUTION: Atlantic, Indian, and Pacific oceans

HABITAT: Marine, occasionally in freshwater; tropical to temperate; continental shelf to continental slope, usually benthic on or in soft bottoms, with some species occuring in mid-waters of the mesopelagic

REMARKS: The Ophichthidae is a large family of eels with considerable morphological diversity. For example, the pectoral, dorsal, anal, and caudal fins can be either present or absent, and the origin of the dorsal fin, when present, can be over the pectoral fin or well posterior. The unusual basket-like structure formed by the branchiostegal rays often involves rays that are detached from any other bone. Some species in the Ophichthinae use their hard, pointed tails to burrow backward, and their burrowing lifestyle implies that the bulk of their diet is likely benthic invertebrates. They are abundant in certain areas, caught by hook-and-line, and occasionally consumed by humans.

REFERENCES: Böhlke and McCosker, in Carpenter and Niem, 1999; McCosker, 1977, 2010; McCosker et al., 1989; McCosker and Rosenblatt, 1998.

(account continued)

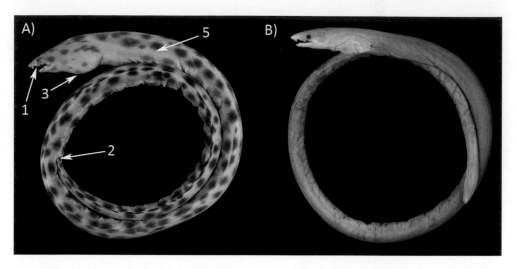

OPHICHTHID CHARACTERISTICS:

1) posterior (excurrent) nostril on upper lip (usually) or inside mouth
2) tail usually hard, pointed; caudal fin absent in most species (present in Myrophinae)
3) branchiostegal rays numerous (15–49), overlapping at ventral midline to form a basket
4) scales absent
5) trunk lateral-line canal complete, left and right sides connected by a canal across nape

ILLUSTRATED SPECIMENS:

A) *Callechelys eristigma,* SIO 65–263, 503 mm TL, holotype
B) *Scolecenchelys chilensis,* SIO 65–645, 284 mm TL, holotype

---

ANGUILLIFORMES : CONGRIDAE—Conger Eels

DIVERSITY: 32 genera, 196 species

REPRESENTATIVE GENERA: *Conger, Heteroconger, Paraconger*

DISTRIBUTION: Atlantic, Indian, and Pacific oceans

HABITAT: Marine; tropical to temperate; continental shelf to abyssal plain, benthic on soft bottoms, often burrowing into substrates

REMARKS: Like the Ophichthidae, the Congridae is a speciose family exhibiting significant morphological and ecological variation, and is difficult to characterize. Conger eels generally have well-developed pectoral fins and large eyes for use in visual predation, but there are exceptions to both traits. They live in association with the benthos and range from depths of less than 5 m to more than 2,000 m. Most species actively feed on small fishes and invertebrates at night, but the Heterocongrinae form vast "gardens" in sands adjacent to reefs and are visual plankton pickers. Conger eels are important in the fish leather industry (Grey et al., 2006).

REFERENCES: Castle and Randall, 1999; Grey et al., 2006; Smith, 1989b; Smith, in Fischer et al., 1995.

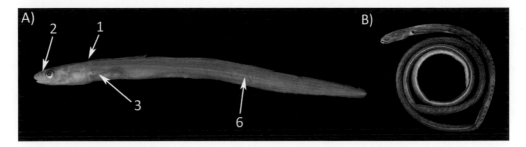

CONGRID CHARACTERISTICS:

1) dorsal-fin origin over or just posterior to pectoral-fin insertion
2) posterior nostril just anterior to eye
3) pectoral fins usually present and well-developed (except in the Heterocongrinae)
4) scales absent
5) eight to twenty-two branchiostegal rays
6) trunk lateral-line canal complete

ILLUSTRATED SPECIMENS:

A) *Ariosoma gilberti,* SIO 69–235, 165 mm TL (Congrinae)
B) *Heteroconger canabus,* SIO 61–261, 760 mm TL, holotype (Heterocongrinae—garden eels)

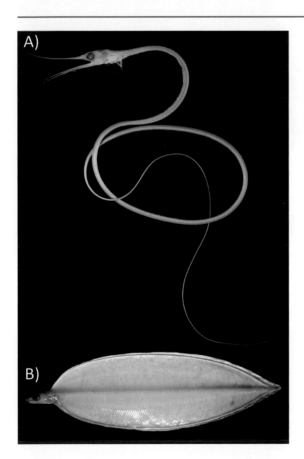

ANGUILLIFORM DIVERSITY:

A) NEMICHTHYIDAE—snipe eels:
   Nemichthys scolopaceus, SIO 88–55, 706 mm TL
B) COLOCONGRIDAE—shorttail eels:
   Thalassenchelys foliaceus, SIO 70–333, 228 mm TL, leptocephalus larva

SACCOPHARYNGIFORMES—Swallowers and Gulper Eels

DIVERSITY: 4 families, 5 genera, 28 species

REPRESENTATIVE GENERA: *Cyema, Eurypharynx, Monognathus, Neocyema, Saccopharynx*

DISTRIBUTION: Atlantic, Indian, and Pacific oceans

HABITAT: Marine; tropical to temperate; midwater, lower mesopelagic to bathypelagic

REMARKS: The deep-sea swallowers and gulper eels are among the most distinctive and bizarre fishes, considered by some (e.g., Nelson, 2006) to be among the most morphologically modified vertebrates. With their huge mouths and highly distensible pharynx, they are clearly adapted for capturing and ingesting large prey items in the food-poor environment of the deep sea. In addition to the features listed below, they lack opercular bones and ribs, and have long, posterior extensions of both jaws. Swallowers and gulpers feed primarily on other fishes, and can swallow items at least as large as their body size. The tail of the swallowers ends in a luminous organ that might be used to attract prey.

REFERENCES: Nelson, 2006; Nielsen et al., 1989; Tighe and Nielsen, 2000.

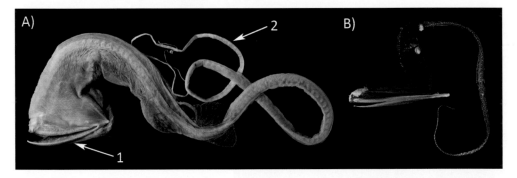

SACCOPHARYNGIFORM CHARACTERISTICS:
1) mouth greatly enlarged
2) tail extremely elongate, caudal fin absent
3) abdominal portion of body much deeper than tail
4) body quite flaccid
5) scales absent
6) branchiostegal rays absent

ILLUSTRATED SPECIMENS:
A) *Saccopharynx lavenbergi,* SIO 75–272, 602 mm TL (Saccopharyngidae—swallowers)
B) *Eurypharynx pelecanoides,* SIO 72–180, 430 mm TL (Eurypharyngidae—pelican eels)

OTOCEPHALA

This group, called the Otomorpha by Wiley and Johnson (2010), includes the herrings and anchovies (Clupeiformes) and their sister group, the Ostariophysi (Arriata, 1999). Its members share several osteological features of the skull, vertebral column, and caudal fin, as well as a unique silvery area associated with the gas bladder (Lecointre and Nelson, 1996; Wiley and Johnson, 2010).

CLUPEIFORMES—Herrings, Anchovies, and Relatives

The Clupeiformes includes five families, 84 genera, and 364 species. They are characterized by a laterally compressed body, one or more unpaired scutes crossing the abdominal midline, and a unique otophysic connection between the gas bladder and the otic region of the brain. This group includes the well-known herrings and anchovies, both primarily planktivores, as well as predators such as the wolfherrings (Chirocentridae). With many species occurring in vast schools, clupeiforms are one of the most important groups to worldwide fisheries (FAO, 2012) and to the ecology of areas where they occur.

REFERENCES: Grande, 1985; Whitehead, 1963.

CLUPEIFORMES : ENGRAULIDAE—Anchovies

DIVERSITY: 16 genera, 147 species

REPRESENTATIVE GENERA: *Anchoa, Anchovetta, Cetengraulis, Engraulis*

DISTRIBUTION: Atlantic, Indian, and Pacific oceans

HABITAT: Marine, estuarine, freshwater; tropical to temperate; neritic and epipelagic

REMARKS: Anchovies are easily recognized by their prominent snout, which houses a unique "rostral organ" (Nelson, 1984), and their large, inferior mouth, with the maxilla reaching well beyond the eye and, in some species, to the posterior end of the head. Anchovies generally are filter feeders and use their large eyes (and perhaps their rostal organ) to seek out planktonic prey. Many species form enormous schools that are targeted by industrial fisheries and form an important ecological link to higher trophic levels. Anchovies are the most heavily harvested fish family both by number of individuals and by weight. A single species, the Peruvian Anchoveta (*Engraulis ringens*), accounts for a large percentage of the world's commercial catch (FAO, 2012).

REFERENCES: FAO, 2012; Nelson, 1984; Nizinski and Munroe, in Carpenter, 2003; Whitehead et al., 1988; Whitehead and Rodriguez-Sanchez, in Fischer et al., 1995; Wongratana et al., in Carpenter and Niem, 1999.

(account continued)

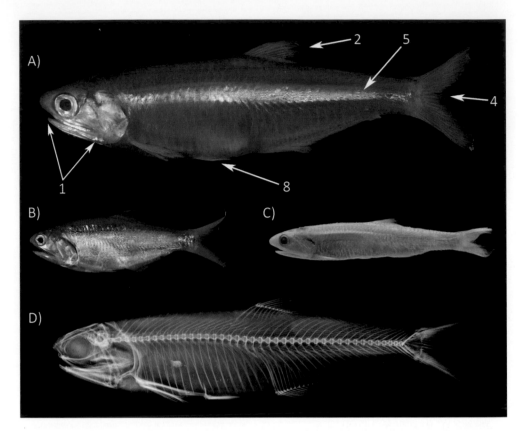

ENGRAULID CHARACTERISTICS:

1) mouth inferior, upper jaw extending well behind eye in most species
2) dorsal fin single, short, usually near midbody
3) body compressed, somewhat elongate
4) caudal fin forked
5) body often translucent, often with silver stripe
6) cephalic lateral-line canals prominent, trunk lateral-line canal absent
7) seven to nineteen branchiostegal rays
8) pelvic fins with seven to ten rays

ILLUSTRATED SPECIMENS:

A) *Anchoa compressa,* SIO 48–29, 104 mm SL
B) *Cetengraulis mysticetus,* SIO 60–487, 107 mm SL
C) *Engraulis mordax,* SIO 59–66, 103 mm SL
D) *Anchoa compressa,* SIO 46–67, radiograph

CLUPEIFORMES : CLUPEIDAE—Herrings

DIVERSITY: 57 genera, 194 species

REPRESENTATIVE GENERA: *Alosa, Clupea, Dorosoma, Harengula, Opisthonema, Sardinops*

DISTRIBUTION: Arctic, Atlantic, Indian, and Pacific oceans

CLUPEID CHARACTERISTICS:

1) abdominal scutes well developed in most species
2) dorsal fin single, usually short and near midbody
3) body compressed, somewhat elongate
4) caudal fin forked
5) mouth usually terminal (subterminal in the Dorosomatinae)
6) cephalic lateral-line canals prominent, trunk lateral-line canal absent
7) five to ten branchiostegal rays
8) teeth small or absent

ILLUSTRATED SPECIMENS:

A) *Opisthonema libertate,* SIO 77-95, 94 mm SL
B) *Dorosoma cepedianum,* SIO 62-317, 143 mm SL
C) *Brevoortia patronus*, SIO 88-95, radiograph

(account continued)

HABITAT: Primarily coastal marine with some anadromous species and some restricted to freshwater; tropical to temperate; neritic and epipelagic

REMARKS: Many herring species form enormous schools. They are generally visual feeders and either pick or filter zooplankton, especially copepods. Herrings exhibit variation in reproductive mode, with most species having pelagic eggs but some attaching eggs to the substrate or floating objects. Many species (e.g., *Brevoortia* spp. and *Sardinops* spp.) are targeted by fisheries for reduction to fish oil and fish meal for use in agriculture and aquaculture, while other species are targeted for human consumption (FAO, 2012). As a result of their abundance in coastal ecosystems, they are an important ecological link to higher trophic levels.

REFERENCES: FAO, 2012; Munroe and Nizinski, in Carpenter, 2003; Munroe et al., in Carpenter and Niem, 1999; Whitehead, 1985; Whitehead and Rodriguez-Sanchez, in Fischer et al., 1995.

OSTARIOPHYSI

The Ostariophysi is a vast lineage of over 8,000 species that dominates freshwater fish faunas around the world, but also includes well over 100 marine species. The lineage comprises five orders: the Gonorynchiformes or milkfishes, the Gymnotiformes or American

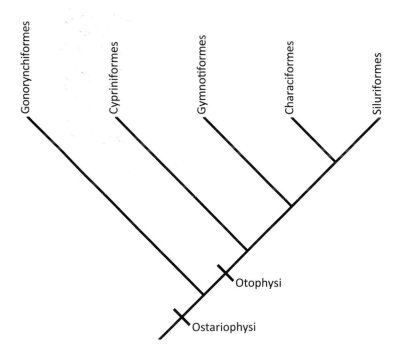

Hypothesized phylogenetic relationships of the Ostariophysi after Nakatani et al. (2011); that study found a portion of the Characiformes to be more closely related to the Siluriformes than to other characiforms.

knifefishes, the Characiformes or characins, the Cypriniformes or carps and minnows, and the Siluriformes or catfishes. The composition and features of the Ostariophysi have been discussed by a variety of authors (e.g., Fink and Fink, 1981, 1996; Rosen and Greenwood, 1970). These are summarized by Wiley and Johnson (2010) who list 13 synapomorphies, most of which are complex internal features such as absence of the basisphenoid and a portion of the palatine bones, and presence of a physostomous gas bladder (bladder lost in some lineages). Many ostariophysans also have keratinous nuptial tubercles in breeding males (Wiley and Collette, 1970; not observed in gymnotiforms), and a unique alarm substance known as Schreckstoff in the epidermis (also absent in the gymnotiforms). The Weberian apparatus, a modification of the anterior-most vertebral elements that mechanically links the gas bladder to the inner ear, often associated with the Ostariophysi, is not found in the Gonorhynchiformes, and instead characterizes the Otophysi, a lineage that includes all other ostariophysans (Britz and Hoffman, 2006; Chardon et al., 2003; Nakatani et al., 2011).

GONORYNCHIFORMES—Milkfishes and Relatives

DIVERSITY: 4 families, 7 genera, 37 species

REPRESENTATIVE GENERA: *Chanos, Gonorynchus, Kneria, Parakneria, Phractolaemus*

DISTRIBUTION: Africa, and Pacific, Indian, and southern Atlantic oceans

HABITAT: Freshwater, and shallow coastal and estuarine areas; tropical to temperate; usually over soft bottoms

REMARKS: Most gonorynchiform species are found in the freshwaters of Africa, while two groups, the Gonorynchidae (beaked sandfishes) and the better-known Chanidae (Milkfish) are mostly marine. Although difficult to characterize superficially, their monophyly is well established, with a number of internal synapomorphies (e.g., a primitive Weberian apparatus and distinctive epibranchial organ). The single chanid species, *Chanos chanos,* is an elongate, silvery fish with a terminal mouth and large, forked caudal fin. Adult *Chanos* feed on algae and spawn in marine waters. *Chanos* is prized as a food fish in Southeast Asia where it is of great importance to the aquaculture industry (Bagarinao, in Carpenter and Niem, 1999; Poyato-Ariza, 1996; Grande and Poyota-Ariza, 1999). The monotypic Phractolaemidae (*Phractolaemus ansorgii*) has an upturned (superior) mouth and is found in tropical Africa. The Kneriidae comprises most species in this group (about 30), is characterized by a subterminal or inferior mouth (also typical of the Gonorynchidae), and includes some transparent species.

REFERENCES: Bagarinao, in Carpenter and Niem, 1999; Ferraris, in Carpenter and Niem, 1999; Fink and Fink, 1981, 1996; Grande and Poyota-Ariza, 1999; Grande et al., 2010; Lecointre, 2010; Poyato-Ariza, 1996; Wiley and Johnson, 2010.

(account continued)

GONORYNCHIFORM CHARACTERISTICS:
1) origin of anal fin usually posterior to insertion of dorsal fin
2) mouth small, jaws without teeth
3) pelvic fins abdominal, at midbody or far posterior
4) three to five branchiostegal rays
5) mouth usually subterminal or inferior
6) pelvic-fin rays typically 6–12

ILLUSTRATED SPECIMENS:
A) *Chanos chanos,* 83–177, 765 mm SL (Chanidae—Milkfish)
B) *Kneria* sp., SIO 01–20, 53 mm SL (Kneriidae—shellears)

CYPRINIFORMES—Carps and Relatives

The Cypriniformes is characterized by a protrusible upper jaw (present in most), and a unique bone, the kinethmoid, present between the ascending processes of the premaxillae. Cypriniforms lack teeth on the oral jaws, but have unique tooth plates in the pharyngeal jaws that oppose a process on the basioccipital bone of the neurocranium. This lineage includes over 4,000 species that are classified in as many as 11 families of minnows, suckers, and loaches. Two of these, the Cyprinidae and Catastomidae, are covered in more detail below. Recent research efforts focusing on reconstructing the cypriniform tree of life (http://bio.slu.edu/mayden/cypriniformes/home.html) have generated a number of new insights into the evolutionary history of this important and diverse clade (e.g., Chen and Mayden, 2009; Mayden et al., 2008, 2009). Cypriniforms are found in the freshwaters of Asia, Africa, Europe, and North America, but are absent from South America and Australia (Berra, 2001).

REFERENCES: Berra, 2001; Chen and Mayden, 2009; Fink and Fink, 1981, 1996; Mabee et al., 2011; Mayden, 1992; Mayden et al., 2008, 2009.

CYPRINIFORMES : CYPRINIDAE—Carps and Minnows

DIVERSITY: Over 220 genera and 2,914 species

REPRESENTATIVE GENERA: *Barbus, Carassius, Cyprinus, Danio, Nocomis, Notropis, Pimephales*

DISTRIBUTION: North America, Eurasia, and Africa

HABITAT: Freshwater streams, rivers, lakes and swamps; tropical to temperate; pelagic to demersal

REMARKS: This lineage dominates the freshwaters of much of the world with the exception of South America and Australia. Traditionally divided into ten or more subfamilies, recent studies have recommended elevation of some of these to family status. This includes the Leucosinae, with over 560 species that are restricted to North America and northern Eurasia (Bufalino and Mayden, 2010; Chen and Mayden, 2009; Mayden et al., 2009). While most cyprinids are microcarnivores, feeding on insects and crustaceans, they exhibit a vast array of feeding behaviors ranging from herbivory to piscivory. Cyprinids include a number of well-known aquarium fishes such as the barbs and the Goldfish (*Carassius auratus*),

CYPRINID CHARACTERISTICS:
1) pharyngeal teeth in one to three rows, always with eight or fewer teeth per row
2) lips usually thin
3) upper jaw usually protrusible and bordered only by the premaxilla
4) fin rays flexible in most species, occasionally with spine-like rays in the dorsal fin
5) small barbels sometimes present
6) breeding tubercles (keratinous structures on the skin) often present

ILLUSTRATED SPECIMENS:
A) *Nocomis micropogon,* SIO 75–390, 135 mm SL (Leuciscinae)
B) *Gila orcuttii,* SIO 62–509, 130 mm SL (Leuciscinae)
C) *Agosia chrysogaster,* SIO 68–144, 54 mm SL (Leuciscinae)

as well as several species that are important in aquaculture and other breeding efforts, both as food fishes and as ornamentals (e.g., the koi, a domesticated form of the common carp, *Cyprinus carpio*). In addition, this group includes the zebrafish (*Danio rerio*), an important model organism for developmental biology, developmental genetics, and genomics (www.zfin.org). Because of their use in aquatic weed control and as bait for larger fishes, many cyprinids have become established well beyond their native ranges, posing a variety of threats to native fishes.

REFERENCES: Arai and Kato, 2003; Boschung and Mayden, 2004; Bufalino and Mayden, 2010; Cavender and Coburn, 1992; Chen and Mayden, 2009; Mayden et al., 2008, 2009; Schönhuth et al., 2012; Wiley and Collette, 1970; Winfield and Nelson, 1991.

CYPRINIFORMES : CATOSTOMIDAE—Suckers

DIVERSITY: 11 genera, 83 species

REPRESENTATIVE GENERA: *Catostomus, Erimyzon, Ictiobus, Moxostoma*

DISTRIBUTION: North America and northeastern Asia (China and Siberia)

HABITAT: Freshwater streams, rivers, lakes, and ponds; tropical to temperate; benthic to demersal

REMARKS: Suckers are so named because of their common behavior of mouthing the substrate with their large lips. They feed on a variety of food items including algae, detritus, insects, and mollusks. Catostomids are one of the few groups of tetraploid fishes (Ferris and Whitt, 1978). Some species support small artisanal fisheries. Several species are considered threatened, most notably the Razorback Sucker (*Catostomus texanus*), which is restricted to the Colorado River basin (Minckley and Deacon, 1991).

CATOSTOMID CHARACTERISTICS:
1) mouth usually subterminal
2) lips usually thick and fleshy with obvious folds or papillae
3) premaxilla and maxilla included in the gape in most species
4) pharyngeal teeth, 16 or more, in a single row
5) anal fin inserted relatively far posteriorly, with slightly elongate rays

ILLUSTRATED SPECIMEN:
Catostomus commersonii, SIO 61–448, 101 mm SL

REFERENCES: Boschung and Mayden, 2004; Ferris and Whitt, 1978; Harris and Mayden, 2001; Minckley and Deacon, 1991; Smith, 1992.

CYPRINIFORM DIVERSITY:

A) CYPRINIDAE—minnows and carps: *Barbonymus schwanenfeldii,* SIO 72–282, 108 mm SL

B) CYPRINIDAE: *Carassius auratus,* SIO 64–228, 47 mm SL

C) COBITIDAE—loaches: *Misgurnus anguillicaudatus,* SIO 69–375, 85 mm SL

D) and E) BALITORIDAE—river loaches: *Sewellia elongata,* SIO 97–62, 55 mm SL (dorsal and ventral views)

CHARACIFORMES—Characins

DIVERSITY: 18 families, 270 genera, over 1,700 species

REPRESENTATIVE GENERA: *Astyanax, Cynodon, Gasteropelecus, Hydrocynus, Serrasalmus*

DISTRIBUTION: Southern North America, Central America, South America, and Africa

HABITAT: Freshwater; tropical to temperate; pelagic to benthic, usually over soft bottoms

REMARKS: The Characiformes is an extraordinarily diverse lineage of freshwater fishes. Its monophyly is well established morphologically, and there is support from some molecular studies (e.g., Lavoue et al., 2005; Li et al., 2008; Olivera et al., 2011), though others do not support the monophyly of this group (e.g, Ortí and Meyer, 1997; Nakatani et al., 2011; Peng et al., 2006). Although members occur in Africa, most of the families and species in this lineage are found in the New World. The largest family, the Characidae, has well over 1,140 species. Some species are air-breathers, most are carnivores (some are fruit-eaters),

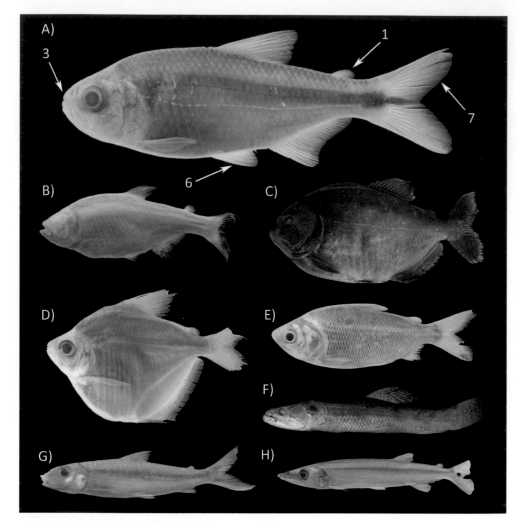

CHARACIFORM CHARACTERISTICS:

1) adipose fin present in most species
2) jaw teeth multicuspid, generally well developed
3) upper jaw usually nonprotrusible
4) body scaled
5) three to five branchiostegal rays
6) pelvic-fin rays five to seven
7) principal caudal-fin rays usually 19

ILLUSTRATED SPECIMENS:

A) *Astyanax mexicanus,* SIO 53-33, 57 mm SL (Characidae—characins)
B) *Astyanax mexicanus* (blind form), SIO 64-228, 62 mm SL (Characidae)
C) *Pygocentrus nattereri,* SIO 64-228, 240 mm SL (Serrasalmidae—piranhas and pacus)
D) *Stethaprion crenatum,* SIO 70-280, 54 mm SL (Characidae)
E) *Psectrogaster rutiloides,* SIO 03-135, 87 mm SL (Curimatidae—toothless characins)
F) *Hoplias microlepis,* SIO 82-92, 323 mm SL (Erythrinidae—trahiras)
G) *Bivibranchia fowleri,* SIO 03-133, 123 mm SL (Hemiodontidae—halftooths)
H) *Acestrorhynchus falcirostris,* SIO 03-136, 172 mm SL (Acestrorhynchidae—smallscale pike characins)

and some apparently have internal fertilization. The group includes numerous popular species of aquarium fishes and many larger species are utilized for food. Also included are the occasionally dangerous piranhas (Serrasalmidae) and the freshwater hatchetfishes (Gasteropelecidae), which are capable of short flights. One species, the Mexican Tetra (*Astyanax mexicanus*), occurs from southwestern Texas to central Mexico and includes 29 cave populations that lack eyes and pigment (Gross, 2012).

REFERENCES: Buckup, 1998; Calcagnotto et al., 2005; Fink and Fink, 1981, 1996; Géry, 1977; Graham, 1997; Gross, 2012; Li et al., 2008; Nakatani et al., 2011; Ortí and Meyer, 1997; Peng et al., 2006; Roberts, 1969; Vari, 1989; Weitzman, 1962.

SILURIFORMES—Catfishes

Catfishes are easily recognized by their prominent barbels, up to four pairs, including one nasal, one maxillary, and two on the chin. These fishes almost always have a rudimentary maxilla that lacks teeth, usually an adipose fin, and spine-like rays at the anterior dorsal and pectoral fins. All members of this lineage also have an electroreceptive sense, and some are venomous. The body may be either naked or covered with bony plates, and the vomer and palatines usually have teeth. Siluriforms lack the symplectic, subopercular, basihyal, and intermuscular bones. Air-breathing has been documented in seven families and more are likely. Some species are seen frequently in the aquarium trade, and the larger species support both recreational and commercial fisheries. Their dietary habits range from being algae eaters to carnivores to parasites, including the infamous Candiru (*Vandellia cirrhosa*) known to enter the human urethra. Catfishes generally occur in freshwater and are circumtropical, but some are found in temperate areas, and some are in coastal marine and estuarine areas. This hugely diverse group includes over 3,000 species classified in approximately 450 genera and 35 families. A focused effort is underway to describe all species of catfishes (http://silurus.acnatsci.org/). The monophyly of the Siluriformes is supported by a large number of internal, morphological characters (de Pinna, 1998; Diogo and Peng, 2010), and by extensive molecular evidence (e.g., Lavoué et al., 2005; Nakatani et al., 2011; Ortí and Meyer, 1997; Sullivan et al., 2006).

REFERENCES: de Pinna, 1998; Diogo and Peng, 2010; Ferraris, 2007; Ferraris and de Pinna, 1999; Graham, 1997; Lavoué et al., 2005; Nakatani et al., 2011; Ortí and Meyer, 1997; Sullivan et al., 2006.

SILURIFORMES : ICTALURIDAE—North American Catfishes

DIVERSITY: 7 genera, 51 species

REPRESENTATIVE GENERA: *Ameiurus, Ictalurus, Noturus, Pylodictis, Satan*

DISTRIBUTION: North America, from Canada to Guatemala; introduced widely

HABITAT: Freshwater streams, rivers, lakes and ponds; tropical to temperate; benthic to demersal on soft bottoms

REMARKS: Ictalurids are important predatory fishes in the freshwaters of North America. Some grow as large as 1.5 m in length and 60 kg in weight. These largely nocturnal fishes feed on a variety of prey including insects, crustaceans, and small fishes. Males guard eggs and young in many species. The madtoms (genus *Noturus*) have venomous cells associated with their pectoral-fin spines (Birkhead, 1972; Wright, 2009). The Ictaluridae includes some the most important aquaculture species in the world and many have been introduced well beyond their native range.

REFERENCES: Birkhead, 1972; Boschung and Mayden, 2004; Lundberg, 1992; Taylor, 1969; Wright, 2009.

ICTALURID CHARACTERISTICS:
1) four pairs of barbels
2) skin thick, scales and bony plates absent
3) robust spine present in dorsal and pectoral fins
4) dorsal fin usually with six soft rays
5) no teeth on palate

ILLUSTRATED SPECIMEN:
Noturus flavus, SIO 89–25, 151 mm SL

SILURIFORMES : ARIIDAE—Sea Catfishes

DIVERSITY: 21 genera, 149 species

REPRESENTATIVE GENERA: *Arius, Bagre, Cathorops, Galeichthys*

DISTRIBUTION: Atlantic, Indian, and Pacific oceans

HABITAT: Marine, many species in estuaries and freshwater; tropical to warm temperate; demersal and benthic on soft bottoms

REMARKS: Sea catfishes feed on a wide variety of organisms including invertebrates and fishes and occasionally detritus. Females lay extremely large eggs that are orally brooded by males, and in some the young are guarded well after hatching. Some species grow to a large size (up to 1.5 m in length) and some are known to make croaking sounds with their gas bladder, and rasping noises with their pectoral-fin spines. Like the madtoms, ariids have venom associated with their serrated fin spines (Wright, 2009). Ariids are food fishes nearly everywhere they occur.

REFERENCES: Acero, in Carpenter, 2003; Acero and Betancur, 2007; Betancur, 2009; Birkhead, 1972; de Pinna, 1998; Kailola, 2004; Kailola, in Carpenter and Niem, 1999; Kailola and Bussing, in Fischer et al., 1985; Marceniuk and Ferraris, 2003; Marceniuk and Menezes, 2007; Wright, 2009.

ARIID CHARACTERISTICS:
1) caudal fin forked
2) adipose fin prominent
3) usually three pairs of barbels (nasal barbels absent)
4) prominent bony plates on head near dorsal-fin origin
5) dorsal and pectoral fins with spines
6) gill membranes attached to isthmus
7) anterior and posterior nostrils closely spaced

ILLUSTRATED SPECIMEN:
Bagre panamensis, SIO 62–81, 172 mm SL

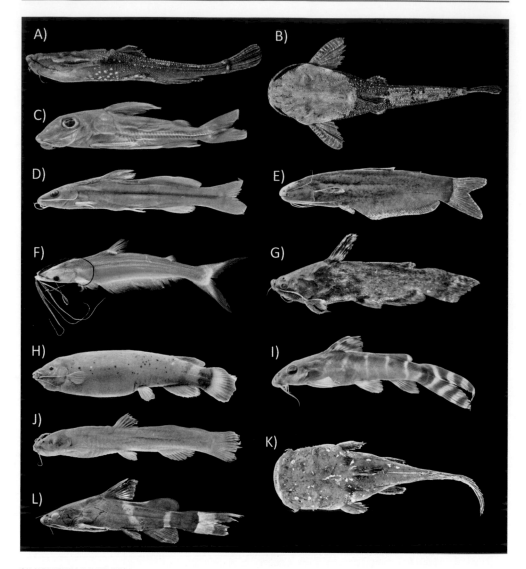

SILURIFORM DIVERSITY:

A and B) ASPERDINIDAE—banjo catfishes (neotropical): *Bunocephalus amaurus,* ANSP 160240, 107 mm SL

C) DORADIDAE—thorny catfishes (neotropical): *Doras carinatus,* SIO 83–4, 99 mm SL

D) HEPTAPTERIDAE—heptapterids (neotropical): *Rhamdia* sp., SIO 63–153, 60 mm SL

E) CETOPSIDAE—whale-like catfishes (neotropical): *Helogenes marmoratus,* SIO 03–132, 52 mm SL

F) PIMELODIDAE—long-whiskered catfishes (neotropical): *Hypophthalmus marginatus,* SIO 03–135, 202 mm SL

G) AUCHENIPTERIDAE—driftwood catfishes (neotropical): *Trachelyopterus galeatus,* SIO 03–138, 62 mm SL

H) MALAPTERURIDAE—electric catfishes (African): *Malapterurus electricus,* SIO 64–228, 147 mm SL

I) MOCHOKIDAE—squeakers (African): *Synodontis brichardi,* SIO 64–228, 93 mm SL

J) AMPHILIIDAE—loach catfishes (African): *Paramphilius trichomycteroides,* SIO 01–23, 41 mm SL

K) CHACIDAE—frogmouth catfishes (Southeast Asian): *Chaca bankanensis,* SIO 64–228, 104 mm SL

L) BAGRIDAE—bagrid catfishes (Southeast Asian): *Pseudomystus siamensis,* SIO 03–140, 52 mm SL

GYMNOTIFORMES—American Knifefishes

DIVERSITY: 5 families, 30 genera, 208 species

REPRESENTATIVE GENERA: *Apteronotus, Brachyhypopomus, Eigenmannia, Electrophorus, Gymnotus, Rhamphichthys*

DISTRIBUTION: South and Central America

HABITAT: Freshwater; tropical; demersal, usually over soft bottoms

REMARKS: The Gymnotiformes, or American knifefishes, are elongate fishes with an extremely long anal-fin base. The anal fin aids in precise forward and backward movement. They are nocturnal and well known for possessing electric organs used to hunt prey at night and in murky waters, for intraspecific communication, and in the case of the Electric Eel (*Electrophorus electricus*), to ward off would-be predators with discharges of up to 600 volts (Albert, 2001; Albert and Crampton, 2005). In addition to the characters listed below, the caudal fin is absent or reduced, and the palatines are not ossified. There also are many internal characters as well as molecular data supporting the well-established monophyly of this group (Fink and Fink, 1981, 1996; Lavoué et al., 2005; Nakatani et al., 2011).

REFERENCES: Albert, 2001; Albert and Crampton, 2005; Albert et al., 1998; Carr and Maler, 1986; Fink and Fink, 1981, 1996; Lavoué et al., 2005; Nakatani et al., 2011.

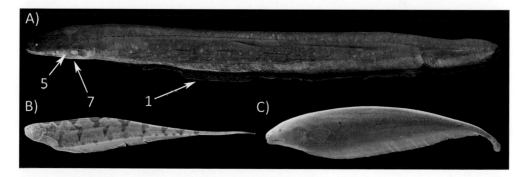

GYMNOTIFORM CHARACTERISTICS:
1) anal fin extremely long, with more than 150 rays
2) dorsal fin absent
3) pelvic fins and pelvic girdle absent
4) maxillary teeth absent
5) gill openings small
6) gill rakers reduced
7) anus far forward, under head or pectoral fins
8) electroreceptive organs

ILLUSTRATED SPECIMENS:
A) *Electrophorus electricus,* SIO 77-375, 900 mm SL (Gymnotidae—naked-back knifefishes)
B) *Steatogenys* sp., SIO 64-228, 177 mm SL (Hypopomidae—bluntnose knifefishes)
C) *Parapteronotus* sp., SIO 64-228, 115 mm SL (Apteronotidae—ghost knifefishes)

EUTELEOSTEI

All remaining teleosts are included within the Eutelostei (Rosen, 1985), a diverse group of fishes diagnosed by features of the caudal fin and the developmental pattern of the supraneurals (Greenwood et al., 1966; Johnson, 1992; Johnson and Patterson, 1996; Wiley and Johnson, 2010).

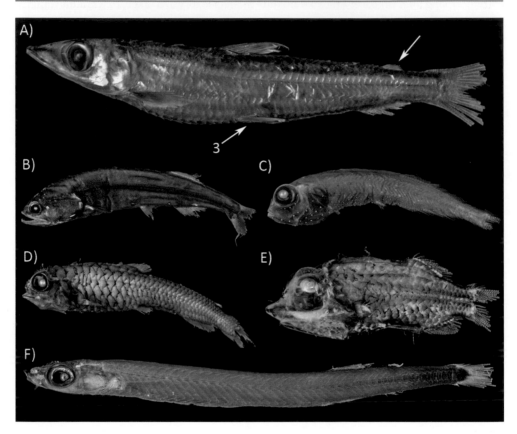

ARGENTINIFORM CHARACTERISTICS:

1) complex "epibranchial" (crumenal) organ present in gill cavity
2) adipose fin present (Argentinoidei) or absent (Alepocephalioidei)
3) pelvic fins abdominal
4) body variable in shape

ILLUSTRATED SPECIMENS:

A) *Argentina sialis,* SIO 55–25, 117 mm SL (Argentinidae—herring smelts)
B) *Sagamichthys abei,* SIO 88–43, 224 mm SL (Platytroctidae—tubeshoulders)
C) *Xenodermichthys copei,* SIO 10–8, 41 mm SL (Alepocephalidae—slickheads)
D) *Bathylagus pacificus,* SIO 87–159, 128 mm SL (Microstomatidae—pencilsmelts)
E) *Macropinna microstoma,* SIO 88–60, 87 mm SL (Opisthoproctidae—barreleyes)
F) *Microstoma microstoma,* SIO 63–372, 115 mm SL (Microstomatidae—pencilsmelts)

ARGENTINIFORMES—Marine Smelts

DIVERSITY: 6 families, 57 genera, 226 species

REPRESENTATIVE GENERA: *Argentina, Bathylagus, Macropinna, Microstoma, Sagamichthys*

DISTRIBUTION: Worldwide in Atlantic, Indian, and Pacific oceans

HABITAT: Marine; tropical to temperate; continental shelves, mesopelagic and bathypelagic

REMARKS: The composition of the Argentiniformes remains controversial. We follow Wiley and Johnson (2010) who include two lineages, the Argentinoidei (Argentinidae, Microstomatidae, and Opisthoproctidae) and the Alepocephaloidei (Alepocephalidae, Bathylaconidae, and Platytroctidae). This highly variable, worldwide lineage of marine fishes is mostly found in the mesopelagic and bathypelagic zones, although some members (e.g., Argentinidae) occur over continental shelves. The unique crumenal organ is a modification of the dorsal portion of the posterior two gill arches and serves to trap food particles. The deep-sea members have a number of unique features such as the upwardly directed eyes of the barreleyes (Opisthoproctidae) and a variety of light-producing organs including one capable of ejecting a luminous fluid from a tube behind the pectoral girdle in the so-called tubeshoulders (Platytroctidae).

REFERENCES: Johnson and Patterson, 1996; Kobyliansky, 1990; Patterson and Johnson, 1997; Poulsen et al., 2009; Wiley and Johnson, 2010.

OSMERIFORMES—Smelts and Relatives

DIVERSITY: 3 families, 22 genera, 92 species

REPRESENTATIVE GENERA: *Galaxias, Lepidogalaxias, Osmerus, Plecoglossus, Retropinna, Salanx*

DISTRIBUTION: Arctic, Atlantic, and Pacific oceans

HABITAT: Freshwater, marine, anadromous, estuarine; temperate (rarely tropical) to polar; pelagic to demersal, usually over soft bottoms

REMARKS: The composition and phylogenetic relationships of the Osmeriformes have been controversial (Wiley and Johnson, 2010). Molecular evidence allies them with the Salmoniformes (Ramsden et al., 2003), but that result is not supported by all data sets (e.g., Near et al., 2012). The wide range in counts for the branchiostegal and caudal-fin rays reflects the diversity in morphology among the three families included herein. These are the "true" smelts (Osmeridae), the New Zealand smelts (Retropinnidae), and the galaxiids (Galaxiidae). Smelts are frequently caught commercially and recreationally nearly everywhere they occur. These fishes consume a variety of prey, some species foraging in the water column for euphausiids, copepods, and other plankton, and also on the bottom for other small invertebrates. At least one species is known to breathe air and more are hypothesized to do so (Graham, 1997). The salamanderfish (*Lepidogalaxias salamandroides*) has the extraordinary ability (for a fish) to bend its neck (Berra et al., 1997).

REFERENCES: Berra et al., 1997; Graham, 1997; Johnson and Patterson, 1996; McAllister,

1963; Mecklenburg et al., 2002; Near et al., 2012; Ramsden et al., 2003; Rosen, 1974; Wiley and Johnson, 2010.

OSMERIFORM CHARACTERISTICS:
1) body elongate
2) maxilla included in gape (few exceptions)
3) scales without radii
4) nuptial tubercles
5) gill rakers without teeth
6) three to ten branchistegal rays
7) adipose fin present or absent
8) caudal fin with 12–19 branched rays

ILLUSTRATED SPECIMENS:
A) *Osmerus mordax*, SIO 89–29, 96 mm SL (Osmeridae—smelts)
B) *Galaxias brevipinnis*, SIO 80–93, 88 mm SL (Galaxiidae—galaxiids)

SALMONIFORMES—Salmons, Trouts, and Relatives

Although the Salmoniformes includes some of the most well-known fishes in the world, its composition and relationships remain unresolved. We follow Nelson (2006), who included only the Salmonidae. Wiley and Johnson (2010) also included the Esocidae (pikes and pickerels) and Umbridae (mudminnows), herein considered under the Esociformes, as well as the Osmeridae (smelts), Retropinnidae (New Zealand smelts), and the Galaxiidae (galaxiids), herein considered under the Osmeriformes. This expanded group lacks unique morphological synapomorphies (Wiley and Johnson, 2010), but its monophyly is supported by some molecular studies (e.g., López et al., 2004), though not by others (e.g., Near et al., 2012). Relationships among these lineages are also unresolved: Ramsden et al. (2003) and López et al. (2004) hypothesized salmonids and esocids as sister groups, while Johnson and Patterson (1996) reported several morphological synapomorphies supporting the sister group relationship of salmonids and osmerids (Wiley and Johnson, 2010).

REFERENCES: Johnson and Patterson, 1996; López et al., 2004; Near et al., 2012; Nelson, 2006; Ramsden et al., 2003; Sanford, 1990, 2000; Stearley and Smith, 1993; Wiley and Johnson, 2010.

SALMONIFORMES : SALMONIDAE—Salmons and Trouts

DIVERSITY: 10 genera, 217 species

REPRESENTATIVE GENERA: *Coregonus, Oncorhynchus, Salmo, Thymallus*

DISTRIBUTION: Northern Hemisphere, introduced widely

HABITAT: Freshwater and anadromous; generally temperate; pelagic to demersal over soft and hard bottoms

REMARKS: The Salmonidae includes a variety of well-known and commercially important species, and some of the most sought-after gamefish. The taxonomy of trouts and salmon has undergone considerable modification in recent years (Smith and Stearly, 1989) and many populations have been elevated to species status, raising the recognized species diversity from 66 (Nelson, 2006) to 217 species (Eschmeyer and Fong, 2013). Salmonids, like the catostomids, are tetraploid (Allendorf and Thorgaard, 1984; Le Comber and Smith, 2004). Their phylogenetic relationships have been studied by a variety of researchers including Sanford (2000) and Crespi and Fulton (2004). These fishes all spawn demersal eggs in freshwaters, with some making long migrations from their feeding grounds

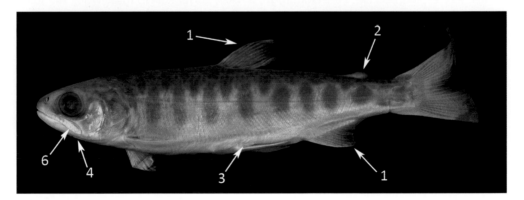

SALMONID CHARACTERISTICS:
1) fins comprised only of soft rays
2) adipose fin distinct
3) pelvic fins thoracic with axillary process
4) gill membranes extending far anteriorly, not fused with isthmus
5) scales small, cycloid
6) toothed maxilla forming part of gape

ILLUSTRATED SPECIMEN:
Oncorhynchus kisutch (juvenile), SIO 60–503, 115 mm SL

in marine waters. These anadromous species use both a magnetic compass and olfactory cues to return to their natal streams, and many are semelparous (dying after one breeding season). Habitat destruction and erection of dams have eliminated or threatened many populations (IUCN, 2013). The Atlantic Salmon (*Salmo salar*) is one of the most intensively aquacultured fishes in the world.

REFERENCES: Allendorf and Thorgaard, 1984; Eschmeyer and Fong, 2013; Crespi and Fulton, 2004; Le Comber and Smith, 2004; López et al., 2004; Mecklenburg et al., 2002; Nelson, 2006; Sanford, 1990, 2000; Smith and Stearley, 1989; Stearley and Smith, 1993.

ESOCIFORMES—Pikes and Mudminnows

Esociformes comprises two families, four genera, and 13 species, and its monophyly is well supported (e.g., Johnson and Patterson, 1996; Rosen, 1974), although its relationships remain controversial (see Salmoniformes and Osmeriformes; Nelson, 2006; Wiley and Johnson, 2010). These are relatively elongate fishes with the anal fin positioned far posteriorly. The operculum and cheek have scales and the maxilla is included in the gape, but it lacks teeth. Unlike their closest relatives, these species have no adipose fin or nuptial tubercles. This group occurs in the freshwaters of temperate North America and Europe. The pikes (Esocidae) are piscivores, while the mudminnows (Umbridae) are omnivores. The Umbridae (seven species) differs from the Esocidae (six species, treated below) in having a round caudal fin with 20–30 total rays, 5–8 branchiostegal rays, an obscure or absent lateral line, and a more rounded snout. Maximum reported length in the umbrids is 35 cm.

REFERENCES: Johnson and Patterson, 1996; Nelson, 2006; Rosen, 1974; Wiley and Johnson, 2010; Wilson and Veilleux, 1982.

ESOCIFORMES : ESOCIDAE—Pikes and Pickerels

DIVERSITY: 1 genus, 6 species

REPRESENTATIVE GENUS: *Esox*

DISTRIBUTION: Northern Hemisphere

HABITAT: Freshwater streams, lakes and ponds; temperate to subpolar; demersal over soft bottoms to pelagic

REMARKS: Pikes are important predators in freshwater systems of the Northern Hemisphere. Like many stalking predators, they are elongate and have their dorsal and anal fins placed far posteriorly on the body. Pikes are visual predators and primarily eat other fishes, but also frogs, snakes, and crayfishes. Their systematics has been studied by a variety of researchers (e.g., Crossman, 1978; Grande et al., 2004; López et al., 2004). While not targeted commercially, pikes are important sport fishes in temperate lakes and rivers, and the Muskellunge (*Esox masquinongy*) commonly reaches lengths of over 1 m and has been reported to attain nearly 2 m.

REFERENCES: Crossman, 1978; Crossman and Casselman, 1987; Grande et al., 2004; López et al., 2004; Nelson, 1972.

ESOCID CHARACTERISTICS:

1) body elongate
2) mouth large, snout produced, resembling a duck's bill
3) palatine, dentary, and vomer with canine teeth
4) single dorsal fin and anal fin located far posteriorly
5) toothless maxilla included in gape
6) trunk lateral-line canal complete
7) ten to twenty branchiostegal rays
8) adipose fin absent

ILLUSTRATED SPECIMEN:
Esox americanus, SIO 62–321, 162 mm SL

STOMIIFORMES—Dragonfishes

The dragonfishes are a bizarre and morphologically diverse group of mesopelagic and bathypelagic fishes. They range in shape from deep bodied and laterally compressed to greatly elongate. The upper jaw includes a toothed premaxilla and maxilla. A variety of luminescent organs are present, including those on the body and on the tip of the chin barbel (in the barbeled dragonfishes), and large light organs under the eyes. Scales, when present, are cycloid and deciduous, and the overall color is usually dark brown or black, occasionally silvery. The group comprises four families (only the Phosichthyidae are not covered here), 53 genera, and 426 mostly deep-sea species.

REFERENCES: Ahlstrom et al., 1984; Fink, 1985; Fink and Weitzman, 1982; Harold and Weitzman, 1996; Weitzman, 1974.

STOMIIFORMES : GONOSTOMATIDAE—Bristlemouths

DIVERSITY: 8 genera, 33 species

REPRESENTATIVE GENERA: *Cyclothone, Diplophos, Gonostoma*

DISTRIBUTION: Worldwide in Atlantic, Indian, and Pacific oceans

HABITAT: Marine; usually mesopelagic, some bathypelagic

REMARKS: The bristlemouths are common and abundant fishes of the mesopelagic zone; in fact, the species of the genus *Cyclothone* together are reputed to be the most abundant vertebrates on earth. Harold (1998) and Miya and Nishida (1996, 2000) report on the phylogenetic relationships of bristlemouths. Unlike that of Nelson (2006), the current classi-fication includes the Diplophidae within the Gonostomatidae. Some widespread species of *Cyclothone* have been shown to include multiple cryptic species (Miya and Nishida, 1996). The diet of gonostomatids consists of small crustaceans and fishes. Some species are sexu-ally dimorphic and reportedly hermaphroditic.

REFERENCES: Harold, 1998; Harold, in Carpenter and Niem, 1999; Harold, in Carpenter, 2003; Miya and Nishida, 1996, 1997, 2000; Nelson, 2010.

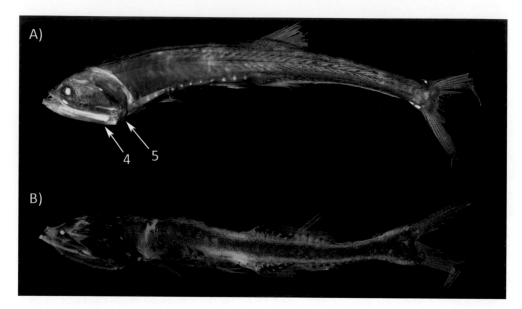

GONOSTOMATID CHARACTERISTICS:

1) body elongate
2) adipose fin usually absent
3) twelve to sixteen branchiostegal rays
4) eight to sixteen photophores on branchiostegal membranes
5) photophores usually present on isthmus
6) postorbital light organ absent
7) gill rakers well developed
8) teeth in upper jaw usually short, bristle-like

ILLUSTRATED SPECIMENS:

A) *Gonostoma elongatum,* SIO 01–174, 194 mm SL
B) *Cyclothone pallida,* SIO 07–45, 50 mm SL

DIVERSITY: 10 genera, 75 species

REPRESENTATIVE GENERA: *Argyropelecus, Danaphos, Polyipnus, Sternoptyx*

DISTRIBUTION: Worldwide in Atlantic, Indian, and Pacific oceans

HABITAT: Marine; mesopelagic, occasionally bathypelagic or benthopelagic (e.g., *Polyipnus*).

REMARKS: The marine hatchetfishes are conspicuous members of mesopelagic fish communities around the world. The anatomy, systematics, and phylogenetic relationships of sternoptychids have been studied by a number of authors (e.g., Baird, 1971; Weitzman, 1974; Harold, 1993; Miya and Nishida, 1998), and their various body shapes have intrigued evolutionary biologists for decades (e.g., Thompson, 1945). Their large, clustered, ventral photophores probably function in crypsis, eliminating their outline (via counter-illumination) when viewed by predators from below. Their diet consists of small fishes and a variety of zooplankton including polychaetes, arrow worms, and small crustaceans.

REFERENCES: Baird, 1971; Harold, 1993, 1994; Harold, in Carpenter and Niem, 1999; Harold, in Carpenter, 2003; Miya and Nishida, 1998; Thompson, 1945; Weitzman, 1974.

STERNOPTYCHID CHARACTERISTICS:
1) body deep, laterally compressed (Sternoptychinae) or somewhat elongate and not extremely compressed (Maurolicinae)
2) mouth oblique to nearly vertical
3) eyes large, dorsally directed in some
4) six to ten branchiostegal rays
5) three to seven photophores on branchiostegal membranes
6) ventral photophores generally in clusters
7) gill rakers well developed
8) pseudobranch present

ILLUSTRATED SPECIMEN:
Argyropelecus sladeni, SIO 77–325, 58 mm SL.

STOMIIFORMES : STOMIIDAE—Barbeled Dragonfishes

DIVERSITY: 28 genera, 294 species

REPRESENTATIVE GENERA: *Astronesthes, Chauliodus, Idiacanthus, Malacosteus, Stomias*

DISTRIBUTION: Worldwide in Atlantic, Indian, and Pacific oceans

HABITAT: Marine; mesopelagic and bathypelagic

REMARKS: Traditionally, the barbeled dragonfishes were divided into six families: Stomiidae (scaly dragonfishes), Astronesthidae (snaggletooths), Chauliodontidae (viperfishes), Melanostomiidae (scaleless black dragonfishes), Idiacanthidae (black dragonfishes), and the Malacosteidae (loosejaws). More recently, these have been considered various lineages (sub-

STOMIID CHARACTERISTICS:
1) body somewhat to greatly elongate
2) mouth large, often with fang-like teeth
3) mental barbel present in most species, with luminous tissue at tip
4) coloration typically black
5) no gill rakers in adults
6) one infraorbital bone

ILLUSTRATED SPECIMENS:
A) *Idiacanthus antrostomus,* SIO 01–47, 313 mm SL (Idiacanthinae—barbeled dragonfishes)
B) *Pachystomias microdon,* SIO 71–296, 183 mm SL (Melanostomiinae—scaleless black dragonfishes)
C) *Stomias atriventer,* SIO 00–163, 205 mm SL (Stomiinae—scaly dragonfishes)
D) *Malacosteus niger,* SIO 73–159, 127 mm SL (Malacosteinae—loosejaws)

families and tribes) within an expanded Stomiidae (Fink, 1985; Nelson, 2006). Kenaley (2010) analyzed relationships of the dragonfishes based on morphology primarily from Fink (1985) and supported the monophyly of the Stomiidae and Malacosteidae, but not of the other lineages. Schnell et al. (2010) described the unusual features associated with the "neck" of these fishes and Kenaley (2010) described their unique photophores. Stomiids are important predators in the mesopelagic and bathypelagic zones of the world's oceans, feeding mainly on fishes and crustaceans. They exhibit a bewildering array of adaptations including a variety of light organs for attracting and searching for prey, the unique ability to produce and detect red light (Partridge and Douglas, 1995), huge mouths and teeth for ingesting large prey, larvae with stalked eyes, and dwarf males.

REFERENCES: Fink, 1985; Harold, in Carpenter and Niem, 1999; Harold, in Carpenter, 2003; Kenaley, 2007, 2009, 2010; Partridge and Douglas, 1995; Schnell et al., 2010.

ATELEOPODIFORMES : ATELEOPODIDAE—Jellynose Fishes

DIVERSITY: 1 family, 4 genera, 13 species

REPRESENTATIVE GENERA: *Ateleopus, Guentherus, Ijimaia, Parateleopus*

DISTRIBUTION: Atlantic, Indian, and Pacific oceans

HABITAT: Marine; tropical to temperate; outer continental shelf to slope, demersal or benthopelagic, presumably over soft bottoms

REMARKS: Although a small group, the jellynose fishes, sometimes called tadpole fishes, are distinguished by a gelatinous snout, elongate, flabby body (up to 2 m), and long anal fin.

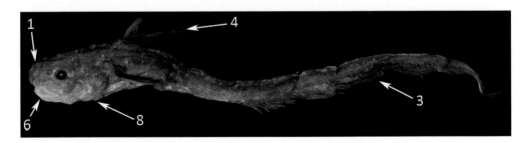

ATELEOPODIFORM CHARACTERISTICS:
1) snout soft and gelatinous
2) body flaccid, elongate
3) base of anal fin very long, confluent with caudal fin in most
4) dorsal fin located just posterior to head, with short base
5) fin spines absent
6) mouth subterminal, protrusible
7) seven branchiostegal rays
8) pelvic fins tiny, jugular
9) skeleton mostly cartilaginous

ILLUSTRATED SPECIMEN:
Ijimaia antillarum, SIO 77-6, 309 mm TL

Although once considered closely aligned with the Lampridiformes, recent morphological evidence tends to support an aulopiform/ateleopodiform relationship (Wiley and Johnson, 2010), while more recent molecular evidence places them as the sister group to aulopiforms plus all other teleosts (Near et al., 2012). They are worldwide in distribution, but their biology remains poorly known.

REFERENCES: Miya et al., 2003; Moore, in Carpenter and Niem, 1999; Moore, in Carpenter, 2003; Near et al., 2012; Olney et al., 1993; Wiley and Johnson, 2010.

AULOPIFORMES—Lizardfishes and Relatives

The lizardfishes and allies are a diverse group of 16 families, over 44 genera, and over 260 species. They are characterized by a moderately elongate body that is rounded in cross-section, a single, soft-rayed dorsal fin, an adipose fin (in most species), and a large mouth. The group lacks a gas bladder and has a series of unique gill-arch specializations (Baldwin and Johnson, 1996). The majority of aulopiforms are deep-sea fishes associated with the bottom or the mesopelagic zone. However, shallow-water representatives include the well-known lizardfishes (Synodontidae), which are found in coastal waters on soft bottoms and coral reefs. Other aulopiforms include the lancetfishes (Alepisauridae), voracious mesopelagic predators that grow to over 1 m in length, and the tripod fishes (Ipnopidae), which are among the deepest occurring vertebrates and have characteristic elongate pectoral-, pelvic-, and caudal-fin rays. Simultaneous hermaphroditism evolved once in the group, and is associated with deep-sea habitats (Davis and Felitz, 2010).

REFERENCES: Baldwin and Johnson, 1996; Davis and Felitz, 2010; Johnson, 1982; Johnson et al., 1996.

AULOPIFORMES : AULOPIDAE—Flagfins

DIVERSITY: 3 genera, 9 species

REPRESENTATIVE GENERA: *Aulopus, Hime, Latropiscis*

DISTRIBUTION: Atlantic and Pacific oceans

HABITAT: Marine; tropical to subtropical; continental shelf to continental slope, demersal over soft bottoms

REMARKS: Flagfins are notable in having relatively large dorsal, anal, and pelvic fins and are known to exhibit sexual dimorphism in fin shape (Gomon et al., 2013). This small group of predatory, benthic fishes feeds primarily on other fishes and crustaceans. There are currently nine recognized species, but a number of undescribed species are known. Flagfins are of minor commercial importance in some areas.

REFERENCES: Gomon et al., 2013; Thompson, 1998; Thompson, in Carpenter, 2003.

AULOPID CHARACTERISTICS:
1) body elongate
2) caudal fin forked
3) dorsal-fin origin in first third of body
4) adipose fin prominent
5) teeth present on tongue
6) maxilla expanded posteriorly but not passing beyond posterior margin of eye
7) fulcral scales present anterior to caudal-fin rays
8) pelvic fins thoracic, with nine rays

ILLUSTRATED SPECIMEN:
Aulopus bajacali, SIO 08–68, 170 mm SL

AULOPIFORMES : SYNODONTIDAE—Lizardfishes

DIVERSITY: 4 genera, 71 species

REPRESENTATIVE GENERA: *Harpadon, Saurida, Synodus, Trachinocephalus*

DISTRIBUTION: Atlantic, Indian, and Pacific oceans

HABITAT: Coastal marine, rarely in brackish waters; tropical to temperate; continental shelves, demersal to benthic

REMARKS: Two lineages are evident within the Synodontidae: the Synodontinae, with 47 species of *Synodus* and *Trachinocephalus,* and the Harpadontinae, with 24 species of *Harpadon* and *Saurida.* Lizardfishes are well-known sit-and-wait ambush predators on coral reefs. Some species launch such attacks while partially buried in the sand. The Bombay Duck (*Harpadon nehereus*) supports a fishery in the South China Sea where this reputedly aromatic fish is often sold dried; other lizardfishes are made into fish balls or cakes because the flesh is not highly regarded.

REFERENCES: Davis and Felitz, 2010; Russell, in Carpenter and Niem, 1999; Russell, in Carpenter, 2003.

(account continued)

SYNODONTID CHARACTERISTICS:
1) body elongate
2) caudal fin forked
3) dorsal fin small, near midbody
4) adipose fin usually present
5) pelvic fins abdominal
6) mouth large, supramaxilla small or absent
7) 8–26 branchiostegals

ILLUSTRATED SPECIMEN:
Synodus lucioceps, SIO 59–103, 204 mm SL

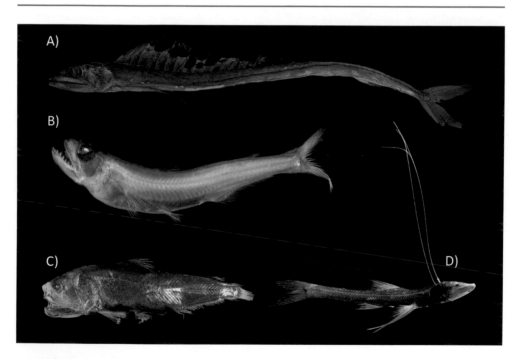

AULOPIFORM DIVERSITY:
A) ALEPISAURIDAE—lancetfishes: *Alepisaurus ferox,* SIO 91–110, 1,020 mm SL
B) SCOPELARCHIDAE—pearleyes: *Scopelarchus analis,* SIO 76–358, 91 mm SL
C) EVERMANNELLIDAE—sabertooth fishes: *Coccorella atrata,* SIO 94–102, 95 mm SL
D) IPNOPIDAE—tripod fishes: *Bathypterois longipes,* SIO 88–177, 94 mm SL

MYCTOPHIFORMES—Lanternfishes and Blackchins

DIVERSITY: 2 families, 35 genera, and 257 species

REPRESENTATIVE GENERA: *Ceratoscopelus, Diaphus, Gonichthys, Myctophum, Notoscopelus, Scopelengys*

DISTRIBUTION: Arctic, Atlantic, Indian, and Pacific oceans

HABITAT: Marine; tropical to polar; mesopelagic to bathypelagic, rarely benthopelagic

REMARKS: The lanternfishes (Myctophidae), with 251 species, and the blackchins (Neoscopelidae), with six species, are well-known mesopelagic and bathypelagic fishes. They are dominant inhabitants of the mesopelagic zone, one of the largest ecosystems on the planet. Lanternfishes and blackchins are opportunistic predators, feeding on pelagic crustaceans, mollusks, and smaller fishes. In turn, these fishes are important prey for larger fishes and for some marine tetrapods. They are important members of the deep scattering layer, with

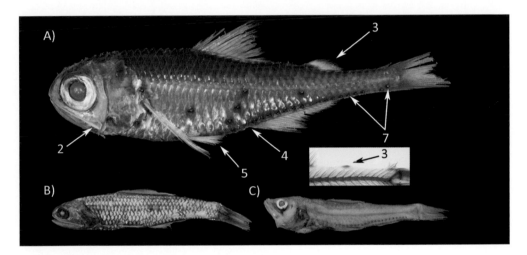

MYCTOPHIFORM CHARACTERISTICS:

1) head and body compressed
2) mouth terminal, typically large, extending well past posterior margin of orbit
3) adipose fin supported by a cartilaginous plate uniquely inserted into underlying muscle
4) anal fin origin under or just behind dorsal-fin base (Myctophidae) or more posteriorly (Neoscopelidae)
5) pelvic fin abdominal, with eight rays
6) scales typically cycloid, deciduous, often lost during capture
7) small photophores typically arranged in species-specific rows and groups on head and body

ILLUSTRATED SPECIMENS:

A) *Myctophum lychnobium,* SIO 60–269, 98 mm SL (Myctophidae—lanternfishes)
B) *Notoscopelus resplendens,* SIO 77–230, 98 mm SL (Myctophidae)
C) *Neoscopelus microchir,* SIO 09–390, 83 mm SL (Neoscopelidae—blackchins).

INSET: Cleared-and-stained *Nannobrachium idostigma* (SIO 65–253) showing support cartilage (blue) of adipose fin.

many species making daily vertical migrations to surface waters at night to feed and returning to the mesopelagic zone during the day to escape predators, and by doing so, transferring a significant amount of surface productivity to the deep sea (Davison et al., 2013). The species-specific patterns of photophores may be important in mate recognition. A few species are targeted commercially by reduction fisheries.

REFERENCES: Bekker, 1983; Craddock and Hartel, in Carpenter, 2003; Davison et al., 2013; Nafpaktitis, 1977, 1978; Paxton, 1972, 1979; Paxton and Hulley, in Carpenter and Niem, 1999; Stiassny, 1996; Wisner, 1974.

Actinopterygii II

ACANTHOMORPHA—Spiny-rayed Fishes

This large and diverse group of teleost fishes, recognized by Rosen (1973) and evaluated in detail by Stiassny (1986) and Johnson and Patterson (1993), includes all remaining fishes covered in this book. Most members have true spines in the dorsal, anal, and pelvic fins (Nelson, 2006). Wiley and Johnson (2010) reviewed the compelling morphological evidence for its monophyly, and this has been corroborated by a variety of molecular studies (e.g., Betancur et al., 2013; Miya et al., 2003; Near et al., 2012, 2013; Wiley et al., 2000). The phylogenetic relationships of acanthomorph fishes are currently under intensive study by a number of reseachers (see http://acanthoweb.fr/en).

LAMPRIDIFORMES—Opahs and Relatives

DIVERSITY: 7 families, 12 genera, 25 species

REPRESENTATIVE GENERA: *Lampris, Lophotus, Radiicephelus, Regalecus, Stylephorus, Trachipterus, Velifer, Zu*

DISTRIBUTION: Atlantic, Indian, and Pacific oceans

HABITAT: Marine; tropical to temperate; epipelagic to abyssopelagic, occasionally neritic

REMARKS: The Lampridiformes is a highly variable group of fishes ranging from deep-bodied to elongate species. This variability is exemplified by the extreme range of pelvic-fin rays, 0–17. Phylogenetic studies (Miya et al., 2007; Olney et al., 1993; Wiley et al., 1998) imply that the elongate morphology is derived within the group. Their morphological diversity is also reflected in their classification: the 25 species are allocated among 12 genera and seven families. They include the deep-bodied opahs (Lampridae), and the elongate ribbonfishes

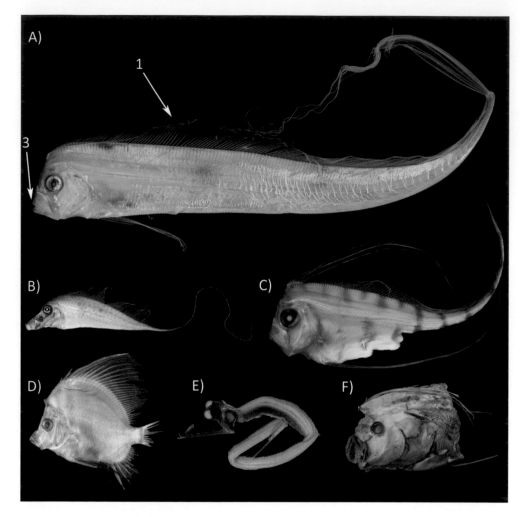

LAMPRIDIFORM CHARACTERISTICS:
1) dorsal fin with extremely long base
2) maxilla excluded from gape by premaxilla
3) upper jaw uniquely protrusible
4) teeth very small or absent
5) fins lacking true spines

ILLUSTRATED SPECIMENS:
A) *Trachipterus altivelis,* SIO 66–56, 268 mm SL (Trachipteridae—ribbonfishes)
B) *Desmodema lorum,* SIO 88–74, 370 mm SL (Trachipteridae)
C) *Zu cristatus,* SIO 76–210, 135 mm SL (Trachipteridae)
D) *Velifer hypselopterus,* SIO 90–152, 49 mm SL (Veliferidae—sailfin moonfishes)
E) *Stylephorus chordatus,* SIO 88–172, 283 mm SL (Stylephoridae—tube-eyes)
F) *Regalecus russelii* (head only), SIO 96–82, ~7,300 mm SL (Regalecidae—oarfishes)

(Trachipteridae) and oarfishes (Regalecidae). The latter grow to over 8 m, the longest bony fish, likely responsible for sea serpent legends (Roberts, 2012). The Opah (*Lampris guttatus*) supports a commercial fishery, and it has a brain heater that facilitates foraging in cold, deep waters below the thermocline (Runcie et al., 2009). The crestfishes (Lophotidae) and the tapertails (Radiicephalidae) have a gland over the hindgut capable of discharging an ink-like fluid as an alarm response. Molecular data indicate that the deep-sea Tube-eye (Stylephoridae), long considered a member of this group, may be more closely related to the Gadiformes (Miya et al., 2007). The Lampridiformes has been hypothesized to be the sister group to all other acanthomorphs (Johnson and Patterson, 1993). The young and adults of some lampridiform species differ dramatically in appearance. These pelagic fishes often wander into shallow waters (occasionally beaching) where they sometimes attract public attention. They are known to eat crustaceans, squids, and small fishes.

REFERENCES: Johnson and Patterson, 1993; Miya et al., 2007; Olney, in Carpenter and Niem, 1999; Olney, in Carpenter, 2003; Olney et al., 1993; Roberts, 2012; Runcie et al., 2009; Smith and Heemstra, 1986; Wiley et al., 1998.

POLYMIXIIFORMES : POLYMIXIIDAE—Beardfishes

DIVERSITY: 1 family, 1 genus, about 10 species

REPRESENTATIVE GENUS: *Polymixia*

DISTRIBUTION: Atlantic, Indian, and western Pacific oceans

HABITAT: Marine; tropical to subtropical; demersal, usually over soft bottoms, outer continental shelf to depths of approximately 800 m

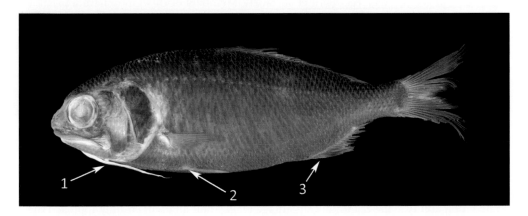

POLYMIXINIFORM CHARACTERISTICS:
1) pair of long barbels on chin, originating well posterior of lower jaw symphysis
2) pelvic fins with one spine and six soft rays, inserted well behind base of pectoral fin
3) anal-fin origin far posterior to dorsal-fin origin
4) seven branchiostegal rays (posterior four apparent; anterior three tiny, supporting the barbel)

ILLUSTRATED SPECIMEN:
Polymixia nobilis, SIO 04–68, 143 mm SL

REMARKS: The beardfishes are known for their long barbels, which are positioned well posterior to the tip of the chin. They have been observed swimming with their barbels in continuous contact with the bottom. Long considered members of the Beryciformes (e.g., Greenwood et al., 1966), recent evidence shows their independence from that group and suggests that they are near the base of the acanthomorphs, perhaps sister to the Percopsiformes (Miya et al., 2007; Near et al., 2012; Wiley and Johnson, 2010). Beardfishes feed on benthic invertebrates, small fishes, and squids, and some species are regarded as food fishes.

REFERENCES: Greenwood et al., 1966; Johnson and Patterson, 1993; Moore, in Carpenter, 2003; Paxton, in Carpenter and Niem, 1999; Stiassny, 1986; Wiley and Johnson, 2010.

PERCOPSIFORMES—Trout-perches and Relatives

DIVERSITY: 3 families, 7 genera, 9 species

REPRESENTATIVE GENERA: *Amblyopsis, Aphredoderus, Chologaster, Percopsis, Typhlichthys*

DISTRIBUTION: Eastern North America

HABITAT: Freshwater; temperate; benthic to demersal over soft and hard bottoms

REMARKS: The Percopsiformes is a small but distinctive group of North American freshwater fishes that is difficult to characterize with external features alone. Both the Pirate Perch (*Aphredoderus*) and the amblyopsids (Amblyopsidae) have the anus positioned far forward between the gill membranes in adults, while the trout-perches (Percopsidae) are the only members with an adipose fin. Percopsiform monophyly was questioned by Murray and Wilson (1999) but has since been supported by both morphological (Springer and Johnson, 2004) and molecular data (Dillman et al., 2011). Members of this group have a proliferation of superficial neuromasts on the skin that are especially prominent in the blind and unpigmented cave-dwelling fishes (*Amblyopsis* and *Typhlichthys;* Poulson, 1963). The phylogenetic hypothesis of Dillman et al. (2011) implies that eyes and pigment re-evolved in the genus *Chologaster* from a blind and unpigmented ancestor. The Pirate Perch (*Aphredoderus sayanus*) is a small, somewhat enigmatic fish typically found in slow-moving freshwaters of eastern North America. The location of the anus under the throat in adults is related to its unique and amazing spawning behavior in which both sexes pass gametes from the genital pore through the branchial and buccal cavities, and out the mouth, depositing them within fine, underwater tree roots (Poly and Wetzel, 2003). The Pirate Perch grows to 14 cm and feeds primarily on insect larvae, as well as eggs of amphibians. A recent study demonstrated their ability to use chemical camouflage to resemble and more readily approach their prey (Resetarits and Binckley, 2013).

REFERENCES: Dillman et al., 2011; Murray and Wilson, 1999; Poulson, 1963; Resetarits and Binckley, 2013; Springer and Johnson, 2004.

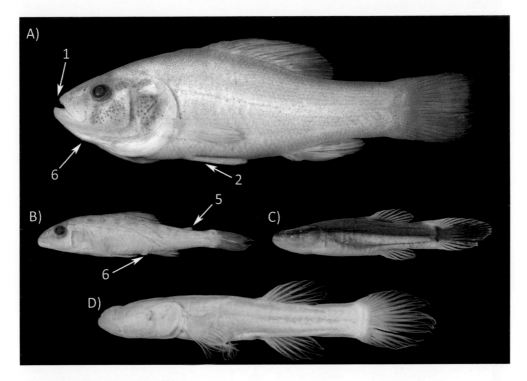

PERCOPSIFORM CHARACTERISTICS:

1) premaxilla nonprotrusible
2) pelvic fins, when present, posterior to pectoral-fin insertion
3) scales ctenoid
4) six branchiostegal rays
5) adipose fin present (Percopsidae) or absent (Aphredoderidae and Amblyopsidae)
6) anus in normal position (Percopsidae) or under throat in adults (Aphredoderidae and Amblyopsidae)

ILLUSTRATED SPECIMENS:

A) *Aphredoderus sayanus,* SIO 82-30, 53 mm SL (Aphredoderidae—pirate perches)
B) *Percopsis omiscomaycus,* SIO 62–457, 92 mm SL (Percopsidae—trout-perches)
C) *Chologaster cornuta,* UMMZ 211256, 32 mm SL (Amblyopsidae—cavefishes)
D) *Typhlichthys subterraneus,* UMMZ 103473, 42 mm SL, lateral (Amblyopsidae)

GADIFORMES—Cods and Relatives

The cods are a diverse group of over 600 species allocated among ten families. They are characterized by a relatively elongate body, absence of true fin spines, long dorsal and anal fins in most, pelvic fins thoracic to jugular (absent in some), pelvic fins with up to 11 rays, cycloid scales in most, and the maxilla excluded from the gape. The Gadiformes includes many commercially important species that inhabit continental shelf areas, especially in the north Atlantic and north Pacific. Relationships within this group are not well resolved. A variety of studies based on morphology and/or molecular data (e.g., Endo, 2002; Howes,

1989; Markle, 1989; Roa-Varón and Ortí, 2009; Teletchea et al., 2006; von der Heyden and Matthee, 2008) provide conflicting hypotheses. We cover two families below.

REFERENCES: Cohen, 1989; Cohen et al., 1990; Endo, 2002; Howes, 1989, 1991; Lloris et al., 2005; Markle, 1989; Roa-Varón and Ortí, 2009; Teletchea et al., 2006; von der Heyden and Matthee, 2008.

GADIFORMES : MACROURIDAE—Grenadiers and Rattails

DIVERSITY: 27 genera, 400 species
REPRESENTATIVE GENERA: *Caelorinchus, Coryphaenoides, Nezumia*
DISTRIBUTION: Atlantic, Indian, and Pacific oceans

MACROURID CHARACTERISTICS:
1) body elongate, greatly tapering posteriorly
2) second dorsal fin and anal fin confluent with tail, caudal fin absent, tail usually ending in a point
3) chin barbel usually present
4) anal-fin rays generally longer than rays in second dorsal fin
5) some species with a light organ on abdominal midline

ILLUSTRATED SPECIMENS:
A) *Nezumia stelgidolepis,* SIO 91-72, 185 mm SL
B) *Mesobius berryi,* SIO 73-170, 344 mm TL, holotype
C) *Coryphaenoides rudis,* SIO 68-479, 1,200 mm TL (tail broken)

HABITAT: Marine; tropical to polar; lower continental shelf to abyssal plain, demersal or benthopelagic over soft bottoms

REMARKS: The grenadiers are some of the deepest living fishes in the world, rarely occurring shallower than 250 m. They are predatory, and feed primarily on benthic and pelagic invertebrates and fishes. Many have very large eyes, consistent with their deep-dwelling habits. Some species reach large sizes and can be important to fisheries. The phylogenetic relationships have been studied based on morphology (Iwamoto, 1989) and molecular data (e.g., Satoh et al., 2006; Wilson and Attia, 2003).

REFERENCES: Cohen et al., 1990; Iwamoto, 1989, 2008; Iwamoto, in Carpenter, 2003; Morita, 1999; Orlov and Iwamoto, 2008; Satoh et al., 2006; Wilson, 1994; Wilson and Attia, 2003; Wilson et al., 1991.

GADIFORMES : GADIDAE—Cods

DIVERSITY: 12 genera, 23 species

REPRESENTATIVE GENERA: *Arctogadus, Gadus, Microgadus, Raniceps*

DISTRIBUTION: Arctic, Atlantic, and Pacific oceans

HABITAT: Marine; temperate to polar; continental shelf to continental slope, demersal and benthopelagic over soft bottoms

REMARKS: Cods are relatively large (reaching nearly 2 m), predatory fishes that live mostly in cold waters of the Northern Hemisphere. Cods feed on a variety of invertebrates including crabs, squid, and shrimp, as well as fishes. Many species are important for commercial fisheries across their entire geographic range. The once abundant Atlantic cod (*Gadus morhua*) supported decades of intensive fisheries but has since collapsed (Kurlansky, 1997).

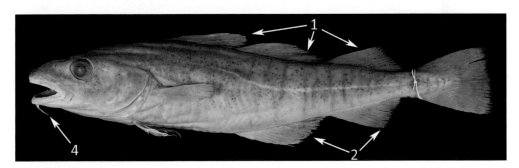

GADID CHARACTERISTICS:
1) three dorsal fins in most species
2) two anal fins in most species
3) teeth present on vomer
4) small chin barbel in most species
5) caudal fin with only one hypural bone

ILLUSTRATED SPECIMEN:
Gadus morhua, SIO 64–164, 290 mm SL

(account continued)

Despite high fecundity (millions of eggs per spawning) and more than a decade of greatly reduced fisheries catch in Canada and the United States, the species has not recovered.

REFERENCES: Bakke and Johansen, 2005; Cohen et al., 1990; Coulson et al., 2006; Dunn, 1989; Kurlansky, 1997; Mecklenburg et al., 2002; Teletchea et al., 2006; von der Heyden and Matthee, 2008.

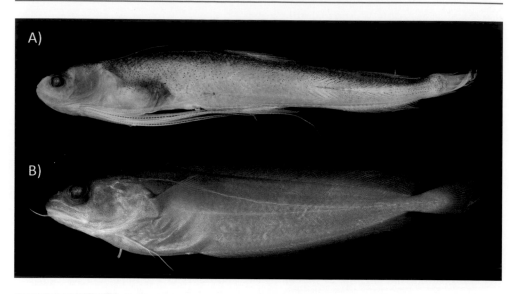

GADIFORM DIVERSITY:
A) BREGMACEROTIDAE—codlets: *Bregmaceros bathymaster,* SIO 91–177, 58 mm SL
B) MORIDAE—deep-sea cods: *Lotella fernandeziana,* SIO 64–645, 93 mm SL

ZEIFORMES—Dories

DIVERSITY: 6 families, 16 genera, 33 species

REPRESENTATIVE GENERA: *Allocyttus, Cyttus, Zeus*

DISTRIBUTION: Atlantic, Indian, and Pacific oceans

HABITAT: Marine; tropical to temperate; pelagic to demersal, above continental shelves, slopes, and seamounts, usually over soft bottoms

REMARKS: Most species of dories occur in less than 600-m depths, but the Oreosomatidae (oreos) are found as deep as 1,000 m. The John Dory (*Zeus faber*), with its distinctive elongate dorsal-fin spines and prominent ocellus on the body, has been a prized food fish since ancient times, although caught in relatively small numbers. The anatomy, systematics, and relationships of the group have been studied by Tyler et al. (2003), Tyler and Santini (2005), and Nolf and Tyler (2006).

REFERENCES: Nolf and Tyler, 2006; Rosen, 1984; Tyler and Santini, 2005; Tyler et al., 2003.

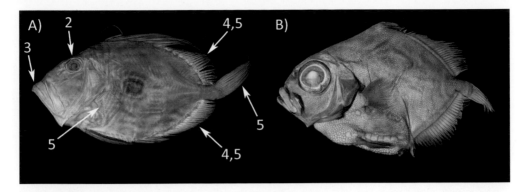

ZEIFORM CHARACTERISTICS:

1) body laterally compressed, extremely so in some species
2) eyes positioned high on head
3) jaws oblique, usually strongly protrusible
4) dorsal and anal fins with long bases
5) dorsal-, anal-, and pectoral-fin rays unbranched; caudal-fin rays branched
6) vomer with teeth, palatines without teeth
7) three and one-half gills

ILLUSTRATED SPECIMENS:

A) *Zeus faber,* SIO 84–282, 206 mm SL (Zeidae—dories)
B) *Allocyttus verrucosus,* SIO 85–6, 115 mm SL (Oreosomatidae—oreos)

STEPHANOBERYCIFORMES—Pricklefishes

DIVERSITY: 7 families, 29 genera, 87 species

REPRESENTATIVE GENERA: *Barbourisia, Melamphaes, Rondeletia*

DISTRIBUTION: Atlantic, Indian, and Pacific oceans

HABITAT: Marine; tropical to temperate; mesopelagic to bathypelagic

REMARKS: The Stephanoberyciformes is difficult to characterize, having the body variously shaped (rounded in most species) and bones of the skull generally thin. They typically have obvious, widened lateral-line canals on the head and sometimes on the trunk (Webb, 1989). Pricklefishes are found in all oceans of the world and include an array of mesopelagic and bathypelagic species, ranging from the mesopelagic "bigscale fishes" (Melamphaidae) to the bathypelagic whalefishes (Barbourisiidae, Rondeletiidae, and Cetomimidae). Some authors (e.g., Eschmeyer and Fong, 2013) place the whalefishes in their own order, while others (e.g., Betancur et al., 2013; Near et al., 2012) include them within the Beryciformes. We follow Wiley and Johnson (2010) in treating this group separate from the Beryciformes. In an extraordinary study, Johnson et al. (2009) demonstrated that two previously recognized families, the Megalomycteridae (largenose fishes) and the Mirapinnidae (tapertails), are in fact different life stages of the flabby whalefishes (Cetomimidae).

REFERENCES: Ebeling and Weed, 1973; Johnson et al., 2009; Kotlyar, 1996; Moore, 1993; Paxton, 1989; Paxton et al., 2001; Webb, 1989.

(account continued)

STEPHANOBERYCIFORM CHARACTERISTICS:
1) body variably shaped, rounded in many
2) bones of skull thin
3) no teeth on palate
4) cephalic lateral-line canals wide

ILLUSTRATED SPECIMENS:
A) *Melamphaes acanthomus,* SIO 87–150, 101 mm SL (Melamphaidae—bigscales)
B) *Barbourisia rufa,* SIO 73–336, 166 mm SL (Barbourisiidae—velvet whalefishes)
C) *Rondeletia loricata,* SIO 76–110, 88 mm SL (Rondeletiidae—redmouth whalefishes)

BERYCIFORMES—Alfonso Squirrelfishes

Traditionally the Beryciformes included a diverse array of over 160 species classified in seven families and 29 genera. Some recent molecular hypotheses (e.g., Near et al., 2012) have concluded that the Beryciformes is paraphyletic without the inclusion of the Stephanoberyciformes, while others (e.g., Betancur et al., 2013) suggest that one lineage, the Holocentridae, may be sister to percomorph fishes independent of other beryciforms. The traditional beryciform fishes are found from coastal marine areas to the deep sea. They are characterized by Jakubowski's organ, a unique cluster of sensory cells including the terminal supraorbital neuromasts located between the nasal and lacrimal bones (Freihofer, 1978; Jakubowski, 1974; Johnson and Patterson, 1993; Zehren, 1979). They also have more than five pelvic-fin rays, 16–17 branched caudal-fin rays, and in some species, the maxilla is partially included in the gape. The shallow-water species generally have strong, stout fin spines. The group includes the flashlightfishes (Anomalopidae), unique for having an organ with bioluminescent bacteria in the lower portion of the eye (Johnson and Rosenblatt, 1988); the pinecone fishes (Monocentridae), with their body encased in heavy scales; and the slimeheads or roughies (Trachichthyidae), which are often abundant near seamounts. The latter group includes extremely long-lived species (well over 100 years in the case of the

Orange Roughy). Unsustainable fishing practices have resulted in precipitous declines in that species (Branch, 2001).

REFERENCES: Branch, 2001; Freihofer, 1978; Jakubowski, 1974; Johnson and Patterson, 1993; Johnson and Rosenblatt, 1988; Kotlyar, 1996; Moore, 1993; Paxton, 1989; Paxton et al., 2001; Zehren, 1979.

BERYCIFORMES : ANOPLOGASTRIDAE—Fangtooths

DIVERSITY: 1 genus, 2 species

REPRESENTATIVE GENUS: *Anoplogaster*

DISTRIBUTION: Atlantic, Indian, and Pacific oceans

HABITAT: Marine; tropical to temperate; mesopelagic to abyssopelagic

REMARKS: Fangtooths, with their laterally compressed bodies, large heads, large mouths, and long, fang-like teeth, are distinctive predators in the mesopelagic and bathypelagic zones. These fishes are known to live in a very broad depth range of 75 to 5,000 m, and unlike most beryciform fishes, they lack fin spines. The two known species are indistinguishable

ANOPLOGASTRID CHARACTERISTICS:
1) mouth large, upper and lower jaws with numerous fang-like teeth
2) eyes small
3) scales small, nonoverlapping
4) spines absent from all fins
5) pelvic fins with seven segmented rays
6) anal fin short and posteriorly positioned
7) trunk lateral-line canal relatively open, covered by soft tissue only

ILLUSTRATED SPECIMEN:
Anoplogaster cornuta, SIO 07–165, 119 mm SL

as adults but are easily differentiated as larvae and juveniles by differences in their head spines (Kotlyar, 1987).

REFERENCES: Kotlyar, 1987, 1996, 2003; Moore, in Carpenter, 2003.

BERYCIFORMES : HOLOCENTRIDAE—Squirrelfishes

DIVERSITY: 8 genera, 84 species

REPRESENTATIVE GENERA: *Holocentrus, Myripristis, Plectrypops, Sargocentron*

DISTRIBUTION: Atlantic, Indian, and Pacific oceans

HABITAT: Marine; tropical to warm temperate; coastal to upper continental slope, demersal, typically associated with reefs

REMARKS: Squirrelfishes are conspicuous, nocturnal predators on coral reefs where they are often observed beneath ledges during the day. Most species are reddish in color, with large eyes, prominent ridges and mucous canals on the head, and a well-developed otophysic connection. Squirrelfishes have especially distinctive larvae with prominent spines on the head and a relatively long pelagic phase (Tyler et al., 1993). Two lineages are recognized, the Holocentrinae with a strong preopercular spine and large anal-fin spine, and the Myripristinae, most members of which lack a spine at the angle of the preopercle and have a smaller anal-fin spine (Nelson, 2006). In contrast to most lineages of coastal fishes, the

HOLOCENTRID CHARACTERISTICS:
1) pelvic fins with strong spine
2) dorsal fins separated by deep notch
3) caudal fin forked
4) anal fin with four spines
5) scales large and strongly ctenoid
6) superficial bones of head often serrate or spinous
7) eyes large

ILLUSTRATED SPECIMEN:
Neoniphon sammara, SIO 92–159, 133 mm SL

shallow-water squirrelfishes apparently are derived from deep-dwelling ancestors (Randall and Greenfield, in Carpenter and Niem, 1999).

REFERENCES: Greenfield, in Carpenter, 2003; Randall, 1998; Randall and Greenfield, 1996; Randall and Greenfield, in Carpenter and Niem, 1999; Tyler et al., 1993.

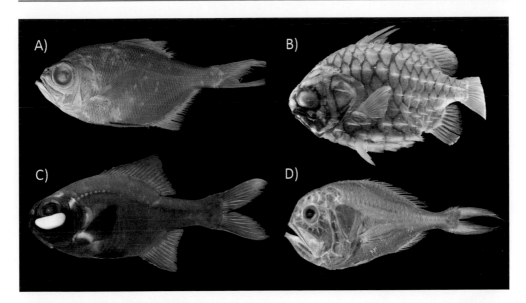

BERYCIFORM DIVERSITY:
A) BERYCIDAE—alfonsinos: *Beryx decadactylus*, SIO 85-77, 289 mm SL
B) MONOCENTRIDAE—pinecone fishes: *Cleidopus gloriamaris*, SIO 75-415, 103 mm SL
C) ANOMALOPIDAE—flashlightfishes: *Photoblepharon palpebratum*, SIO 92-125, 65 mm SL
D) TRACHICHTHYIDAE—slimeheads/roughies: *Hoplostethus atlanticus*, SIO 72-195, 97 mm SL

PERCOMORPHA

The limits and relationships of percomorph fishes are poorly understood and remain under intense study by a variety of researchers. Current concepts of the group differ from traditional concepts in part by the inclusion of the mullets (Mugiliformes), silversides and relatives (Atherinomorpha), sticklebacks (Gasterosteiformes), pipefishes and relatives (Syngnathiformes), scorpionfishes (Scorpaeniformes), sculpins and allies (Cottiformes), cuskeels (Ophidiiformes), and anglerfishes (Lophiiformes). In addition, the distinctive flat-fishes (Pleuronectiformes) and the puffers and relatives (Tetraodontiformes), long classified separately from the percomorphs, are clearly nested within the group. It is not possible to diagnose the group at the present time, but in general it includes perch-like fishes with well-developed fin spines and anteriorly inserted pelvic fins (thoracic to jugular), as well as a wide variety of lineages derived from them. These features are not unique to this group and they are lost or modified in many included lineages. Any arrangement of taxa within the percomorphs is necessarily tentative. We largely follow the sequence presented in Helfman and Collette (2011), which is based on Nelson (2006) and Wiley and Johnson (2010), with additional modifications where warranted. We follow Wiley and Johnson (2010) in recognizing many included lineages as orders rather than suborders of the Perciformes, as in Nelson (2006) and Helfman and Collette (2011). This is done not to reflect the relative importance or distinctiveness of a group, but to emphasize the current lack of consensus regarding the higher-level classification of percomorphs.

MUGILIFORMES : MUGILIDAE—Mullets

DIVERSITY: 1 family, 17 genera, 72 species

REPRESENTATIVE GENERA: *Agonostomus, Joturus, Liza, Mugil*

DISTRIBUTION: Atlantic, Indian, and Pacific oceans

HABITAT: Marine to freshwater; tropical to temperate; generally neritic and demersal over soft bottoms

MUGILIFORM CHARACTERISTICS:
1) two widely separated dorsal fins, first with four spines
2) pelvic fins inserted well behind pectoral fins
3) scales typically ctenoid
4) trunk lateral-line canal poorly developed or absent
5) mouth small, triangular with small or absent teeth
6) gill rakers long
7) stomach muscular; intestine extremely long

ILLUSTRATED SPECIMENS:
A) *Liza falcipinnis,* SIO 63–563, 133 mm SL
B) *Agonostomus monticola,* SIO 67–237, 106 mm SL
C) *Chaenomugil proboscideus,* SIO 73–69, radiograph

REMARKS: Most mullets are schooling, usually surface-dwelling fishes, well known for their curious behavior of leaping from the water. With their small mouth, muscular stomach, and greatly elongate intestine, mullets form an important ecological link between detritus and higher trophic levels within coastal ecosystems (Odum, 1970). Most mullets are euryhaline and migrate to spawn, with coastal species spawning offshore and freshwater species generally spawning in estuaries (Harrison, in Carpenter, 2003). The relationships of mullets have been controversial (Johnson and Patterson, 1993; Stiassny, 1993), but recent molecular studies have included them in the Ovalentaria, a recently recognized lineage of percomorphs (e.g., Li et al., 2009; Near et al., 2012; Wainwright et al., 2012). A recent study on relationships within the Mugilidae (Durand et al., 2012) implies that many of the currently recognized genera are not monophyletic. Mullets support artisanal fisheries around the world, and some species are important in aquaculture.

REFERENCES: Durand et al., 2012; Harrison, in Fischer et al., 1995; Harrison, in Carpenter, 2003; Harrison and Senou, in Carpenter and Niem, 1999; Li et al., 2009; Odum, 1970; Stiassny, 1993; Thomson, 1997; Wainwright et al., 2012.

ATHERINOMORPHA

The Atherinomorpha includes the Atheriniformes, Beloniformes, and Cyprinodontiformes (Parenti, 1993, 2005). Wiley and Johnson (2010) list 17 synapomorphies for the group, sev-

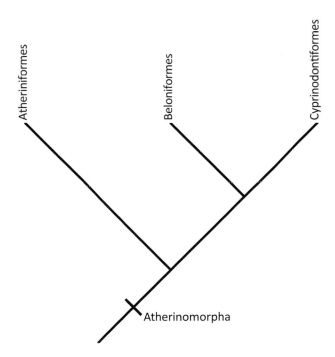

Hypothesized phylogenetic relationships of the Atherinomorpha after Parenti (2005).

eral of which are related to reproduction and early development. Relationships of the group have been studied in detail by Parenti (1993, 2005). Beloniformes and Cyprinodontiformes are sister lineages, sharing derived conditions of four characters, including absence of a stomach and pyloric caecae (Parenti, 2005; Rosen and Parenti, 1981; Wiley and Johnson, 2010).

ATHERINIFORMES—Silversides and Relatives

The silversides and their relatives are characterized by a relatively elongate body, two dorsal fins in most species, with the first dorsal fin, if present, comprising flexible spines. They also usually have a single, weak anal-fin spine, a weak or absent trunk lateral-line canal, cycloid scales, and paired nostril openings. The position of the pelvic fins varies from abdominal to thoracic. The ten families, about 50 genera, and 343 species constitute the exclusively marine surf sardines (Notocheiridae) and several groups found in both marine and freshwaters, including the New World silversides (Atherinopsidae, covered below), the Old World silversides (Atherinidae), the rainbowfishes (Melanotaeniidae) from Madagascar, New Guinea, and Australia, and the unusual priapiumfishes (Phallostethidae) from southeast Asia.

REFERENCES: Dyer and Chernoff, 1996; Parenti, 2005; Rosen, 1964.

ATHERINIFORMES : ATHERINOPSIDAE—New World Silversides

DIVERSITY: 11 genera, 109 species

REPRESENTATIVE GENERA: *Atherinopsis, Labidesthes, Leuresthes, Menidia*

DISTRIBUTION: North and South America.

HABITAT: Coastal marine and brackish waters, as well as freshwater streams and lakes; tropical to temperate; usually over soft bottoms.

REMARKS: This lineage of silversides is restricted to the New World and can be distinguished from all other silversides (e.g., the Atherinidae *sensu stricto*) by the presence of a protrusible premaxillary. Two subfamilies, the Atherinopsinae and the Menidiinae, are recognized, and Dyer (1997, 1998) and Setiamarga et al. (2008) have studied their relationships. New World silversides feed on zooplankton and small benthic invertebrates. This group includes the grunions (*Leuresthes*), well known for predictable spawning events in which they temporarily leave the water to bury their eggs in the intertidal, where the eggs remain until subsequent high tides reach them, stimulating hatching. Some species of silversides have been shown to exhibit temperature-dependent sex determination (Conover and Kynard, 1981), while others are parthenogenetic (Echelle et al., 1983). The two grunion species support recreational fisheries in southern California and northwestern Mexico.

REFERENCES: Chernoff, in Carpenter, 2003; Conover and Kynard, 1981; Dyer, 1997, 1998; Echelle et al., 1983; Setiamarga et al., 2008.

ATHERINOPSID CHARACTERISTICS:
1) two widely separated dorsal fins, the first with two to nine flexible spines
2) mouth small and terminal
3) premaxillary protrusible
4) pectoral fins inserted high on body
5) pelvic fins usually abdominal
6) trunk lateral-line canal absent
7) broad, silvery lateral stripe present in life

ILLUSTRATED SPECIMENS:
A) *Odontesthes gracilis*, SIO 65–634, 123 mm SL (Atherinopsinae)
B) *Atherinops affinis*, SIO uncataloged, 79 mm SL (Atherinopsinae)
C) *Labidesthes sicculus*, SIO 80–61, 46 mm SL (Menidiinae)

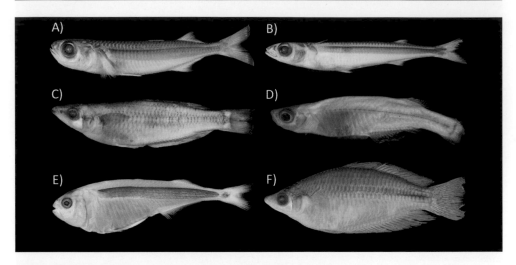

ATHERINIFORM DIVERSITY:
A) ATHERINIDAE—Old World silversides: *Atherinomorus lacunosus,* SIO 77-349, 76 mm SL
B) ATHERINIDAE—*Hypoatherina temminckii,* SIO 56-126, 81 mm SL
C) BEDOTIIDAE—Madagascar rainbowfishes: *Bedotia geayi,* UMMZ 223575, 87 mm SL
D) PHALLOSTETHIDAE—priapiumfishes: *Gulaphallus mirabilis,* SIO 80-102, 22 mm SL
E) NOTOCHEIRIDAE—surf sardines: *Iso rhothophilus,* UMMZ 217631, 49 mm SL
F) MELANOTAENIIDAE—rainbowfishes: *Melanotaenia splendida,* SIO 84-286, 79 mm SL

BELONIFORMES—Needlefishes and Relatives

The needlefishes, flyingfishes, and their relatives are a diverse group that includes six families, 36 genera, and nearly 260 species. They are characterized by having the lower pharyngeal bones fused into a triangular plate, the upper jaw fixed or nonprotrusible, the dorsal and anal fins located on the rear half of the body, abdominal pelvic fins with six soft rays, and the lower caudal-fin lobe with more principal rays than the upper lobe. They also lack fin spines and an interhyal bone. Included in this group are the flyingfishes (Exocoetidae), the halfbeaks (Hemiramphidae), and the needlefishes (Belonidae), all described further below, as well as the ricefishes/duckbilled fishes (Adrianichthyidae), the sauries (Scomberesocidae), and the predominately freshwater halfbeaks (Zenarchopteridae).

REFERENCES: Parenti, 2005, 2008; Rosen and Parenti, 1981.

BELONIFORMES : BELONIDAE—Needlefishes

DIVERSITY: 10 genera, 38 species

REPRESENTATIVE GENERA: *Ablennes, Strongylura, Tylosurus*

DISTRIBUTION: Atlantic, Indian, and Pacific oceans

HABITAT: Marine, occasionally in freshwater; tropical to warm temperate; neritic and epipelagic, coastal to oceanic

REMARKS: Needlefishes are long, slender fishes (reaching 2 m in some species), noted for their elongate beaks, which are armed with sharp teeth. They feed primarily on small fishes at the surface and capture their prey with a fast, lateral motion of the head. They are known to leap from the water when startled, posing a danger to fishers. Some species are fished heavily, and their flesh is generally highly regarded, though their sometimes green bones may deter some people from eating them.

REFERENCES: Collette, in Carpenter and Niem, 1999; Collette, in Carpenter, 2003; Collette, 2003a; Lovejoy, 2000; Lovejoy and Collette, 2001; Lovejoy et al., 2004.

BELONID CHARACTERISTICS:
1) mouth large, both upper and lower jaws elongate
2) jaws with numerous needle-like teeth
3) body long and slender
4) scales small, cycloid
5) trunk lateral-line canal low on body
6) nostrils located in depression anterior to eyes

ILLUSTRATED SPECIMEN:
Strongylura exilis, SIO 49–139, 318 mm SL

BELONIFORMES : HEMIRAMPHIDAE—Halfbeaks

DIVERSITY: 7 genera, 61 species

REPRESENTATIVE GENERA: *Euleptorhamphus, Hemiramphus, Hyporhamphus*

DISTRIBUTION: Atlantic, Indian, and Pacific oceans

HABITAT: Marine, occasionally in freshwater; tropical to warm temperate; neritic to epipelagic, coastal to oceanic

REMARKS: Halfbeaks are distinguished from the needlefishes by having only the lower jaw elongate. They are omnivorous with a modified pharyngeal mill (Tibbetts and Carseldine, 2003) and are known to feed on floating vegetation, as well as small crustaceans and fishes. Some species leap from the water when startled, and one species can glide long distances much like a flyingfish. Halfbeaks are eaten in many parts of the world and are particularly important as bait when fishing for large pelagic piscivores.

REFERENCES: Banford and Collette, 2001; Collette, in Carpenter and Niem, 1999; Collette, in Carpenter, 2003; Collette, 2004; Tibbetts and Carseldine, 2003.

(account continued)

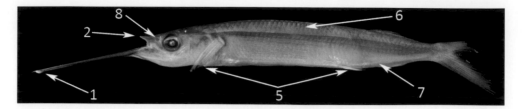

HEMIRAMPHID CHARACTERISTICS:

1) lower jaw elongate, upper jaw short
2) mouth small
3) premaxillae pointed anteriorly
4) teeth minute or absent
5) pectoral fins and pelvic fins short
6) scales relatively large, cycloid
7) trunk lateral-line canal low on body
8) nostrils located in depression anterior to eyes

ILLUSTRATED SPECIMEN:
Hemiramphus brasiliensis, SIO 63–23, 97 mm SL

BELONIFORMES : EXOCOETIDAE—Flyingfishes

DIVERSITY: 8 genera, 66 species

REPRESENTATIVE GENERA: *Cheilopogon, Exocoetus, Fodiator, Hirundichthys*

DISTRIBUTION: Atlantic, Indian, and Pacific oceans

HABITAT: Marine; tropical to warm temperate; neritic and epipelagic, coastal to oceanic

REMARKS: Flyingfishes have incredibly long pectoral fins that they use to glide, sometimes for extremely long distances, reportedly as much as 400 m. Some species, known as the "four wing" flyingfishes, also have elongate pelvic fins that aid in gliding. This adaptation is clearly a means of avoiding predation, as flyingfishes are the favorite prey of several pelagic predators. Their highly deciduous scales may also be a means of escaping predation. They feed primarily on zooplankton and small fishes and can reach up to 45 cm in length. Although many species school, they do not undertake long migrations. Though flyingfishes are epipelagic, they are not broadcast spawners and instead attach their eggs, via sticky filaments, to floating debris. These fishes are pursued with gillnets and purse seines and are used for human consumption and as bait. Lewallen et al. (2011) studied their phylogenetic relationships, recognizing four subfamilies.

REFERENCES: Dasilao and Sasaki, 1998; Lewallen et al., 2011; Parin, in Carpenter, 2003; Parin, in Carpenter and Niem, 1999.

EXOCOETID CHARACTERISTICS:

1) pectoral fins large and wing-like
2) pelvic fins large in some species
3) mouth small, upper and lower jaws short
4) teeth minute or absent
5) trunk lateral-line canal low on body
6) scales relatively large and cycloid
7) lower lobe of caudal fin longer than upper lobe
8) gas bladder extended posteriorly into haemal canal

ILLUSTRATED SPECIMEN:
Hirundichthys marginatus, SIO 52–357, 134 mm SL (lateral and dorsal views).

CYPRINODONTIFORMES—Killifishes

The Cyprinodontiformes comprises ten families, well over 100 genera, and 1,213 species, many of which are especially well known. They are characterized by a lateral line with pores on the head and pitted scales on the body; paired nostril openings; a protrusible upper jaw bordered by the premaxilla; a symmetrical, truncate, or rounded caudal fin supported by a single bony element; and pectoral fins usually set low on the body. The systematics and taxonomy of the killifishes have been studied extensively. Hertwig (2008), using features of the head musculature, largely corroborated the results of other studies (Costa, 1998; Parenti, 1981, 1993), supporting the monophyly of the Cyprinodontiformes as including most notably the Rivulidae (New World rivulines), Fundulidae (topminnows), Goodeidae (goodeids), Cyprinodontidae (pupfishes), and Poeciliidae (livebearers). The early diversification of the group appears to be related to the breakup of Gondwana (Hertwig, 2008).

Killifishes are small and found mostly in freshwater and coastal marine areas, especially marshes. The group includes several popular aquarium fishes (e.g., guppies, mollies, and swordtails), and the behavior and ecology of many species have been studied extensively. Certain members of the Cyprinodontiformes exhibit internal fertilization and give birth to well-developed young (= embryoparity). Phylogenetic reconstructions indicate that these reproductive features evolved independently three times within the group (Parenti, 2005).

REFERENCES: Costa, 1998; Hertwig, 2008; Parenti, 1981, 1993, 2005.

CYPRINODONTIFORMES : FUNDULIDAE—Topminnows

DIVERSITY: 3 genera, 41 species

REPRESENTATIVE GENERA: *Fundulus, Leptolucania, Lucania*

DISTRIBUTION: North America, Cuba, and Bermuda

HABITAT: Freshwater to brackish waters and coastal marine; tropical to temperate; neritic to demersal in slow-moving, shallow waters and in estuaries, generally over soft bottoms

REMARKS: Topminnows are omnivorous, feeding on insects and other invertebrates, vegetation, and other small fishes. They live from the water's surface to the shallow substrate. As the name implies, most species can be observed or captured near the surface. Relationships within the Fundulidae have been studied by several researchers (e.g., Wiley, 1986; Bernardi and Powers, 1995; Bernardi, 1997; Ghedotti and Davis, 2013). While the majority of topminnow species live in freshwater, Whitehead (2010) concluded that the ancestral condition was marine, and that freshwaters were invaded repeatedly during the group's evolutionary history. The reverse transition from freshwater to marine waters has apparently not occurred in the group. Species of the genus *Fundulus* are considered a "model organism" for studies of fish physiology (Burnett et al., 2007).

REFERENCES: Bernardi, 1997; Bernardi and Powers, 1995; Burnett et al., 2007; Ghedotti and Davis, 2013; Whitehead, 2010; Wiley, 1986.

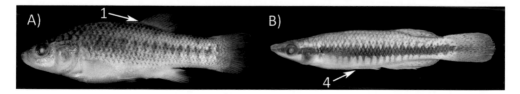

FUNDULID CHARACTERISTICS:
1) dorsal-fin origin at midbody, anterior to or above anal-fin origin
2) ventral arm of maxillae often with hooks
3) maxillae twisted
4) pelvic fins abdominal
5) teeth conical

ILLUSTRATED SPECIMENS:
A) *Fundulus parvipinnis,* SIO 09–234, 65 mm SL
B) *Fundulus notatus,* SIO 04–184, 42 mm SL

CYPRINODONTIFORMES : CYPRINODONTIDAE—Pupfishes

DIVERSITY: 9 genera, 123 species

REPRESENTATIVE GENERA: *Aphanius, Cubanichthys, Cyprinodon, Floridichthys, Jordanella*

DISTRIBUTION: North America to northern South America, and the Mediterranean basin including northern Africa

HABITAT: Freshwater to brackish waters and coastal marine; tropical to temperate; neritic to demersal in slow-moving, shallow waters and in estuaries, usually over soft bottoms

REMARKS: The generally small pupfishes are, like the topminnows, omnivorous, feeding on insects and other invertebrates, vegetation, and other small fishes. Some species are remarkable in that they are completely restricted to very small (< 10 m^2), isolated bodies of water in desert oases. As a consequence, many species are considered threatened or endangered (Minckley and Deacon, 1991; IUCN, 2013). The two species of *Cubanichthys* from Cuba and Jamaica are sisters to the remaining cyprinodontids and are the only two species of pupfishes without tricuspid teeth. A number of species are used as bait fishes and some feature in the aquarium trade.

REFERENCES: Costa, 2003; Echelle et al., 2005; Fuller et al., 2007; Hrbek and Meyer, 2003; Parenti, 1981; Wiley, 1986.

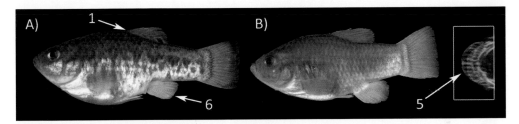

CYPRINODONTID CHARACTERISTICS:
1) dorsal-fin origin anterior to anal-fin origin
2) maxillae with lateral process
3) left and right maxillae nearly joined medially
4) body moderately deep
5) teeth usually tricuspid
6) all anal-fin rays branched

ILLUSTRATED SPECIMENS:
A) *Cyprinodon macularius,* SIO 62–161, 41 mm SL (female)
B) *Cyprinodon macularius,* SIO 62–161, 39 mm SL (male)

INSET: *Cyprinodon variegatus,* SIO 62–25, radiograph showing tricuspid teeth

CYPRINODONTIFORMES : POECILIIDAE—Livebearers

DIVERSITY: 37 genera, 353 species

REPRESENTATIVE GENERA: *Gambusia, Heterandria, Poecilia, Poeciliopsis, Xiphophorus*

DISTRIBUTION: Eastern North America to South America and Africa (including Madagascar)

HABITAT: Freshwater to brackish waters; tropical to temperate; generally demersal in slow-moving, shallow waters and in estuaries

REMARKS: The behavior and ecology of poeciliids is especially well known (Evans et al. 2011). As the common name implies, livebearers are embryoparous, and males have a pronounced gonopodium, used for internal fertilization. Sexual selection and mate choice have been studied extensively in several species, most notably the Guppy (*Poecilia reticulata*) and several species of swordtails (*Xiphophorus*). The group also includes a number of parthenogenic species (Schlupp, 2005), including the Amazon Molly (*Poecilia formosa*). The phylo-

POECILIID CHARACTERISTICS:
1) pectoral fins positioned high on body
2) pelvic fins inserted relatively far anteriorly
3) males with intromittent organ derived from modified anal-fin elements
4) head flattened, scaled dorsally
5) pleural ribs on first several haemal arches

ILLUSTRATED SPECIMENS:
A) *Gambusia affinis,* SIO 52–224, 44 mm SL (female)
B) *Poecilia latipinna,* SIO 69–166, 50 mm SL, radiograph (male)
C) *Lamprichthys tanganicanus,* SIO 64–242, 75 mm SL

INSET: Close-up of gonopodium of (B)

genetic relationships of poeciliids have been studied extensively (Ghedotti, 2000; Hertwig, 2008; Hrbek et al., 2007; Lucinda and Reiss, 2005; Parenti and Rauchenberger, 1989; Rosen and Bailey, 1963). The Mosquitofish, *Gambusia affinis,* has been widely introduced to control mosquito populations (Vidal et al., 2010), and several species of livebearers (e.g., guppies, mollies, and swordtails) are prominent in the aquarium trade.

REFERENCES: Evans et al., 2011; Ghedotti, 2000; Hertwig, 2008; Hrbek et al., 2007; Lucinda and Reiss, 2005; Meffe and Snelson, 1989; Parenti and Rauchenberger, 1989; Rauchenberger, 1989; Rosen and Bailey, 1963; Schlupp, 2005; Vidal et al., 2010.

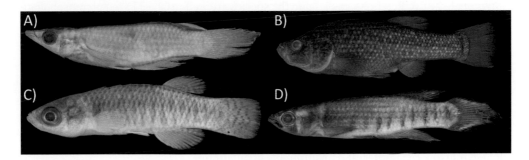

CYPRINODONTIFORM DIVERSITY:

A) APLOCHEILIDAE—Asian killifishes: *Aplocheilus panchax,* UMMZ 187859, 38 mm SL
B) GOODEIDAE—splitfins: *Goodea luitpoldii,* SIO 53-34, 126 mm SL
C) VALENCIIDAE—toothcarps: *Valencia letourneuxi,* UMMZ 213902, 25 mm SL
D) NOTHOBRANCHIIDAE—African rivulines: *Epiplatys sexfasciatus,* UMMZ 220252, 56 mm SL

GASTEROSTEIFORMES—Sticklebacks

The recently redefined Gasterosteiformes is a small group of interesting fishes found in coastal marine waters as well as freshwater streams and lakes. They are characterized by a long caudal peduncle, a series of isolated dorsal-fin spines (absent in Hypoptychidae), a posteriorly located soft dorsal fin, and a protrusible upper jaw. Considered by many (e.g., Britz and Johnson, 2002; Nelson, 2006; Wiley and Johnson, 2010) to be closely related to the pipefishes and relatives (Syngnathiformes), recent molecular studies (e.g., Betancur et al., 2013; Kawahara et al., 2008; Li et al., 2009) have shown the two groups to be unrelated. As currently construed, the group includes seven genera and 21 species allocated among three families: the monotypic Aulorhynchidae (Tubesnout), the Hypoptychidae, with two species of so-called sand eels, and the Gasterosteidae (sticklebacks), covered below. The Indostomidae (armored sticklebacks), pictured below and included in this group by some (e.g., Wiley and Johnson, 2010), may be more closely related to synbranchids (Near et al., 2012).

REFERENCES: Bowne, 1994; Britz and Johnson, 2002; Kawahara et al., 2008; Keivany and Nelson, 2000, 2006; Pietsch, 1978.

GASTEROSTEIFORMES : GASTEROSTEIDAE—Sticklebacks

DIVERSITY: 5 genera, 18 species

REPRESENTATIVE GENERA: *Apeltes, Culea, Gasterosteus, Pungitius, Spinachia*

DISTRIBUTION: Coastal and freshwaters of the Northern Hemisphere

HABITAT: Marine, brackish waters and freshwater lakes and streams; temperate; neritic to demersal usually over soft bottoms

REMARKS: Sticklebacks, especially the Threespine Stickleback (*Gasterosteus aculeatus*), are among the most well-studied fishes, serving as model systems for a variety of ecological, ethological, and evolutionary studies (Wooten, 1984), as well as for developmental genetics (Jones et al., 2012). Relationships among the species are well resolved based on morphological, behavioral, and genetic data (Mattern and McLennan, 2004). The Threespine Stickleback exhibits an extraordinary amount of genetic variation among populations, with clear evidence of repeated, parallel divergence of freshwater populations from a common marine ancestor (Rundle et al., 2000). Sticklebacks show evidence of rapid responses to predation pressures, including changes in the fin spines and number and size of the lateral plates (modified scales), in both fossil and extant populations (Bell, 1974; Schluter et al., 2010). Sticklebacks are omnivorous and males guard eggs deposited in their nests. Some species include marine, freshwater, and anadromous populations.

REFERENCES: Bell, 1974; Bell and Foster, 1994; Bowne, 1994; Jones et al., 2012; Keivany and Nelson, 2000, 2006; Mattern, 2004; Mattern and McLennan, 2004; Rundle et al., 2000; Schluter et al., 2010; Wootton, 1984.

GASTEROSTEID CHARACTERISTICS:
1) body typically elongate, with long, narrow caudal peduncle
2) dorsal fin comprises 2–16 well-developed isolated spines followed by a soft dorsal with 6–14 rays
3) pelvic fin (when present) with one spine and two soft rays
4) mouth small, oblique
5) three branchiostegal rays

ILLUSTRATED SPECIMEN:
Gasterosteus aculeatus, SIO 62-797, 56 mm SL

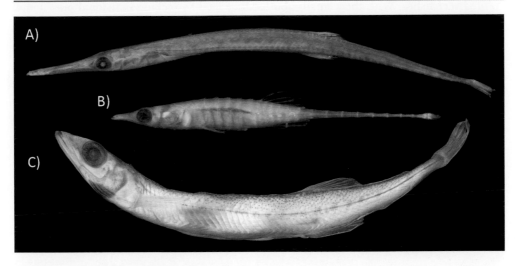

GASTEROSTEIFORM DIVERSITY:
A) AULORHYNCHIDAE—Tubesnout: *Aulorhynchus flavidus,* SIO 76–327, 125 mm SL
B) INDOSTOMIDAE—armored sticklebacks: *Indostomus paradoxus,* SIO 70–383, 25 mm SL
C) HYPOPTYCHIDAE—sand eels: *Hypoptychus dybowskii,* SIO 52–179, 72 mm SL

SYNGNATHIFORMES—Pipefishes and Relatives

The pipefishes and relatives constitute an unusual group of six families, 62 genera, and over 360 species. They include the well-known pipefishes and seahorses (covered below), as well as some obscure and odd groups such as the seamoths (Pegasidae), ghost pipefishes (Solenostomidae), shrimpfishes and snipefishes (Centriscidae), trumpetfishes (Aulostomidae), and cornetfishes (Fistulariidae). Most members of this group have a tube-shaped snout, a small mouth with a non-protrusible upper jaw, and abdominal pelvic fins (absent in some; Pietsch, 1978). Their phylogenetic relationships have been studied most recently by Kawahara et al. (2008), Wilson and Rouse (2010), and Wilson and Orr (2011).

REFERENCES: Kawahara et al., 2008; Pietsch, 1978; Wilson and Orr, 2011; Wilson and Rouse, 2010.

SYNGNATHIFORMES : SYNGNATHIDAE—Pipefishes and Seahorses

DIVERSITY: 52 genera, 338 species

REPRESENTATIVE GENERA: *Corythoichthys, Cosmocampus, Hippocampus, Syngnathus*

DISTRIBUTION: Atlantic, Indian, and Pacific oceans

HABITAT: Coastal areas, brackish waters, with a few species in freshwater; tropical to temperate; benthic to demersal over soft and hard bottoms, often associated with seaweeds and algae

REMARKS: Pipefishes are common members of shallow-water coastal communities. They include both highly cryptic species that resemble seagrass blades as well as conspicuously colored reef fishes. With their small mouths, they feed by picking individual prey items, usually small crustaceans, from the water column or substrate. They are most well known for the incubation behavior of males: eggs are deposited on the abdomen or on the tail of males, sometimes within a pouch, where they are retained until hatching. The distinctive seahorses (*Hippocampus*) have a bent neck and a prehensile tail with no caudal fin, adaptations reputedly associated with the diversification of seagrass beds (Teske and Beheregaray, 2009). Relationships have been studied by Wilson and Rouse (2010) and Wilson and

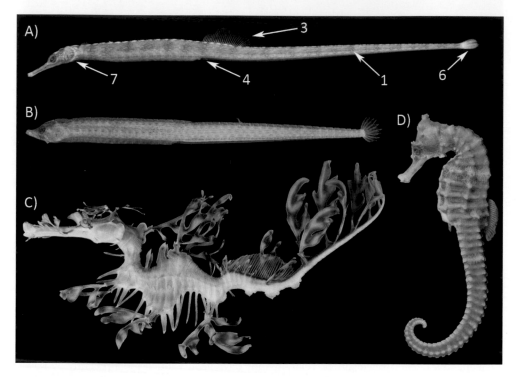

SYNGNATHID CHARACTERISTICS:
1) body elongate, encased in bony rings
2) dorsal-fin spines absent
3) soft dorsal fin variable in size, with 15–60 rays
4) anal fin small to minute
5) pelvic fins absent
6) caudal fin present or absent
7) gill openings restricted

ILLUSTRATED SPECIMENS:
A) *Corythoichthys intestinalis*, SIO 61–132, 100 mm SL
B) *Cosmocampus brachycephalus*, SIO 71–274, 70.5 mm SL
C) *Phycodurus eques*, SIO 04-28, 275 mm TL
D) *Hippocampus ingens*, SIO 00–74, 247 mm TL

Orr (2011). Many syngnathids are important in the aquarium trade. Unfortunately, these most interesting fishes have fallen victim to myths alluding to their medicinal properties and, as a consequence, many species are endangered (Foster and Vincent, 2004).

REFERENCES: Ahnesjö and Craig, 2011; Dawson, 1985; Foster and Vincent, 2004; Fritzsche, 1980; Fritzsche and Vincent, in Carpenter, 2003; Kuiter, 2001; Paulus, in Carpenter and Niem, 1999; Teske and Beheregaray, 2009; Wilson and Orr, 2011; Wilson and Rouse, 2010.

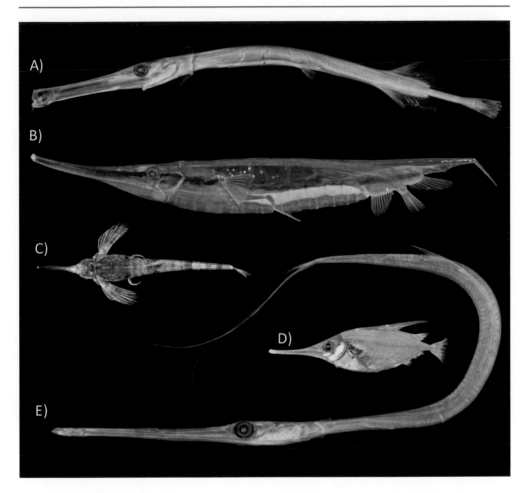

SYNGNATHIFORM DIVERSITY:
A) AULOSTOMIDAE—trumpetfishes: *Aulostomus chinensis,* SIO 61-83, 138 mm SL
B) CENTRISCIDAE—shrimpfishes: *Aeoliscus strigatus,* SIO 74-102, 98 mm SL
C) PEGASIDAE—seamoths: *Pegasus volitans,* SIO 66-525, 110 mm SL (dorsal view)
D) CENTRISCIDAE: *Macroramphosus scolopax,* SIO 65-649, 156 mm SL
E) FISTULARIIDAE—cornetfishes: *Fistularia corneta,* SIO 65-163, 150 mm SL

SYNBRANCHIFORMES—Swamp Eels

The swamp eels are characterized by their elongate bodies and their lack of pelvic fins. The gill openings are restricted to the ventral half of the body and the premaxillae are non-protrusible. Some species are well known as air-breathers, with an assortment of specialized organs, and one species survived in a drying laboratory burrow for nine months (Graham, 1997). Long known to be percomorphs and not true eels (Anguilliformes), recent molecular studies (e.g, Betancur et al., 2013; Near et al., 2012) place them near the Anabantiformes. The highly unusual Synbranchidae have vestigial dorsal and anal fins, no pectoral fins, a small (or non-existent) caudal fin, and some species are hermaphroditic (Liem, 1968). The earthworm eels (Chaudhuriidae) do have dorsal and anal fins, but without spines, and do not have a rostral appendage, in contrast with the spiny eels (Mastacembelidae), which are treated in more detail below. Synbranchiformes comprises three families, approximately 15 genera, and 121 species known mostly from tropical and subtropical freshwaters of the world. Their systematics and phylogenetic relationships were studied by Travers (1984a, 1984b).

REFERENCES: Gosline 1983; Graham, 1997; Liem, 1968; Travers, 1984a, 1984b.

SYNBRANCHIFORMES : MASTACEMBELIDAE—Spiny Eels

DIVERSITY: 5 genera, 88 species

REPRESENTATIVE GENERA: *Aethiomastacembelus, Macrognathus, Mastacembelus*

DISTRIBUTION: Africa to southern Asia

HABITAT: Freshwater and brackish water; tropical; benthic, usually in or on soft bottoms

REMARKS: The spiny eels, occasionally called tire-tread eels in the aquarium trade, are

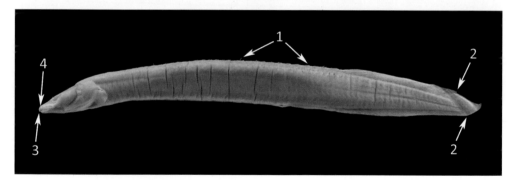

MASTACEMBELID CHARACTERISTICS:
1) dorsal-fin spines isolated, usually in long series
2) dorsal and anal fins long, either contiguous or continuous with caudal fin
3) fleshy appendage on rostrum
4) anterior nostrils tube-like
5) specialized air-breathing organs absent
6) gas swimbladder physoclistous

ILLUSTRATED SPECIMEN:
Sinobdella sinensis, SIO 69–379, 170 mm SL

known to occur in very shallow water. Although lacking an air-breathing organ, some species gulp air, and some will burrow into the mud of drying ponds or streams for considerable periods (Graham, 1997). Mastacembelids are thought to be generalist predators and are considered fine food fishes. Some species are found in the aquarium trade.

REFERENCES: Britz and Kottelat, 2003; Graham, 1997; Gosline, 1983; Travers, 1984a, 1984b.

DACTYLOPTERIFORMES : DACTYLOPTERIDAE—Flying Gurnards

DIVERSITY: 1 family, 2 genera, 7 species

REPRESENTATIVE GENERA: *Dactyloptena, Dactylopterus*

DISTRIBUTION: Atlantic, Indian, and Pacific oceans (excluding the eastern Pacific)

HABITAT: Marine; tropical to temperate; benthic, usually on soft bottoms, continental shelf to upper slope

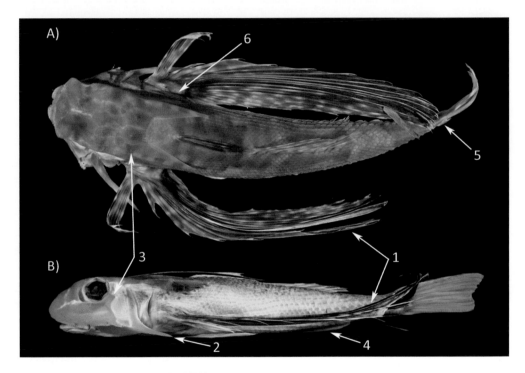

DACTYLOPTERIFORM CHARACTERISTICS:
1) pectoral fins extremely long, reaching at least to base of caudal fin in adults
2) angle of preopercle with prominent spine
3) head with bony armor
4) anal fin with six to seven rays, spines absent
5) base of caudal fin with two keels (enlarged scales)
6) base of pectoral fins horizontal
7) pelvic fins with one spine and four soft rays

ILLUSTRATED SPECIMENS:
A) *Dactylopterus volitans,* SIO 71–278, 195 mm SL (dorsal view)
B) *Dactyloptena peterseni,* SIO 02–92, 360 mm SL

REMARKS: Because of their long pectoral fins, flying gurnards were once thought capable of short bouts of flight above the water surface, but those stories have been discounted as they are seen almost exclusively on or just over the bottom. Dactylopterids were traditionally a part of the Scorpaeniformes, but most recent authors place them in a separate order. Imamura (2000) hypothesized a close relationship of this group with the malacanthids, but this has not been confirmed by other authors. Some molecular studies (e.g., Betancur et al., 2013; Chen et al., 2003; Smith and Wheeler, 2004, 2006) have indicated a relationship with a diverse group of fishes that includes the syngnathids. Flying gurnards feed on benthic crustaceans, clams, and small fishes, and some species occasionally are utilized as food.

REFERENCES: Chen et al., 2003; Eschmeyer, 1997; Imamura, 2000; Poss and Eschmeyer, in Carpenter and Niem, 1999; Smith and Wheeler, 2004, 2006; Smith-Vaniz, in Carpenter, 2003; Springer and Johnson, 2004.

SCORPAENIFORMES—Scorpionfishes, Seabasses, and Relatives

Historically the Scorpaeniformes included a variety of fish lineages characterized by a suborbital stay, a bony process connecting the infraorbital series with the preopercle. Recent morphological (Imamura, 1996; Imamura and Yabe, 2002) and molecular studies (Smith and Wheeler, 2004, 2006) have revealed that two unrelated lineages have independently evolved this feature. We follow these authors and Wiley and Johnson (2010) in restricting the Scorpaeniformes to the scorpionfishes and flatheads (Scorpaenoidei and Platycephaloidei), together with the seabasses and groupers (Serranoidei). Other members of the once broadly defined Scorpaeniformes, notably the sculpins and relatives (Cottoidei), are included instead with the eelpouts (Zoarcoidei) in the Cottiformes. As currently delimited, the Scorpaeniformes is difficult to diagnose, but most members have a posteriorly directed spine on the opercle (Imamura, 1996), unique muscle features, and a single postocular spine in the larvae (Imamura and Yabe, 2002; Moser and Ahlstrom, 1978). These features occur in other percomorph fishes, and resolution of the limits and relationships of this group awaits further study. Within the Scorpaeniformes, the traditional scorpionfishes (Scorpaenoidei) plus the flatheads and relatives (Platycephaloidei) have a suborbital stay on the third infraorbital (Imamura and Yabe, 2002) and a unique extrinsic gas bladder muscle (Wiley and Johnson, 2010), while members of the Serranoidei (Serranidae and Epinephelidae) lack a suborbital stay but have three spines on the opercle (see Johnson, 1983). Other families are likely to be included in the latter group.

REFERENCES: Imamura, 1996; Imamura and Shinohara, 1998; Imamura and Yabe, 2002; Johnson, 1983; Moser and Ahlstrom, 1978; Shinohara and Imamura, 2007; Smith and Wheeler, 2004, 2006; Wiley and Johnson, 2010.

SCORPAENIFORMES : SCORPAENIDAE—Scorpionfishes

DIVERSITY: 56 genera, over 450 species

REPRESENTATIVE GENERA: *Pterois, Scorpaena, Scorpaenodes, Sebastes*

DISTRIBUTION: Atlantic, Indian, and Pacific oceans

HABITAT: Marine, rarely freshwater; tropical to temperate; coastal to continental slope, benthic, demersal, or neritic

REMARKS: Scorpionfishes are a hugely diverse group, perhaps best known for possessing toxic dorsal-fin spines (Smith and Wheeler, 2006). These are especially well developed in the lionfishes (*Pterois*). Many scorpionfishes are sit-and-wait predators, but a variety of

SCORPAENID CHARACTERISTICS:
1) head usually with bony ridges and/or spines
2) body usually somewhat laterally compressed
3) one or two opercular spines and three to five preopercular spines
4) dorsal fin continuous or with a notch
5) anal fin usually with three spines and five soft rays
6) gill membranes free from isthmus
7) suborbital stay usually securely fastened to preopercle

ILLUSTRATED SPECIMENS:
A) *Scorpaenodes xyris,* SIO 83–106, 149 mm SL
B) *Sebastes exsul,* SIO 68–1, 170 mm SL, holotype
C) *Pterois russelii,* SIO 64–225, 226 mm SL

has evolved repeatedly within the group (Erisman et al., 2009). Several species form spawning aggregations where they are especially vulnerable to overfishing (Sadovy et al., 2008). The skin of the curious soapfishes (Grammistinini) releases a chemical when disturbed that may ward off predators (Randall et al., 1971).

REFERENCES: Baldwin and Johnson, 1993; Craig and Hastings, 2007; Craig et al., 2011; Erisman et al., 2009; Heemstra, in Fischer et al., 1995; Heemstra et al., in Carpenter, 2003; Heemstra and Randall, in Carpenter and Niem, 1999; Heemstra and Randall, 1993; Randall et al., 1971; Smith, 1971; Smith and Craig, 2007.

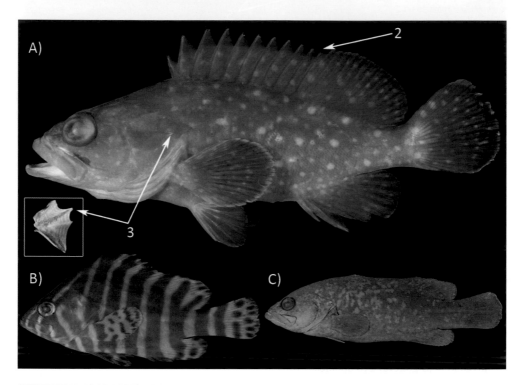

EPINEPHELID CHARACTERISTICS:
1) body laterally compressed
2) dorsal fin continuous
3) three opercular spines
4) scales relatively small in most species
5) larvae with elongate dorsal-fin spine encased in fleshy sheath
6) larvae of most species with elongate pelvic-fin spine

ILLUSTRATED SPECIMENS:
A) *Epinephelus labriformis*, SIO 71–170, 88 mm SL
B) *Dermatolepis dermatolepis*, SIO 70–394, 60 mm SL
C) *Rypticus courtenayi*, SIO 72–67, 119 mm SL, holotype (Grammistini—soapfishes)

INSET: Opercle from *Mycteroperca rosacea*, SIO uncatalogued.

SCORPAENIFORMES : SERRANIDAE—Seabasses and Anthiines

DIVERSITY: 34 genera, 180 species

REPRESENTATIVE GENERA: *Anthias, Centropristis, Diplectrum, Paralabrax, Plectranthias, Pseudanthias, Serranus*

DISTRIBUTION: Atlantic, Indian, and Pacific oceans

SERRANID CHARACTERISTICS:
1) opercle with three spines
2) scales ctenoid
3) dorsal fin with seven to ten spines
4) anal fin with three spines
5) posterior tip of maxilla exposed, not covered by infraorbitals

ILLUSTRATED SPECIMENS:
A) *Cratinus agassizii,* SIO 78–175, 321 mm SL (Serraninae—seabasses)
B) *Pronotogrammus multifasciatus,* SIO 92–12, 154 mm SL (Anthiinae—anthiines)
C) *Pseudanthias squamipinnis,* SIO 04–66, 67 mm SL (Anthiinae)
D) *Diplectrum eumelum,* SIO 65–160, 129 mm SL, holotype (Serraninae)
E) *Hypoplectrus gemma,* SIO 70–197, 88 mm SL (Serraninae)

notably by the trunk lateral-line canal extending to the tip of the caudal fin. Recent studies using molecular data support the inclusion of the Centropomidae within a diverse group of fishes that includes jacks and their relatives, barracudas, flatfishes, and others (Li et al., 2011; Near et al., 2012). The snooks are important predators on crustaceans and small fishes in tropical estuarine and mangrove systems, and at least one species is a protandrous hermaphrodite (Taylor et al., 2000). These fishes support significant recreational fisheries in the southern United States, where their commercial harvest is prohibited. They are, however, commercially important in other parts of the world.

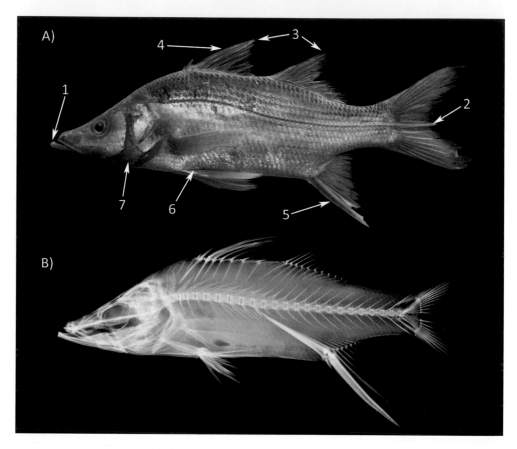

CENTROPOMID CHARACTERISTICS:
1) lower jaw protruding beyond upper jaw
2) trunk lateral-line canal extending to posterior margin of caudal fin
3) two separate dorsal fins
4) first dorsal fin triangular
5) anal fin with three spines, one typically very strong
6) pelvic fin with scaly process in axil
7) margin of preopercle usually serrated
8) seven branchiostegal rays

ILLUSTRATED SPECIMENS:
A) *Centropomus robalito*, SIO 13–238, 186 mm SL
B) *Centropomus robalito*, SIO 13–238, radiograph

REFERENCES: Bussing, in Fischer et al., 1995; Greenwood, 1976; Larson, in Carpenter and Niem, 1999; Li et al., 2011; Mooi and Gill, 1995; Orrell, in Carpenter, 2003; Otero, 2004; Taylor et al., 2000.

PERCIFORMES : MORONIDAE—Temperate Basses

DIVERSITY: 2 genera, 6 species

REPRESENTATIVE GENERA: *Dicentrarchus, Morone*

DISTRIBUTION: North America, Europe, and northern Africa

HABITAT: Freshwater to marine, temperate lakes, rivers, and coastal areas; pelagic to demersal over soft bottoms

REMARKS: The taxonomic placement of the temperate basses has long been a topic of contention among systematic ichthyologists. Once included in the Percichthyidae, these fishes were allocated to the Moronidae by Johnson (1984). Chiba et al. (2009) found evidence for including the moronids with the sparids (porgies and relatives). The temperate basses are predatory, feeding on fishes and invertebrates. The Striped Bass (*Morone saxatilis*) is an important game fish in its native habitat in the temperate western Atlantic, as well as other areas where it has been introduced and become established (Setzler et al., 1980). Some additional species are highly regarded food fishes and are also of increasing importance in the aquaculture industry.

REFERENCES: Chiba et al., 2009; Johnson, 1984; Setzler et al., 1980.

MORONID CHARACTERISTICS:

1) two separate dorsal fins
2) opercle with one or two spines
3) teeth present on glossohyal (tongue bone)
4) lower jaw usually extending beyond upper jaw
5) seven branchiostegal rays

ILLUSTRATED SPECIMEN:
Morone saxatilis, SIO 60–497, 184 mm SL

INSET: Opercle from *Morone saxatilis,* SIO uncatalogued

PERCIFORMES : OPISTOGNATHIDAE—Jawfishes

DIVERSITY: 3 genera, 82 species

REPRESENTATIVE GENERA: *Lonchopisthus, Opistognathus, Stalix*

DISTRIBUTION: Atlantic, Indian, and Pacific oceans

HABITAT: Marine; tropical; benthic, on reefs and surrounding sandy bottoms, occasionally deeper, on soft substrates

REMARKS: Jawfishes are well known for constructing their own burrows in sand and cobble areas, often adjacent to reefs, and usually are seen above or with their head at the entrance of the burrow (Colin, 1973). Their eggs have long filaments, and are orally incubated by males in their large mouths. Recent molecular studies affiliate them with the Ovalentaria, a diverse lineage with eggs bearing filaments (Wainwright et al., 2012). Jawfishes, especially the Yellowhead Jawfish (*Opistognathus aurifrons*) of the Caribbean region, are highly sought by the aquarium industry. While most jawfishes are small, the Giant Jawfish (*Opistognathus rhomaleus*) grows to over 50 cm in length and is consumed locally by humans in the Gulf of California region.

REFERENCES: Colin, 1973; Smith-Vaniz, 1989; Smith-Vaniz, in Carpenter and Niem, 1999; Smith-Vaniz, in Carpenter, 2003.

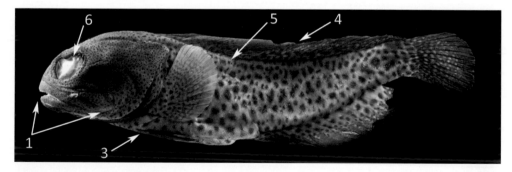

OPISTOGNATHID CHARACTERISTICS:
1) mouth large
2) body with cycloid scales, head naked
3) pelvic fins inserted anterior to pectoral fins
4) dorsal fin continuous, with 9–12 spines
5) trunk lateral-line canal high on body, ending near middle of dorsal fin
6) eyes relatively large and high on head

ILLUSTRATED SPECIMEN:
Opistognathus punctatus, SIO 85–198, 216 mm SL

PERCIFORMES : CENTRARCHIDAE—Sunfishes

DIVERSITY: 8 genera, 38 species

REPRESENTATIVE GENERA: *Ambloplites, Archoplites, Elassoma, Lepomis, Micropterus, Pomoxis*

DISTRIBUTION: North America; widely introduced.

HABITAT: Freshwater rivers, streams, swamps, ponds, and lakes; temperate to warm temperate; pelagic to demersal over soft bottoms

CENTRARCHID CHARACTERISTICS:
1) body laterally compressed and deep in many species
2) pseudobranchs small and hidden
3) dorsal fin continuous, with 5–13 spines
4) anal fin with three to five spines
5) five to seven branchiostegal rays

ILLUSTRATED SPECIMENS:
A) *Micropterus dolomieu,* SIO 62–327, 94 mm SL
B) *Archoplites interruptus,* SIO 77–387, 121 mm SL
C) *Elassoma zonatum,* SIO 79–373, 26 mm SL
D) *Ambloplites rupestris,* SIO 79–367, 40 mm SL
E) *Lepomis cyanellus,* SIO 62–328, 66 mm SL

(account continued)

REMARKS: Although centrarchids are one of the most well-studied groups of freshwater fishes (Cooke and Philipp, 2009; Ross, 2013), their phylogenetic relationships have only recently been studied in detail, and some controversy remains. Historically the enigmatic pygmy sunfishes (*Elassoma*) were considered to be paedomorphic centrarchids, as they are small and lack several bones such as the infraorbital series. While detailed morphological studies (e.g., Johnson, 1984, 1993) have variously placed them outside the percoids (reviewed in Nelson, 2006), recent studies using extensive molecular data (Near, Kassler, et al., 2003; Near, Bolnick and Wainwright, 2004; Near, Sandel, et al., 2012) place them back within the Centrarchidae, a relationship followed herein. This traditional Centrarchidae exhibits two internal synapomorphies (Near, Sandel, et al., 2012): presence of wing-like transverse processes on the first haemal spine, and more than one anal-fin pterygiophore anterior to the first haemal spine. Female centrarchids lay demersal eggs in nests constructed by males who guard the eggs and, in most species, also the young. Due to the variable lifestyles of these fishes, they exhibit considerable dietary range and include planktivores, molluskivores, and piscivores. Several species of sunfishes have been introduced around the world for sportfishing. The Sacramento Perch, *Archoplites interruptus,* is the only extant species naturally occurring west of the Rocky Mountains (Crain and Moyle, 2011).

REFERENCES: Boschung and Mayden, 2004; Cooke and Philipp, 2009; Crain and Moyle, 2011; Johnson, 1984, 1993; Kassler et al., 2002; Mabee, 1993; Near, Bolnick, et al., 2004; Near, Kassler, et al., 2003; Near, Sandel, et al., 2012; Page and Burr, 2011; Roe et al., 2002; Ross, 2013.

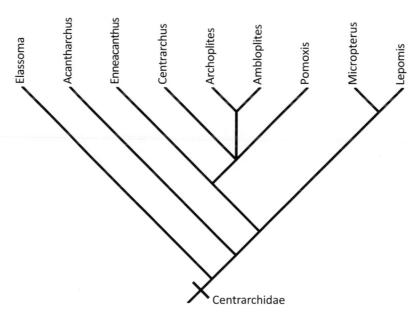

Hypothesized phylogenetic relationships of the genera of the Centrarchidae after Near (2004) and Near, Sandel et al., 2012.

PERCIFORMES : PERCIDAE—Perches

DIVERSITY: 10 genera, 234 species

REPRESENTATIVE GENERA: *Ammocrypta, Etheostoma, Gymnocephalus, Perca, Percina, Sander*

DISTRIBUTION: Northern Hemisphere

HABITAT: Freshwater streams, rivers and lakes; temperate to warm temperate; benthic, demersal, or pelagic, over soft and hard bottoms

REMARKS: Although percids are one of the most diverse and well-studied groups of Northern Hemisphere freshwater fishes, they are difficult to characterize using external features. Three subfamiles are recognized: the Percinae with 12 species, the Luciopercinae with ten species, and the North American Etheostomatinae with 212 species of darters. Numerous phylogenetic studies have addressed relationships within percids in general (Song et al., 1998; Sloss et al., 2004) and within the Etheostomatinae in particular, most notably Page (1981) and Near et al. (2011). The larger-bodied species (e.g., *Sander* and *Perca*) are primarily

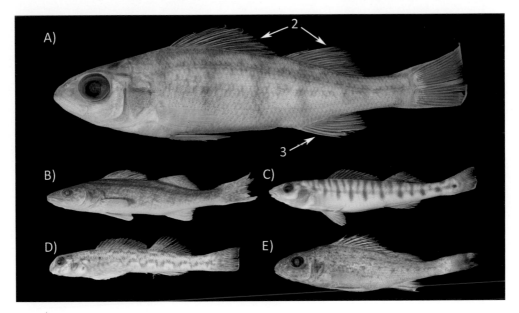

PERCID CHARACTERISTICS:
1) body somewhat elongate
2) two separate (or narrowly joined) dorsal fins
3) one or two anal-fin spines
4) vomer and palatines usually with teeth
5) five to eight branchiostegal rays
6) one or absent predorsal bones

ILLUSTRATED SPECIMENS:
A) *Perca flavescens,* SIO 79–370, 61 mm SL
B) *Sander vitreus,* SIO 79–368, 156 mm SL
C) *Percina caprodes,* SIO 63–257, 65 mm SL
D) *Etheostoma blennioides,* SIO 62–327, 63 mm SL
E) *Gymnocephalus cernua,* SIO 64–165, 74 mm SL

piscivorous and support intensive recreational and small-scale commercial fisheries (Craig, 2000). The biology and ecology of darters, some of the most colorful freshwater fishes in the world, have been studied extensively (e.g., Craig, 2000; Kelly et al., 2012; Page, 1983; Page and Swofford, 1984). These small, benthic species feed primarily on arthropods.

REFERENCES: Bailey and Etnier, 1988; Boschung and Mayden, 2004; Collette and Banarescu, 1977; Craig, 2000; Kelly et al., 2012; Near et al., 2011; Page, 1981, 1983; Page and Burr, 2011; Page and Swofford, 1984; Ross, 2013; Sloss et al., 2004; Song et al., 1998; Wiley, 1992.

PERCIFORMES : PRIACANTHIDAE—Bigeyes

DIVERSITY: 4 genera, 19 species

REPRESENTATIVE GENERA: *Cookeolus, Heteropriacanthus, Priacanthus, Pristigenys*

DISTRIBUTION: Atlantic, Indian, and Pacific oceans

HABITAT: Marine; tropical to subtropical; coastal to upper continental slope, demersal over hard bottoms, often near reefs

PRIACANTHID CHARACTERISTICS:
1) body compressed
2) eyes extremely large
3) mouth large, upturned, lower jaw generally protruding beyond upper
4) dorsal fin continuous
5) pelvic fins large, generally attached to abdomen by membrane
6) pelvic-fin origin anterior to pectoral fin
7) scales modified cycloid with small spines
8) branchiostegal rays covered with scales

ILLUSTRATED SPECIMENS:
A) *Heteropriacanthus cruentatus,* SIO 61–237, 135 mm SL
B) *Pristigenys serrula,* SIO 67–13, 187 mm SL
C) *Pristigenys serrula,* SIO 72–10, radiograph

REMARKS: Bigeyes are typically found on rocky or coral reefs in crevices or under ledges during the day. Most are nocturnal, but some may also feed in daylight. Some species are found in more open areas to depths over 400 m. These predatory fishes feed on invertebrates and other small fishes. Associated with their nocturnal habits, the eyes of priacanthids have a well-developed reflective layer, the tapetum lucidum, which assists their vision in low light. Starnes (1988) reviewed the systematics and evolution of this group. Many species support artisanal fisheries around the world.

REFERENCES: Starnes, 1988; Starnes, in Fischer et al., 1995; Starnes, in Carpenter, 2003; Starnes, in Carpenter and Niem, 1999.

PERCIFORMES : APOGONIDAE—Cardinalfishes

DIVERSITY: Over 20 genera, 347 species

REPRESENTATIVE GENERA: *Apogon, Astrapogon, Foa, Pseudamia*

DISTRIBUTION: Atlantic, Indian, and Pacific oceans

HABITAT: Marine; tropical to subtropical; demersal, common on coral and rocky reefs, with a few species in brackish water and streams of the tropical Pacific

REMARKS: Cardinalfishes are primarily nocturnal, with most species spending the day in caves and crevices and emerging at night to feed on zooplankton. Males incubate fertilized eggs within their buccal cavity, which is significantly larger than that of females (Barnett and Bellwood, 2005). Several studies based on molecular data have found a close relationship between cardinalfishes and the gobies and kurtids (e.g., Betancur et al., 2013; Near et al., 2012; Thacker and Roje, 2009). Thacker and Roje (2009) found support for the monophyly of the Apogonidae with the exclusion of the genus *Pseudamia*. They also document the repeated evolution of bioluminescence associated with the gut of the Apogonidae and its close relatives. The species-level systematics of apogonids has been studied extensively (e.g., Fraser, 1972, 2005, 2013). Some cardinalfishes are prominent in the aquarium trade.

REFERENCES: Allen, in Carpenter and Niem, 1999; Barnett and Bellwood, 2005; Fraser, 1972, 2005, 2013; Gon, in Carpenter, 2003; Thacker and Roje, 2009.

(account continued)

APOGONID CHARACTERISTICS:
1) two separate dorsal fins in nearly all species
2) anal fin with two spines, typically positioned opposite second dorsal fin
3) caudal peduncle long
4) mouth terminal, large, oblique
5) eyes large
6) rear margin of preopercle double-edged
7) seven branchiostegal rays

ILLUSTRATED SPECIMENS:
A) *Apogon atradorsatus*, SIO 64–1004, 69 mm SL
B) *Astrapogon puncticulatus*, SIO 70–179, 39 mm SL
C) *Pseudamia zonata*, SIO 92–121, 85 mm SL

PERCIFORMES : LUTJANIDAE—Snappers

DIVERSITY: 17 genera, 109 species

REPRESENTATIVE GENERA: *Lutjanus, Ocyurus, Pristipomoides, Rhomboplites, Symphorichthys*

DISTRIBUTION: Atlantic, Indian, and Pacific oceans

HABITAT: Marine, brackish, and rarely in freshwater; tropical to temperate; coastal to continental slope, demersal over hard bottoms, common on rocky and coral reefs

REMARKS: Often confused with the grunts (Haemulidae), snappers typically have large canines in the jaws (absent in grunts) and small mandibular lateral-line pores on the lower jaw (in contrast to large pores in grunts). Miller and Cribb (2007) provided evidence from molecular data for the monophyly of an expanded Lutjanidae that includes the fusiliers (Caesionidae), but did not support the monophyly of the genus *Lutjanus* (see Gold et al., 2011). Snappers are conspicuous predators on coral and rocky reefs, although a few (e.g., *Ocyurus*) are largely planktivorous (Davis and Birdsong, 1973). Many snappers are some of the most highly prized food fishes in the world and as a consequence are heavily exploited by artisanal and commercial fisheries (Allen, 1985). Some snappers are responsible for ciguatera poisoning, particularly large individuals of piscivorous species (Randall, 1958).

REFERENCES: Allen, 1985; Allen, in Fischer et al., 1995; Anderson, in Carpenter, 2003; Anderson and Allen, in Carpenter and Niem, 2001; Davis and Birdsong, 1973; Gold et al., 2011; Johnson, 1980; Miller and Cribb, 2007; Randall, 1958.

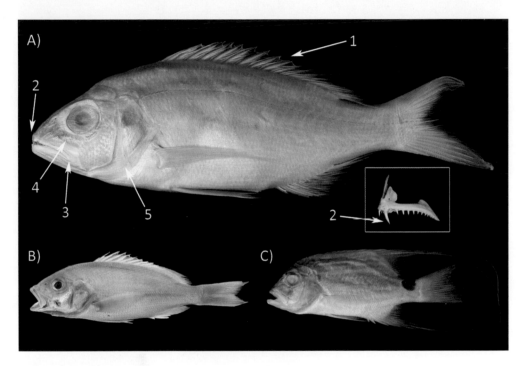

LUTJANID CHARACTERISTICS:
1) dorsal fin continuous or with shallow notch
2) mouth terminal, large, usually with conspicuous canine teeth
3) maxilla covered by premaxilla and infraorbital bones when mouth closed
4) suborbital area between eye and mouth without scales
5) preopercle typically finely serrate
6) seven branchiostegal rays
7) mandibular lateral-line pores (on lower jaw) not enlarged

ILLUSTRATED SPECIMENS:
A) *Pristipomoides zonatus,* SIO 75–372, 190 mm SL
B) *Lutjanus peru,* SIO 60–3, 205 mm SL
C) *Symphorichthys spilurus,* SIO 64–229, 99 mm SL

INSET: Premaxilla of *Lutjanus* sp., SIO uncatalogued

PERCIFORMES : GERREIDAE—Mojarras

DIVERSITY: 8 genera, 54 species

REPRESENTATIVE GENERA: *Diapterus, Eucinostomus, Eugerres, Gerres*

DISTRIBUTION: Atlantic, Indian, and Pacific oceans

HABITAT: Marine to freshwater; tropical to warm temperate; coastal, demersal over soft bottoms

GERREID CHARACTERISTICS:
1) body compressed
2) mouth terminal, highly protrusible, with long premaxillary ascending processes
3) head mostly covered with scales
4) dorsal fin continuous
5) dorsal and anal fins partially covered with sheath of scales
6) caudal fin deeply forked
7) jaws with very small teeth (appearing toothless)
8) vomer and palatines without teeth

ILLUSTRATED SPECIMENS:
A) *Eugerres lineatus*, SIO 64-80, 115 mm SL
B) *Eucinostomus entomelas*, SIO 65-175, 170 mm SL, holotype
C) *Diapterus aureolus*, SIO 62-37, radiograph showing protruding jaws

REMARKS: Mojarras are silvery, schooling fishes with tiny, brush-like teeth. They resemble the ponyfishes (Leiognathidae) but have larger, deciduous scales that at least partially cover the dorsal and anal fins. They feed on buried invertebrates and plant matter by probing the sand with their highly protrusible mouths. Their phylogenetic relationships were studied by Chen et al. (2007). Some species are considered good table fare and are of some commercial importance.

REFERENCES: Bussing, in Fischer et al., 1995; Chen et al., 2007; Deckert and Greenfield, 1987; Gilmore and Greenfield, in Carpenter, 2003; Matheson and McEachran, 1984; Woodland, in Carpenter and Niem, 2001.

PERCIFORMES : HAEMULIDAE—Grunts

DIVERSITY: 17 genera, 132 species

REPRESENTATIVE GENERA: *Anisotremus, Haemulon, Orthopristis, Plectorhinchus, Pomadasys*

DISTRIBUTION: Atlantic, Indian, and Pacific oceans

HABITAT: Marine, occasionally in freshwater; tropical to temperate; demersal on hard and soft bottoms, common on coral and rocky reefs as well as in estuaries

REMARKS: As their common name implies, grunts are vocal fishes, making a variety of sounds by scraping their pharyngeal teeth together (Burkenroad, 1930). Sometimes confused with snappers (Lutjanidae), grunts typically lack large canines in the jaws (canines present in snappers), but have large, mandibular lateral-line pores on the lower jaw (in contrast to small pores in snappers). Adults of some grunts, especially those in the genus *Haemulon,* form loose schools over reefs during the day and disperse to feed on benthic invertebrates in surrounding areas at night, transferring energy to the reef ecosystem (Helfman et al., 1982). Johnson (1980) recognized two lineages, the mostly New World Haemulinae, and the Indo-West Pacific Plectorhinchinae (sweetlips). Rocha et al. (2008) studied the phylogenetic relationships of Caribbean species of *Haemulon* and included *Inermia vittata* (Inermiidae) within that genus, and Sanciango et al. (2011) hypothesized relationships within the family. Most grunts are considered food fishes in many parts of the world, and some larger species are highly prized sport fishes.

REFERENCES: Burkenroad, 1930; Helfman et al., 1982; Johnson, 1980; Lindeman and Toxey, in Carpenter, 2003; McKay, in Carpenter and Niem, 2001; McKay and Schneider, in Fischer et al., 1995; Rocha et al., 2008; Sanciango et al., 2011.

(account continued)

HAEMULID CHARACTERISTICS:

1) mouth usually small, teeth (in jaws) generally small, pointed, and numerous
2) mandibular lateral-line pores (on lower jaw) enlarged
3) scales ctenoid, present in suborbital area
4) margin of preopercle serrate
5) seven branchiostegal rays

ILLUSTRATED SPECIMENS:

A) *Haemulon plumierii,* SIO 70–179, 156 mm SL
B) *Haemulon californiensis,* SIO 65–177, 132 mm SL
C) *Anisotremus virginicus,* SIO 65–175, 114 mm SL
D) *Haemulon flavolineatum,* SIO 70–178, 133 mm SL
E) *Conodon serrifer,* SIO 63–106, 153 mm SL
F) *Plectorhinchus* sp. (juvenile), SIO 69–305, 41 mm SL (Plectorhinchinae—sweetlips)

PERCIFORMES : POLYNEMIDAE—Threadfins

DIVERSITY: 8 genera, 42 species

REPRESENTATIVE GENERA: *Polydactylus, Polynemus*

DISTRIBUTION: Atlantic, Indian, and Pacific oceans

HABITAT: Marine to freshwater rivers; tropical to warm temperate; coastal, demersal over soft bottoms

REMARKS: Threadfins are silvery, grayish fishes that superficially resemble one another and thus are difficult to identify to species (Motomura, 2002, 2004). They appear to be closely related to the drums (Sciaenidae) based on their similar larvae and several osteological characters (Johnson, 1993). Threadfins are found in coastal areas usually over sand or over muddy bottoms where they sometimes form large schools and feed on benthic invertebrates and small fishes. Their free pectoral-fin rays appear to function as sensory organs that are especially useful in their muddy habitats. Some species are protandrous hermaphrodites. Juveniles are commonly encountered in seagrass beds. A few species grow to a large size, over 1.5 m in length, and several species are important targets of fisheries and used in the aquaculture industry (Motomura, 2004).

REFERENCES: Feltes, in Carpenter, 2003; Feltes, in Carpenter and Niem, 2001; Johnson, 1993; Motomura, 2002, 2004.

POLYNEMID CHARACTERISTICS:
1) pectoral fin in two sections, lower portion with 3–16 free rays
2) two separate dorsal fins
3) caudal fin forked
4) pelvic fins inserted posterior to pectoral fins
5) snout conical, extending beyond subterminal mouth
6) eye covered with adipose tissue
7) trunk lateral-line canal extends across caudal fin, occasionally bifurcate posteriorly

ILLUSTRATED SPECIMEN:
Polydactylus approximans, SIO 62–68, 185 mm SL

DIVERSITY: 70 genera, 291 species

REPRESENTATIVE GENERA: *Aplodinotus, Atractoscion, Bairdiella, Cynoscion, Umbrina*

DISTRIBUTION: Atlantic, Indian, and Pacific oceans

HABITAT: Marine to freshwater; tropical to temperate; coastal to continental slopes, neritic to demersal over soft and hard bottoms

REMARKS: Sciaenid fishes are most often found in estuarine settings, but also occur on continental shelves, and around coral and rocky reefs. A few species (e.g., the North American Freshwater Drum, *Aplodinotus grunniens*) are restricted to freshwaters. Their common names reflect their sound-producing behavior, especially evident during mating. The gas bladders of sciaenids are often complex, with secondary chambers, numerous and complex extensions enhancing their auditory abilities, and complex drumming muscles for sound production. These fishes are generalist predators and eat fishes and benthic invertebrates, including crustaceans and even hard-shelled mollusks. Sciaenids were once considered to be closely related to the Haemulidae (grunts) based on similar otoliths and pores in the lower jaw, as well as on their sonorous behaviors (Chao, 1978; Schwarzhans, 1993; Trewavas, 1977). However, Sasaki (1989) argued against this hypothesis but was unable to determine their relationships within the perciforms. Johnson (1993) hypothesized a close relationship with the Polynemidae (threadfins) based on their similar larvae and several osteological characters. Sasaki (1989) recognized four major lineages within the Sciaenidae based on morphological features, and Vergara-Chen et al. (2009) hypothesized relationships of *Cynoscion* and related genera based on molecular data. Many species of drums are extremely flavorful and heavily exploited by commercial and recreational fisheries. The Totoaba (*Totoaba macdonaldi*), restricted to the northern Gulf of California, has been fished to near extinction and is listed as critically endangered by the IUCN (2013).

REFERENCES: Chao, 1978, 1986; Chao, in Carpenter, 2003; Chao, in Fischer et al., 1995; Cui et al., 2009; Johnson, 1993; Sasaki, 1989; Sasaki, in Carpenter and Niem, 2001; Schwarzhans, 1993; Trewavas, 1977; Vergara-Chen et al., 2009.

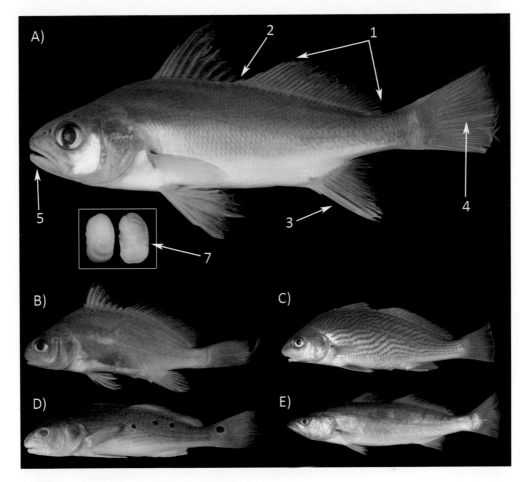

SCIAENID CHARACTERISTICS:

1) soft-ray portion of dorsal fin much longer than soft-ray portion of anal fin
2) spinous and soft-ray portions of dorsal fin usually separated by notch
3) anal fin with two (rarely one) spines
4) trunk lateral-line canal well developed, extending to end of caudal fin
5) cephalic lateral-line canals large and conspicuous, especially those on snout and lower jaw
6) gas bladder large, often with branches or subdivisions
7) otoliths remarkably large
8) vomer and palatines toothless

ILLUSTRATED SPECIMENS:

A) *Bairdiella icistia,* SIO 64–82, 112 mm SL (Stelliferinae—stardrums)
B) *Aplodinotus grunniens,* SIO 79–370, 96 mm SL (Sciaeninae—croakers and drums)
C) *Umbrina wintersteeni,* SIO 60–366, 193 mm SL, holotype (Sciaeninae)
D) *Sciaenops ocellatus,* SIO 88–95, 208 mm SL (Sciaeninae)
E) *Cynoscion parvipinnis,* SIO 62–237, 228 mm SL (Cynoscioninae—weakfishes)

INSET: Otoliths of *Totoaba macdonaldi,* SIO 70–125

PERCIFORMES : MULLIDAE—Goatfishes

DIVERSITY: 6 genera, 84 species

REPRESENTATIVE GENERA: *Mulloidichthys, Mullus, Parupeneus, Upeneus*

DISTRIBUTION: Atlantic, Indian, and Pacific oceans

HABITAT: Marine, rarely in brackish water; tropical to temperate; coastal, demersal over soft bottoms, often around reefs

REMARKS: Goatfishes are readily recognizable by their highly mobile barbels. Gosline (1984) and Kim et al. (2001) described the form and function of these chemosensory barbels that are used to probe sandy areas in search of crustaceans, worms, and other small prey. Other fishes often follow closely behind foraging goatfishes in search of escaping prey. The biology and ecology of goatfishes were reviewed by Uiblein (2007). Some goatfishes have a long-lived pelagic phase (McCormick and Milicich, 1993). The morphology and relationships of goatfishes were studied in detail by Kim (2002) who recognized six genera. Recent analyses of molecular data indicate that goatfishes are related to the syngnathiform and scombriform fishes (Betancur et al., 2013; Near et al., 2013). Certain mullid species are common ingredients in classic southern European fish stews. Several other goatfishes are highly prized food fishes.

REFERENCES: Gosline, 1984; Kim, 2002; Kim et al., 2001; McCormick and Milicich, 1993; Randall, in Carpenter, 2003; Randall, in Carpenter and Niem, 2001; Uiblein, 2007.

MULLID CHARACTERISTICS:
1) two long movable barbels on chin, derived from modified branchiostegal rays
2) median groove on ventral side of head to accommodate barbels
3) dorsal fins well separated
4) anal fin with one or two spines
5) caudal fin forked

ILLUSTRATED SPECIMEN:
Parupeneus multifasciatus, SIO 68–531, 130 mm SL

DIVERSITY: 4 genera, 15 species

REPRESENTATIVE GENERA: *Hermosilla, Kyphosus, Neoscorpis, Sectator*

DISTRIBUTION: Atlantic, Indian, and Pacific oceans

HABITAT: Marine; tropical to temperate; coastal to oceanic, epipelagic and neritic over rocky and coral reefs

REMARKS: The composition of the Kyphosidae is controversial and unresolved. Nelson (2006) included the Kyphosinae (rudderfishes, with the maxillary exposed), the Girellinae (nibblers, with the maxillary concealed beneath the infraorbitals), the Scorpidinae (halfmoons, lacking incisor-like teeth), the Microcanthinae, and the Parascorpidinae (jutjaws, with a prolonged lower jaw). Yagashita et al. (2002) did not support the monophyly of this group, but included with these *Oplegnathus* (knifejaws, Oplegnathidae) and *Kuhlia* (flagtails, Kuhliidae) in a poorly supported clade. Near et al. (2012) recovered a similar clade that included a few additional groups. We have elected to consider the Kyphosidae to include only 15 species of "rudderfishes" in four genera. Typical of other herbivorous fishes, the rudderfishes have incisor-like teeth for scraping algae, and a long gut for digesting plant material. Some species are highly regarded as food fishes.

REFERENCES: Carpenter, in Carpenter, 2003; Johnson and Fritzsche, 1989; Sakai, in Carpenter and Niem, 2001; Sommer, in Fischer et al., 1995; Yagashita et al., 2002.

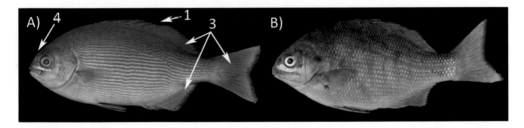

KYPHOSID CHARACTERISTICS:
1) dorsal fin continuous
2) anterior jaw teeth incisiform
3) scales ctenoid, covering soft-ray portions of dorsal, anal, and anterior caudal fins
4) head relatively small, snout short
5) digestive tract exceptionally long

ILLUSTRATED SPECIMENS:
A) *Kyphosus cinerascens,* SIO 60–250, 98 mm SL
B) *Hermosilla azurea,* SIO 59–379, 134 mm SL

PERCIFORMES : CHAETODONTIDAE—Butterflyfishes

DIVERSITY: 11 genera, 131 species

REPRESENTATIVE GENERA: *Chaetodon, Forcipiger, Heniochus, Johnrandallia*

DISTRIBUTION: Atlantic, Indian, and Pacific oceans

HABITAT: Marine; tropical to warm temperate; coastal to lower continental shelf, demersal, common on coral and rocky reefs

REMARKS: Butterflyfishes are conspicuous members of coral-reef communities. Most species are strikingly colored and many swim in pairs or small groups during the day. Their diet consists of coral polyps, other small invertebrates, and algae. Species of *Chaetodon* have a unique connection between the gas bladder and the lateral-line canals of the head (Smith et al., 2003; Webb and Smith, 2000; Webb et al., 2006). Their phylogenetic relationships were studied by Littlewood et al. (2004) and later by Fessler and Westneat (2007), who recovered two major clades, the "bannerfish clade" including several genera such as *Forcipiger, Heniochus,* and *Johnrandallia,* and the "butterflyfish clade" with *Prognathodes* sister to the speciose genus *Chaetodon.* Several species of butterflyfishes are common in the aquarium trade.

REFERENCES: Allen et al., 1998; Burgess, 1978; Burgess, in Carpenter, 2003; Fessler and Westneat, 2007; Littlewood et al., 2004; Pyle, in Carpenter and Niem, 2001; Smith et al., 2003; Webb and Smith, 2000; Webb et al., 2006.

(Opposite)

CHAETODONTID CHARACTERISTICS:
1) body deep, strongly compressed
2) mouth small, upper jaw protrusible
3) dorsal fin usually continuous
4) scaly axillary process at base of pelvic-fin spine
5) no spine at angle of preopercle
6) eye generally obscured by dark coloration on head
7) gas bladder well developed, with two anteriorly directed extensions

ILLUSTRATED SPECIMENS:
A) *Chaetodon ocellatus,* SIO 64–80, 115 mm SL
B) *Prognathodes falcifer,* SIO 82–79, 97 mm SL
C) *Johnrandallia nigrirostris,* SIO 70–135, 96 mm SL
D) *Forcipiger flavissimus,* SIO 74–150, 89 mm SL
E) *Chaetodon ocellatus,* SIO 64–80, radiograph
F) *Chaetodon ocellatus,* SIO 64–80, magnetic resonance image illustrating large gas bladder (from Digital Fish Library; Berquist et al., 2012)

PERCIFORMES : POMACANTHIDAE—Angelfishes

DIVERSITY: 8 genera, 88 species

REPRESENTATIVE GENERA: *Centropyge, Geniacanthus, Holacanthus, Pomacanthus*

DISTRIBUTION: Atlantic, Indian, and Pacific oceans

HABITAT: Marine; tropical to subtropical; coastal, demersal, common on coral and rocky reefs

REMARKS: Angelfishes are among the most conspicuous members of coral-reef fish communities. They are the sister group of the butterflyfishes and are easily recognized by their prominent preopercular spine that is conspicuously colored in some species. While most species of angelfishes are omnivores, picking small organisms such as sponges, corals, and other small invertebrates, herbivory (within *Centropyge*) and planktivory (*Geniacanthus*) have evolved within the group (Bellwood et al., 2004). Juveniles often differ greatly in coloration from adults and several species are cleaners as juveniles. Hybrids are commonly reported among angelfishes, perhaps because they are easily recognized by their intermediate color patterns (Pyle and Randall, 1994). Bellwood et al. (2004) hypothesized relationships of 24 species, supporting the monophyly of all genera except *Centropyge*. Alva-Campbell et al. (2010) studied relationships within the New World genus *Holacanthus*.

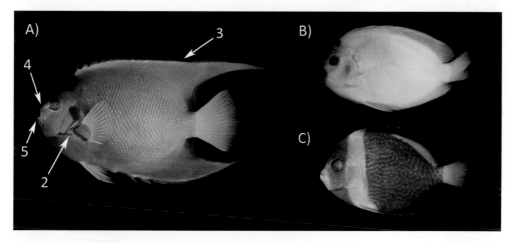

POMACANTHID CHARACTERISTICS:
1) body strongly compressed, disc-like
2) preopercle with one or more strong spines
3) dorsal fin continuous
4) head relatively small, snout short
5) mouth extremely small, terminal
6) jaw teeth usually arranged in brush-like bands
7) no pelvic-fin axillary process

ILLUSTRATED SPECIMENS:
A) *Holacanthus ciliaris,* SIO 70–205, 165 mm SL
B) *Apolemichthys trimaculatus,* SIO 92–144, 111 mm SL
C) *Chaetodontoplus duboulayi,* SIO 64–229, 56 mm SL

Several species of angelfishes are commonly kept in public and private aquaria, and this group is among the most valuable fish families in the aquarium trade.

REFERENCES: Allen et al., 1998; Alva-Campbell et al., 2004; Bellwood et al., 2004; Burgess, 1974; Burgess, in Carpenter, 2003; Pyle, in Carpenter and Niem, 2001; Pyle and Randall, 1994.

PERCIFORMES : CIRRHITIDAE—Hawkfishes

DIVERSITY: 12 genera, 35 species

REPRESENTATIVE GENERA: *Amblycirrhitus, Cirrhitichthys, Cirrhitus, Oxycirrhites, Paracirrhites*

DISTRIBUTION: Atlantic, Indian, and Pacific oceans

HABITAT: Marine; tropical; coastal, benthic on coral and rocky reefs

REMARKS: Hawkfishes are conspicuous benthic inhabitants of coral reefs, often resting

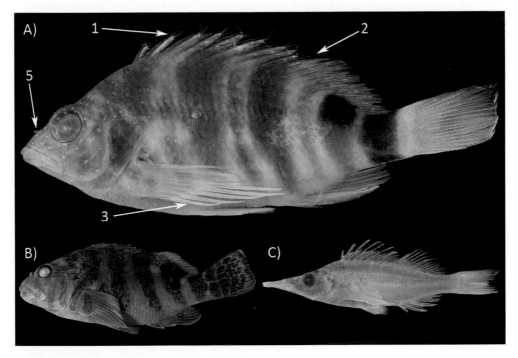

CIRRHITID CHARACTERISTICS:
1) dorsal-fin spines usually incised with tuft of cirri on each tip
2) dorsal fin continuous
3) lower five to seven rays of pectoral fins unbranched, thickened, tips free
4) scales usually cycloid
5) tuft of cirri on posterior margin of anterior nostril
6) gas bladder absent

ILLUSTRATED SPECIMENS:
A) *Amblycirrhitus pinos,* SIO 67–91, 59 mm SL
B) *Cirrhitus rivulatus,* SIO 62–704, 126 mm SL
C) *Oxycirrhites typus,* SIO 65–329, 72 mm SL

within live corals where they prey on invertebrates and small fishes. Some perch atop pinnacles of the reef, giving rise to their common name. Most are small species, but the Giant Hawkfish of the eastern Pacific reaches lengths of over 50 cm (Thomson et al., 2000) and is taken by artisanal fisheries. Males defend territories occupied by one or more females (Donaldson, 1990), and at least some species are protogynous hermaphrodites (Sadovy and Donaldson, 1995). A few species are important in the aquarium trade.

REFERENCES: Donaldson, 1990; Randall, 1963, 2001a; Randall, in Carpenter and Niem, 2001; Sadovy and Donaldson, 1995; Thomson et al., 2000.

PERCIFORMES : SPHYRAENIDAE—Barracudas

DIVERSITY: 1 genus, 29 species

REPRESENTATIVE GENUS: *Sphyraena*

DISTRIBUTION: Atlantic, Indian, and Pacific oceans

HABITAT: Coastal marine, juveniles often in estuaries; tropical to warm temperate; neritic, coastal, including coral reefs

REMARKS: Barracudas, with their elongate body and jutting lower jaw are among the most recognizable predators in coastal marine systems, especially on coral reefs. Many species are small, but some, such as the Great Barracuda (*Sphyraena barracuda*), grow to well over 1.5 m in length and occasionally have been implicated in attacks on humans. While barracudas are consumed by humans in some areas, in others they are well known for causing ciguatera poisoning and are avoided (Randall, 1958). Johnson (1986) hypothesized that barracudas are the sister group to scombriform fishes, while others (e.g., Betancur et al., 2013; Near et al., 2012) hypothesize a relationship with the Carangiformes and related fishes.

REFERENCES: de Sylva, 1975, 1984; Johnson, 1986; Randall, 1958; Senou, in Carpenter and Niem, 2001.

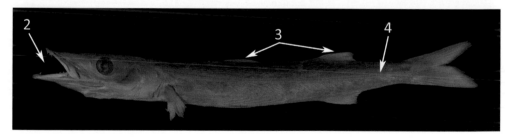

SPHYRAENID CHARACTERISTICS:
1) body elongate
2) mouth large, with protruding lower jaw and strong fang-like teeth
3) two widely separated dorsal fins, second positioned over similar, short-based anal fin
4) trunk lateral-line canal well developed
5) gill rakers vestigial to absent

ILLUSTRATED SPECIMEN:
Sphyraena argentea, SIO 51–17, 193 mm SL

PERCIFORMES : SPARIDAE—Porgies

DIVERSITY: 33 genera, 138 species

REPRESENTATIVE GENERA: *Archosargus, Calamus, Diplodus, Lagodon, Sparus*

DISTRIBUTION: Atlantic, Indian, and Pacific oceans

HABITAT: Marine, rarely in freshwater; tropical to temperate; continental shelf to continental slope, demersal over soft and hard bottoms

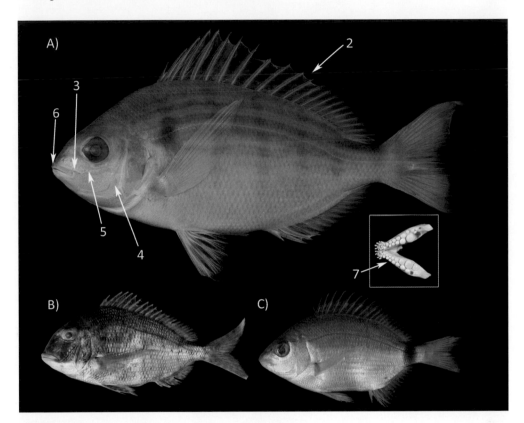

SPARID CHARACTERISTICS:
1) body laterally compressed
2) dorsal fin continuous
3) maxilla covered by premaxilla and infraorbital bones when mouth closed
4) preopercle with scales, smooth along posterior margin
5) suborbital area between eye and mouth without scales
6) mouth relatively small, rarely reaching posteriorly to midline of eye
7) jaw teeth either conical, flat, or round
8) six branchiostegal rays

ILLUSTRATED SPECIMENS:
A) *Lagodon rhomboides,* SIO 70–207, 122 mm SL
B) *Calamus brachysomus,* SIO 62–231, 173 mm SL
C) *Diplodus annularis,* SIO 90–7, 70 mm SL

INSET: Dentary of *Calamus brachysomus,* SIO uncataloged

(Porgies continues on page 172)

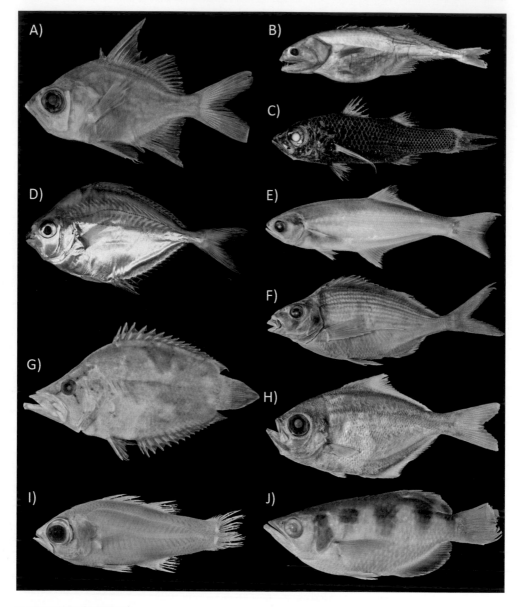

PERCIFORM DIVERSITY 1:

A) AMBASSIDAE—Asiatic glassfishes: *Chanda* sp., SIO 64–228, 61 mm SL

B) CHIASMODONTIDAE—swallowers: *Chiasmodon subniger,* SIO 66–42, 128 mm SL

C) HOWELLIDAE—howellas: *Howella sherborni,* SIO 08–32, 65 mm SL

D) LEIOGNATHIDAE—ponyfishes: *Leiognathus equulus,* AMNH 244305, 85 mm SL

E) LEPTOBRAMIDAE—beachsalmon: *Leptobrama muelleri,* SIO 61–58, 104 mm SL

F) CHEILODACTYLIDAE—morwongs: *Nemadactylus gayi,* SIO 65–647, 185 mm SL

G) POLYCENTRIDAE—Afro-American leaffishes: *Monocirrhus polyacanthus,* SIO 70–294, 58 mm SL

H) MONODACTYLIDAE—moonfishes: *Schuettea* sp., SIO 84–274, 96 mm SL

I) EPIGONIDAE—deepwater cardinalfishes: *Rosenblattia robusta,* SIO 61–45, 92 mm SL, holotype

J) TOXOTIDAE—archerfishes: *Toxotes jaculatrix,* SIO 74–63, 135 mm SL

PERCIFORM DIVERSITY 2:
A) CEPOLIDAE—bandfishes: *Acanthocepola limbata,* SIO 90–151, 300 mm SL
B) APLODACTYLIDAE—marblefishes: *Aplodactylus punctatus,* SIO 87–140, 215 mm SL
C) PSEUDOCHROMIDAE—dottybacks: *Congrogadus subducens,* SIO 87–75, 182 mm SL
D) GLAUCOSOMATIDAE—pearl perches: *Glaucosoma scapulare,* SIO 88–22, 115 mm SL
E) LOBOTIDAE—tripletails: *Lobotes surinamensis,* SIO 71–266, 68 mm SL
F) NEMIPTERIDAE—threadfin breams: *Nemipterus japonicus,* SIO 65–38, 140 mm SL
G) EPHIPPIDAE—spadefishes: *Parapsettus panamensis,* SIO 69–386, 121 mm SL
H) PEMPHERIDAE—sweepers: *Pempheris schomburgkii,* SIO 78–119, 102 mm SL
I) PLESIOPIDAE—roundheads: *Plesiops nigricans,* SIO 61–130, 57 mm SL
J) PENTACEROTIDAE—armorheads: *Pentaceros decacanthus,* SIO 84–275, 105 mm SL

REMARKS: Relationships of the Sparidae have been studied by a variety of researchers, with most supporting their close relationship with the Centracanthidae (picaral porgies, included in the Sparidae by some authors), Nemipteridae (threadfin breams), and Lethrinidae (emperors; Johnson, 1980; Carpenter and Johnson, 2002; Orrell et al., 2002; Chiba et al., 2009; Near et al., 2012). Many species of sparids have large, molar-like teeth and are able to feed on hard-shelled invertebrates, including crabs and mollusks. Day (2002) documented a high degree of trophic and associated morphological convergence within the group. In addition to their high degree of trophic specializations, sparid fishes exhibit an amazing array of sexual patterns including gonochorism, protogyny, and protandry (Buxton and Garratt, 1990). Many porgies are highly regarded food fishes and are commercially important throughout their range.

REFERENCES: Buxton and Garratt, 1990; Carpenter, in Carpenter and Niem, 2001; Carpenter, in Carpenter, 2003; Carpenter and Johnson, 2002; Chiba et al., 2009; Day, 2002; Johnson, 1980; Orrell et al., 2002.

CARANGIFORMES—Jacks and Relatives

The jacks are active predatory fishes characterized by one or two tubular ossifications (prenasals) around an extension of the nasal canal, and small, adherent scales. This group of five families, 38 genera, and 152 species (as treated here) includes the well-known and diverse Carangidae (jacks), the Coryphaenidae (dolphinfishes), the Nematistiidae (Roosterfish), the Rachycentridae (Cobia), and the Echeneidae (remoras). Recent studies indicate that jacks and flatfishes are closely related (Betancur et al., 2013; Little et al., 2010; Near et al., 2012).

REFERENCES: Smith-Vaniz, 1984; Wiley and Johnson, 2010.

CARANGIFORMES : ECHENEIDAE—Remoras

DIVERSITY: 3 genera, 8 species

REPRESENTATIVE GENERA: *Echeneis, Phtheirichthys, Remora*

DISTRIBUTION: Atlantic, Indian, and Pacific oceans

HABITAT: Marine; tropical to temperate; neritic to epipelagic

REMARKS: One of the most recognizable groups of fishes, the remoras are phoretic, able to hitch a ride by temporarily attaching to large vertebrates (including marlins, sharks, rays, sea turtles, cetaceans, and others, including humans) using the disc on the dorsal side of the head. This complex structure includes modified elements of the spinous dorsal fin (Britz and Johnson, 2012; Friedman et al., 2013). As a result of their ecology, the distribution of this group reflects the distribution of the host animals. Remoras feed on the leftover food scraps of their hosts, and some species feed on their hosts' parasitic copepods (Collette, in Carpenter, 2003). Their relationships have been studied based on morphology (O'Toole, 2002) and molecular sequence data (Gray et al., 2009), and the genus *Remorina* recently was found to be nested within, and thus synonymized with, *Remora* (Gray et al., 2009).

REFERENCES: Britz and Johnson, 2012; Collette, in Carpenter, 2003; Friedman et al., 2013; Gray et al., 2009; Miller and Lea, 1972; O'Toole, 2002.

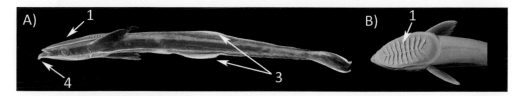

ECHENEID CHARACTERISTICS:

1) flat, sucking disc on head
2) body elongate, head depressed
3) dorsal and anal fins with long bases, lacking spines
4) lower jaw projecting anterior to upper jaw
5) scales small, cycloid
6) gas bladder absent

ILLUSTRATED SPECIMENS:

A) *Echeneis neucratoides,* SIO 71–85, 180 mm SL
B) head of *Phtheirichthys lineatus,* SIO 62–657, 103 mm SL (dorsal view)

CARANGIFORMES : CARANGIDAE—Jacks

DIVERSITY: 32 genera, 148 species

REPRESENTATIVE GENERA: *Caranx, Oligoplites, Seriola, Trachinotus*

DISTRIBUTION: Atlantic, Indian, and Pacific oceans

HABITAT: Marine and rarely brackish waters; tropical to temperate; neritic to epipelagic

REMARKS: Jacks are generally coastal fishes that actively swim in the water column. They are convergent in many ways with the typically more pelagic scombrid fishes (tunas). Many jack species form large schools that occur near reefs where they are important predators, especially on smaller schooling fishes. Juveniles of many species commonly are found in seagrass beds or associated with flotsam. Many are important to both commercial and recreational fisheries. Four lineages are recognized within the Carangidae (Gushiken, 1988; Hilton and Johnson, 2007; Reed et al., 2002; Smith-Vaniz, 1984): Caranginae (e.g., *Caranx, Decapterus, Selene, Trachurus*), Naucratinae (e.g., *Seriola*), Trachinotinae (e.g., *Trachinotus*) and Scomberoidinae (e.g., *Oligoplites*). Various aspects of the osteology of carangids recently have been discussed with respect to the evolution of the group, including the caudal fin (Hilton and Johnson, 2007), dorsal-fin pterygiophores (Springer and Smith-Vaniz, 2008), and gill arch skeleton (Hilton et al., 2010).

REFERENCES: Gushiken, 1988; Hilton and Johnson, 2007; Hilton et al., 2010; Reed et al. 2002; Smith-Vaniz, 1984; Smith-Vaniz, in Fischer et al., 1995; Smith-Vaniz, in Carpenter and Niem, 1999; Smith-Vaniz, in Carpenter, 2003; Springer and Smith-Vaniz, 2008.

(account continued)

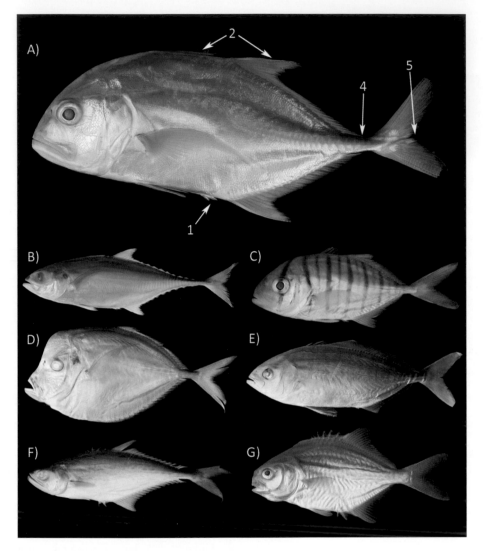

CARANGID CHARACTERISTICS:
1) anal fin with three spines, first two detached from fin
2) two dorsal fins, the second with long base
3) body compressed in most species and very deep in some
4) caudal peduncle slender, sometimes with row of lateral scutes
5) caudal fin forked
6) scales small, cycloid in most species, with some species naked

ILLUSTRATED SPECIMENS:
A) *Caranx ignobilis,* SIO 66–559, 210 mm SL (Caranginae—jacks)
B) *Megalaspis cordyla,* SIO 64–261, 208 mm SL (Caranginae)
C) *Gnathanodon speciosus,* SIO 71–204, 171 mm SL (Caranginae)
D) *Selene peruviana,* SIO 98–09, 111 mm SL (Caranginae)
E) *Seriola rivoliana,* SIO 06–260, 130 mm SL (Naucratinae—pilotfishes)
F) *Oligoplites altus,* SIO 64–235, 180 mm SL (Scomberoidinae—leatherjacks)
G) *Trachinotus paitensis,* SIO 60–367, 67 mm SL (Trachinotinae—pompanos)

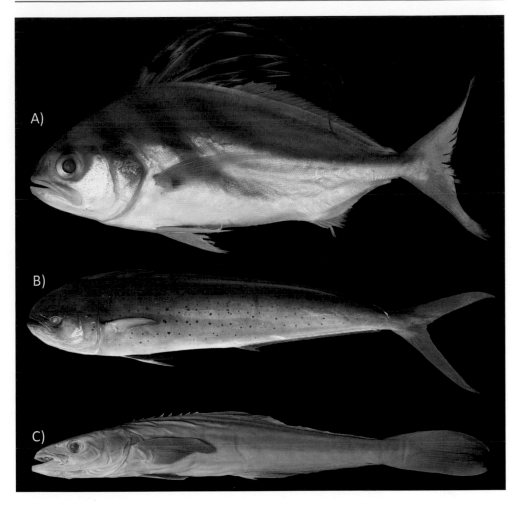

CARANGIFORM DIVERSITY:
A) NEMATISTIIDAE—Roosterfish: *Nematistius pectoralis*, SIO 64–77, 167 mm SL
B) CORYPHAENIDAE—dolphinfishes: *Coryphaena hippurus*, SIO 08–170, 758 mm SL
C) RACHYCENTRIDAE—Cobia: *Rachycentron canadum*, SIO 75–5, 228 mm SL

LABRIFORMES—Wrasses and Relatives

Historically, the Labriformes (*sensu* Kaufman and Liem, 1982) comprised six families, 235 genera, and well over 2,000 species. This included the families Labridae, Scaridae, Odacidae, Embiotocidae, Cichlidae, and Pomacentridae. Wiley and Johnson (2010), noting that its monophyly has been questioned by several authors, list eight morphological synapomorphies, none of which are unique to the included members. The group has been diagnosed primarily by features related to the pharyngeal jaws. Simply stated, the fifth ceratobranchials are united to form a single lower pharyngeal jaw that is suspended in a muscular sling, and the opposing upper pharyngeal jaw articulates by means of a diarthrosis with the basicranium (Nelson, 2006). This pharyngeal jaw apparatus has been considered a "key innovation" that led to the rapid diversification of these fishes (Liem, 1973; Kaufman and Liem, 1982; Stiassny and Jensen, 1987; Wainwright et al., 2012). Using extensive mitogenomic data, however, Mabuchi et al. (2007) did not support labriform monophyly, but instead identified two lineages, one including the Labridae, Scaridae, and Odacidae, and a second including the Cichlidae, Pomacentridae, and Embiotocidae. This result has been supported by several subsequent molecular analyses (e.g., Betancur et al., 2013; Near et al., 2012, 2013; Wainwright et al., 2012) implying that the specialized pharyngeal jaws of these groups evolved independently. Thus, we recognize the Labriformes *sensu stricto* as including the wrasses and related fishes, and a separate clade comprising the Cichlidae, Pomacentridae, and Embiotocidae. These families, sometimes called the "chromides," were recently included in the Ovalentaria (Wainwright et al., 2012).

REFERENCES: Kaufman and Liem, 1982; Liem, 1973; Mabuchi et al., 2007; Stiassny and Jensen, 1987; Wainwright et al., 2012; Wiley and Johnson, 2010.

LABRIFORMES: LABRIDAE—Wrasses and Relatives

DIVERSITY: 82 genera, 633 species

REPRESENTATIVE GENERA: *Bodianus, Halichoeres, Odax, Scarus, Thalassoma*

DISTRIBUTION: Atlantic, Indian, and Pacific oceans

HABITAT: Marine; tropical to temperate; coastal, demersal or neritic, especially common on coral and rocky reefs

REMARKS: Among marine fishes, the Labridae is second only to the Gobiidae in terms of species diversity. A recent study of their phylogenetic relationships (Westneat and Alfaro, 2005) provides strong evidence that the parrotfishes (formerly Scaridae) and the so-called cales (formerly Odacidae) are nested within the large and diverse Labridae. Westneat and Alfaro (2005) identified eight lineages within the Labridae. Parenti and Randall (2000, 2010) list the known species of wrasses and parrotfishes. Wrasses are diverse, abundant, and conspicuous members of tropical reef communities where they have radiated into a wide variety of feeding niches, including parasite picking at "cleaning stations" (Westneat et al., 2005). The group is especially diverse in the Indo-Pacific region. At least five lineages

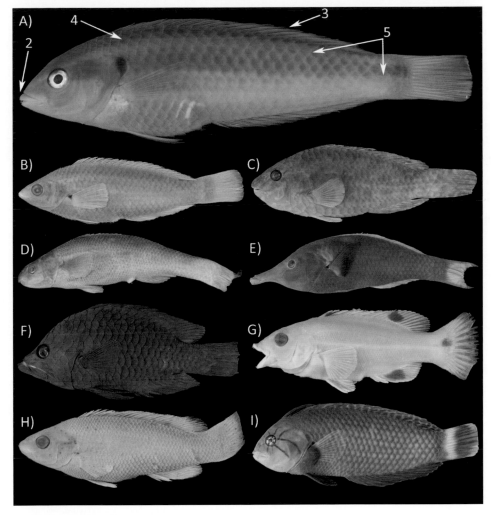

LABRID CHARACTERISTICS:
1) body usually laterally compressed and somewhat elongate
2) upper jaw protrusible in most species (not in parrotfishes and cales)
3) dorsal fin long, continuous
4) scales cycloid
5) trunk lateral-line canal complete or interrupted
6) many species very colorful

ILLUSTRATED SPECIMENS:
A) *Halichoeres melanotis,* SIO 65–185, 89 mm SL (Julidini)
B) *Pseudolabrus gayi,* SIO 65–629, 75 mm SL (Pseudolabrini)
C) *Nicholsina denticulata,* SIO 65–340, 167 mm SL (Scarini)
D) *Olisthops cyanomelas,* SIO 84–296, 152 mm SL (Hypsigeni)
E) *Gomphosus caeruleus,* SIO 60–408, 201 mm SL (Julidini)
F) *Epibulus insidiator,* SIO 61–644, 95 mm SL (Cheilini)
G) *Semicossyphus pulcher* (juvenile), SIO 62–272, 100 mm SL (Hypsigeni)
H) *Tautogolabrus adspersus,* SIO 62–456, 94 mm SL (Labrini)
I) *Novaculichthys taeniourus,* SIO 61–252, 92 mm SL (Novaculini)

(account continued)

have invaded the New World tropics, while others (e.g., the Labrini of the Northern Hemisphere and Odacini of the Southern Hemisphere) have invaded temperate waters (Clements et al., 2004; Westneat and Alfaro, 2005). Two radiations, within the Julidini (*Halichoeres* and related genera) and the Scarini (parrotfishes), are largely responsible for the extraordinary biodiversity of the group (Alfaro et al., 2009). Labrids exhibit a range of reproductive modes: most species are protogynous hermaphrodites, while gonochorism (separate sexes) has evolved repeatedly within the group (Kazancioğlu and Alonzo, 2010). Similarly, most species spawn pelagic eggs, while some lay and guard demersal eggs (e.g., Potts, 1985). Most labrids are highly variable in coloration, with juveniles, females, and males differing dramatically in appearance. Several large-bodied species in tropical and temperate systems are targeted for human consumption, both commercially and recreationally, and a number of small species are included in the aquarium trade.

REFERENCES: Alfaro et al., 2009; Clements et al., 2004; Gomon, in Fischer et al., 1995; Kazancioğlu and Alonzo, 2010; Parenti and Randall, 2000, 2010; Potts, 1985; Schultz, 1958, 1969; Westneat, in Carpenter, 2003; Westneat, in Carpenter and Niem, 2001; Westneat and Alfaro, 2005; Westneat et al., 2005.

"CHROMIDES"— Cichlids, Damselfishes, and Relatives

The following three families, traditionally included in the Labriformes, are treated together as the "chromides." The monophyly of this group has not be demonstrated and it has no official taxonomic standing, but was recently removed from the Labriformes and included in a large and diverse clade, the Ovalentaria, by Wainwright et al. (2012).

REFERENCES: Wainwright et al., 2012

"CHROMIDES" : CICHLIDAE—Cichlids

DIVERSITY: 112 genera; estimates of species diversity vary greatly from 1,350 (Nelson, 2006), to 1,640 (Eschmeyer and Fong, 2013), to 2,200 (Turner, 2007), to over 3,000 (Kocher, 2004)

REPRESENTATIVE GENERA: *Cichlasoma, Haplochromis, Melanochromis, Oreochromis, Tilapia*

DISTRIBUTION: Southern North America to South America, Africa, Madagascar, Middle East, and India; widely introduced into other freshwaters

HABITAT: Freshwater to brackish waters; tropical to warm temperate; demersal to pelagic, various habitats in lakes, rivers, and streams

REMARKS: Cichlids are one of the most well-studied lineages of tropical freshwater fishes because of their broad distribution, extraordinary diversity, and popularity in the aquarium industry (Barlow, 2000; Keenleyside, 1991). The distribution and relationships of cichlids is indicative of a group whose evolution is intimately tied to the breakup of Gondwana (Chakrabarty, 2004; Sparks and Smith, 2004). Sparks and Smith (2004) recovered four primary lineages: Etropilinae and Ptychochrominae, both restricted to Madagascar and southern Asia; Pseudocrenilabrinae from Africa; and Cichlinae from the neotropics. The phylogeny of neotropical cichlids was recently hypothesized by Smith et al. (2008), and that of African cichlids by numerous authors in conjunction with their studies of the remarkable radiations of these fishes. The cichlids were recently identified as one of five rapidly radiating lineages of acanthomorph fishes (Near et al., 2013). The large rift lakes of Africa are home to an astounding diversity of cichlids with over 500 species in Lake Victoria, 600 spe-

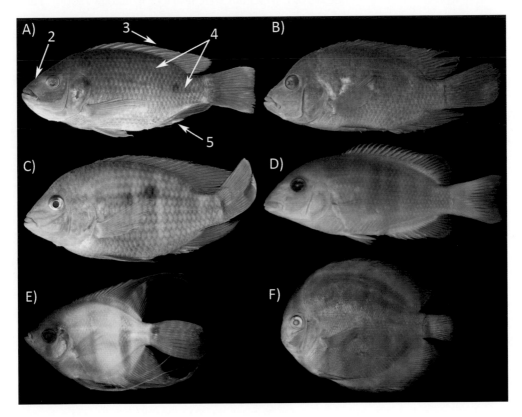

CICHLID CHARACTERISTICS:
1) body shape variable, typically laterally compressed and deep (especially so in some species), slightly elongate in some species
2) single nostril on either side of head
3) dorsal fin continuous, with 7–25 spines (high end of range most typical)
4) trunk lateral-line canal interrupted in most species
5) anal-fin spines variable in number (3–15)

ILLUSTRATED SPECIMENS:
A) *Oreochromis mossambicus,* SIO 64-266, 130 mm SL
B) *Herichthys cyanoguttatus,* SIO 69-188, 110 mm SL
C) *Andinoacara rivulatus,* SIO 63-745, 78 mm SL
D) *Paretroplus nourissati,* AMNH 229555, 83 mm SL
E) *Pterophyllum scalare,* SIO 64-228, 56 mm SL
F) *Symphysodon aequifasciatus,* SIO 64-228, 96 mm SL

cies in Lake Malawi, and perhaps 1,800 species in Lake Tanganyika (Fryer and Iles, 1972; Genner and Turner, 2005; Kocher, 2004; Turner, 2007). It was once widely thought that most of this diversity derived from independent radiations from within each lake, but recent studies of cichlids in Lake Tanganyika (Genner et al., 2007) and Lake Malawi (Joyce et al., 2011) demonstrate that these are derived from multiple invasions of each lake. The genomes of cichlids are under study (see http://cichlid.umd.edu/cichlidlabs/kocherlab/bouillabase.html) with the goal of understanding the genetic features associated with rapid speciation in these fishes. The behavior and biology of cichlids were reviewed by Barlow (2000). Their dietary diversity is immense; the group includes omnivores, planktivores, piscivores, and herbivores (Fryer and Illes, 1972). Many cichlids are important aquarium species and several are important to aquaculture. As a consequence, some species, especially those in the genera *Oreochromis* and *Tilapia,* have been widely introduced beyond their normal range.

REFERENCES: Barlow, 2000; Carpenter, in Carpenter and Niem, 2001; Carpenter, in Carpenter, 2003; Chakrabarty, 2004; Fryer and Iles, 1972; Genner et al., 2007; Genner and Turner, 2005; Joyce et al., 2011; Keenleyside, 1991; Kocher, 2004; Kullander, 1998, 2003; Miller et al., 2005; Smith et al., 2008; Sparks and Smith, 2004; Stiassney, 1991; Turner, 2007; Turner et al., 2001.

"CHROMIDES" : POMACENTRIDAE—Damselfishes

DIVERSITY: 28 genera, 390 species
REPRESENTATIVE GENERA: *Abudefduf, Amphiprion, Chromis, Dascyllus, Pomacentrus, Stegastes*
DISTRIBUTION: Atlantic, Indian, and Pacific oceans
HABITAT: Marine to brackish waters; tropical to temperate; demersal or neritic over hard bottoms, especially common on reefs
REMARKS: Damselfishes are diverse, conspicuous, and abundant members of reef communities, especially on coral reefs (Allen, 1991; Cooper, 2006). Their relationships have been studied by a number of researchers (e.g., Cooper et al., 2009; Tang, 2001). Cooper et al. (2009) recognized five subfamilies and questioned the monophyly of several currently recognized genera. Pomacentrids are in many respects model organisms of the reef-fish community, and the behavior and ecology of many species are well studied. The group includes numerous herbivorous species (e.g., *Stegastes*), some tending their own gardens, as well as planktivorous lineages (e.g., *Abudefduf* and *Chromis*). Many damselfishes are fiercely territorial (Myrberg, 1972), having a significant impact on a variety of other reef species (Ceccarelli et al., 2001). Damselfishes lay demersal eggs that are guarded by males (Petersen, 1995) and a few species, notably *Dascyllus* and the so-called anemonefishes (*Amphiprion and Premnas*), are hermaphroditic (Cooper et al., 2009). The anemonefishes are also well-known symbionts of sea anemones. Most damselfishes are small, and many species are heavily exploited by the aquarium trade.

REFERENCES: Allen, 1991; Allen, in Carpenter and Niem, 2001; Ceccarelli et al., 2001; Cooper, 2006; Cooper et al., 2009; Myrberg, 1972; Petersen, 1995; Schneider and Krupp, in Fischer et al., 1995; Tang, 2001.

POMACENTRID CHARACTERISTICS:
1) body compressed and usually deep
2) mouth small
3) dorsal fin continuous
4) trunk lateral-line canal incomplete or discontinuous
5) scales cteniod
6) anal fin usually with two spines
7) eye diameter usually greater than snout length
8) many species very colorful

ILLUSTRATED SPECIMENS:
A) *Neoglyphidodon nigroris,* SIO 73–196, 72 mm SL (Pomacentrinae)
B) *Amphiprion percula,* SIO 64–1034, 54 mm SL (Pomacentrinae, Amphiprionini)
C) *Stegastes nigricans,* SIO 73–196, 82 mm SL (Stegastinae)
D) *Chromis atrilobata,* SIO 65–290, 95 mm SL (Chrominae)
E) *Abudefduf troschelii,* SIO 61–272, 136 mm SL (Abudefdufinae)
F) *Dascyllus aruanus,* SIO 83–56, 52 mm SL (Chrominae)

"CHROMIDES" : EMBIOTOCIDAE—Surfperches

DIVERSITY: 13 genera, 23 species

REPRESENTATIVE GENERA: *Amphistichus, Embiotica, Hysterocarpus, Zalembius*

DISTRIBUTION: North Pacific Ocean from northern Mexico to Korea

HABITAT: Marine; subtropical to temperate; demersal to neritic, coastal habitats, includ-

EMBIOTOCID CHARACTERISTICS:
1) body laterally compressed
2) dorsal fin continuous
3) scales cycloid
4) caudal fin forked
5) males with modified anal-fin elements, used to inseminate females
6) vomerine and palatine teeth usually absent
7) trunk lateral-line canal complete

ILLUSTRATED SPECIMENS:
A) *Amphistichus koelzi,* SIO 80–20, 112 mm SL
B) *Hysterocarpus traskii,* SIO 77–369, 151 mm SL
C) *Cymatogaster aggregata,* SIO 48–127, 115 mm SL, radiograph of pregnant female

ing bays and estuaries, with one species (the Tule Perch, *Hysterocarpus traskii*) restricted to freshwaters

REMARKS: Embiotocids are one of the few lineages of fishes restricted to the north Pacific Ocean. Their systematics and morphology were studied by Tarp (1952) and intrarelationships by Bernardi and Bucciarelli (1999) and Westphal et al. (2011). They feed on a variety of small benthic invertebrates, as well as plankton. Surfperches are rare among marine, bony fishes for their reproductive pattern, which includes internal fertilization and extended embryoparity, with females giving birth to large, well-developed offspring (Balz, 1983). Remarkably, males of the Dwarf Surfperch (*Micrometrus minimus*) are sexually mature at birth (Warner and Harlan, 1982). Several species of surfperches are targeted by recreational fisheries.

REFERENCES: Balz, 1983; Bernardi and Bucciarelli, 1999; Eschmeyer and Herald, 1983; Liem, 1986; Miller and Lea, 1972; Tarp, 1952; Warner and Harlan, 1982; Westphal et al., 2011.

NOTOTHENIIFORMES—Icefishes and Relatives

DIVERSITY: 8 families, 44 genera, 159 species

REPRESENTATIVE GENERA: *Bovichtus, Chaenocephalus, Channichthys, Notothenia, Pagetopsis*

DISTRIBUTION: Southern Hemisphere, restricted to southern South America and Antarctica

HABITAT: Marine; temperate to polar; coastal areas, sometimes closely associated with ice; benthic, demersal, neritic to pelagic

REMARKS: The Nototheniiformes comprises an impressive ecological radiation of Southern Hemisphere fishes, with most species found south of the Antarctic convergence (Eastman, 2005; Near, Pesavento, and Cheng, 2003). Members of this group range from small benthic fishes, to large, mobile predators, to small schooling planktivores (*Pleurogramma*). Their phylogenetic relationships have been studied by several researchers (e.g., Balushkin, 2000; Eastman, 2005; Eastman and Eakin, 2000; Hastings, 1993; Near, Pesavento, and Cheng, 2003, 2004). They exhibit several unusual adaptations to cold conditions, especially the crocodile icefishes (Channichthyidae) whose blood lacks hemoglobin and muscles lack myoglobin. At least one species (Patagonian Blenny—Eleginopidae) is known to be a protandrous hermaphrodite. Several species are important to commercial fisheries, including the well-known Patagonian Toothfish (*Dissostichus eleginoides*).

REFERENCES: Baluskin, 2000; Eakin, 1981; Eastman, 2005; Eastman and Eakin, 2000; Gon and Heemstra, 1990; Hastings, 1993; Miller, 1993; Montgomery and Clements, 2000; Near, Pesavento, and Cheng, 2003, 2004.

(account continued)

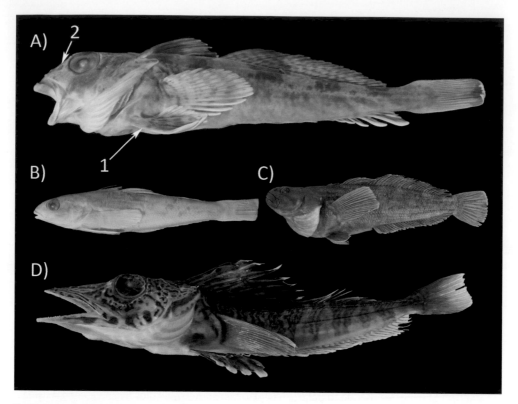

NOTOTHENIIFORM CHARACTERISTICS:
1) pelvic fins jugular
2) single nostril on either side of head
3) multiple trunk lateral-line canals in most species
4) pectoral-fin radials plate-like
5) five to nine branchiostegal rays

ILLUSTRATED SPECIMENS:
A) *Bovichtus argentinus,* SIO 65–670, 87 mm SL (Bovichtidae—thornfishes)
B) *Eleginops maclovinus,* SIO 65–670, 60 mm SL (Eleginopsidae—Patagonian blennies)
C) *Notothenia coriiceps,* SIO 00–163, 330 mm SL (Nototheniidae—cod icefishes)
D) *Pagetopsis macropterus,* SIO 10–104, 255 mm SL (Channichthyidae—crocodile icefishes)

TRACHINIFORMES—Weeverfishes and Relatives

The Trachiniformes is especially difficult to characterize, and its reality and composition have been questioned by a number of authors. This group, as presented by Pietsch (1989) and Pietsch and Zabetian (1990), is characterized by a pelvic spur (a short process on the pelvic bones) and short, wide pectoral radials but is likely not monophyletic (Wiley and Johnson, 2010). Nelson (2006) included 12 families, 53 genera, and 237 species from a variety of habitats. These included the coastal marine stargazers (Uranoscopidae, covered below) and sandburrowers (Creediidae), the freshwater New Zealand torrentfishes (Cheimarrhichthyidae), and the deep-sea swallowers (Chiasmodontidae, here included in Perciformes).

REFERENCES: Johnson, 1993; Mooi and Johnson, 1997; Pietsch, 1989; Pietsch and Zabetian, 1990; Wiley and Johnson, 2010.

TRACHINIFORMES : URANOSCOPIDAE—Stargazers

DIVERSITY: 8 genera, 53 species

REPRESENTATIVE GENERA: *Astroscopus, Kathetostoma, Uranoscopus*

DISTRIBUTION: Atlantic, Indian, and Pacific oceans

HABITAT: Marine, tropical to temperate; coastal to continental slope, benthic on soft bottoms, including estuaries, and adjacent to reefs

REMARKS: Stargazers are sit-and-wait predators that typically bury in the sediment. In addition to their dorsally directed eyes and mouth, fringed lips, and dorsally located trunk lateral-line canal, some have a worm-like appendage in the mouth for attracting prey. Their diet consists of small fishes and probably benthic invertebrates. Members of the genus *Astroscopus* have electric organs derived from eye muscles for stunning prey. These electric organs no doubt also serve to ward off would-be predators, as do the venomous spines on the cleithrum (Smith and Wheeler, 2006).

REFERENCES: Carpenter, in Carpenter, 2003; Pietsch, 1989; Smith and Wheeler, 2006.

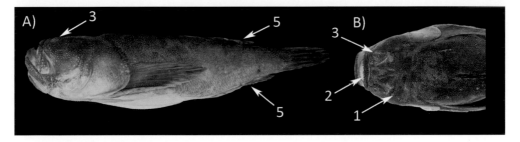

URANOSCOPID CHARACTERISTICS:
1) head large and broad
2) mouth upturned with fringed lips
3) eyes dorsal
4) trunk lateral-line canal high on body
5) dorsal and anal fins with long bases

ILLUSTRATED SPECIMEN:
A) *Astroscopus zephyreus,* SIO 08–66, 312 mm SL
B) head of *Astroscopus zephyreus,* SIO 08–66 (dorsal view)

Trachiniformes   185

PHOLIDICHTHYIFORMES : PHOLIDICHTHYIDAE—Convict Blennies

DIVERSITY: 1 family, 1 genus, 2 species

REPRESENTATIVE GENUS: *Pholidichthys*

DISTRIBUTION: Indo-West Pacific

HABITAT: Marine; tropical; benthic or demersal over reefs or soft bottoms strewn with rocks and shells

REMARKS: The convict blennies or convict fishes have a long, dark stripe as juveniles and a series of dark bars as adults that contrast with their light or white background color. This enigmatic taxon previously has been placed within or associated with the blenniiforms, labriforms, and trachiniforms (Springer and Freihofer, 1976). Most recently, it has been hypothesized to be the sister group to the Cichlidae (Near et al., 2012, 2013; Wainwright et al., 2012). Juveniles often form large schools around coral reefs and frequently enter burrows. Convict blennies occasionally are seen in the aquarium trade.

REFERENCES: Near et al., 2012, 2013; Springer, in Carpenter and Niem, 2001; Springer and Freihofer, 1976; Springer and Larson, 1996; Stiassny and Jensen, 1987; Wainwright et al., 2012.

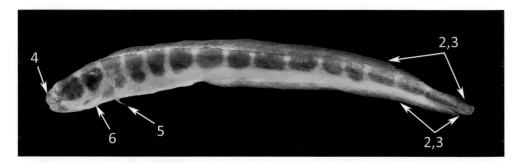

PHOLIDICHTHYIFORM CHARACTERISTICS:
1) body elongate, without scales
2) dorsal and anal fins with segmented rays only
3) dorsal and anal fins long, confluent with caudal fin
4) single nostril on each side of head
5) pelvic fins with two or three rays
6) gill openings attached to isthmus
7) trunk lateral-line canal consisting of small, inconspicuous pits

ILLUSTRATED SPECIMEN:
Pholidichthys leucotaenia, SIO 13–258, 117 mm SL

BLENNIIFORMES—Blennies

The Blenniiformes includes six families, 151 genera, and nearly 900 species of coastal benthic fishes. They are characterized by jugular pelvic fins with one embedded spine and two to four simple soft rays, pelvic bones forming a nut-like pod that is open ventrally, and an anal fin with zero to two spines and only unbranched soft rays (Hastings and Springer, 2009a; Springer, 1993). Recently, their phylogenetic relationships were hypothesized by Lin and Hastings (2013) and they were identified as one of five rapidly radiating lineages of acanthomorph fishes (Near et al., 2013). Blennies are classic examples of cryptobenthic fishes that are often numerically dominant members of reef communities (Depczynski and Bellwood, 2003). Most are microcarnivores, although many blenniids are herbivores (Kotrschal and Thomson, 1986; Patzner et al., 2009). Blennies generally have external fertilization of demersal eggs that are guarded by the male, although internal fertilization has evolved independently in the Clinidae and Labrisomidae (Hastings and Petersen, 2010). Growing evidence from both molecular and morphological data (Lin and Hastings, 2013) indicate that the Blenniiformes is most closely related to another cryptobenthic group, the clingfishes (Gobiesociformes), with both groups recently included in the Ovalentaria (Wainwright et al., 2012). In addition to the four families covered below, the Blenniiformes includes the temperate kelp blennies (Clinidae) and the tropical sand stargazers (Dactyloscopidae).

REFERENCES: Depczynski and Bellwood, 2003; Hastings and Petersen, 2010; Hastings and Springer, 2009a; Kotrschal and Thomson, 1986; Lin and Hastings, 2013; Patzner et al. 2009; Springer, 1993; Wainwright et al., 2012.

BLENNIIFORMES : TRIPTERYGIIDAE—Triplefin Blennies

DIVERSITY: 32 genera, 171 species

REPRESENTATIVE GENERA: *Axoclinus, Enneanectes, Enneapterygius, Helcogramma, Tripterygion*

DISTRIBUTION: Atlantic, Indian, and Pacific oceans, including Antarctica

HABITAT: Marine; tropical to polar; coastal to continental slope, benthic, typically on rocky and coral reefs from the intertidal to over 500 m

REMARKS: The common name triplefin blenny derives from their distinctly divided three-part dorsal fin. Triplefins are small, active fishes that can be quite abundant at some reef sites where they feed on a variety of small invertebrates (Kotrschal and Thomson, 1986; Wilson, 2009). The group is especially diverse in the warm temperate reefs of New Zealand (Hickey et al., 2009), and one species is endemic to Antarctica. Like most other blenniiforms, triplefins lay demersal eggs that are guarded by males (Hastings and Petersen, 2010). Fricke (2009) reviewed the species-level systematics of triplefins and recognized eight lineages within the group.

REFERENCES: Fricke, 2009; Hickey et al., 2009; Kotrschal and Thomson, 1986; Williams, in Carpenter, 2003; Williams and Fricke, in Carpenter and Niem, 2001; Wilson, 2009.

(account continued)

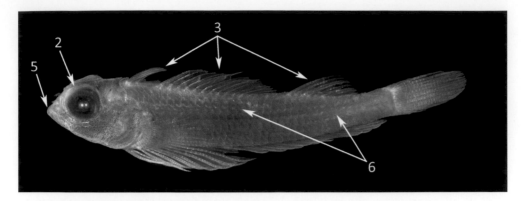

TRIPTERYGIID CHARACTERISTICS:
1) body somewhat elongate in most species, but not strongly compressed
2) eyes relatively large, set high on head
3) dorsal fin in three distinct parts, first two spinous, third with segmented rays
4) no cirri on nape
5) upper jaw protrusible
6) trunk lateral-line canal well developed, typically with anterior series of pored scales and posterior series of notched scales

ILLUSTRATED SPECIMEN:
Enneanectes glendae, SIO 11–394, 27 mm SL, holotype

BLENNIIFORMES : BLENNIIDAE—Combtooth Blennies

DIVERSITY: 56 genera, 404 species

REPRESENTATIVE GENERA: *Blennius, Ecsenius, Hypsoblennius, Ophioblennius, Parablennius, Plagiotremus*

DISTRIBUTION: Atlantic, Indian, and Pacific oceans

HABITAT: Marine, rarely brackish and freshwater; tropical to temperate; benthic in most coastal habitats, including tidepools

REMARKS: The common name, combtooth blennies, comes from this group's characteristic close-set, comb-like row of incisiform teeth in the outer row of both jaws. The systematics and biology of blenniids have been widely studied (Hastings and Springer, 2009b; Hundt et al., 2013; Lin and Hastings, 2013). Blenniids have flexible dorsal-fin spines and adults of most species, in keeping with their benthic habits, usually lose the gas bladder seen in larvae. The nemophin blennies are aggressive mimics of cleaner wrasses, and, being more active swimmers, retain a gas bladder as adults (Smith-Vaniz, 1976). Many blenniids use their comb-like teeth to scrape algae and other food items from the substrate, but the group includes a variety of microcarnivores and even some skin- and scale-eaters (Wilson, 2009).

REFERENCES: Hastings, in Fischer et al., 1995; Hastings and Springer, 2009b; Hastings and Petersen, 2010; Hundt et al., 2013; Lin and Hastings, 2013; Smith-Vaniz, 1976; Springer, 1968; Springer, in Carpenter and Niem, 2001; Williams, in Carpenter, 2003; Wilson, 2009.

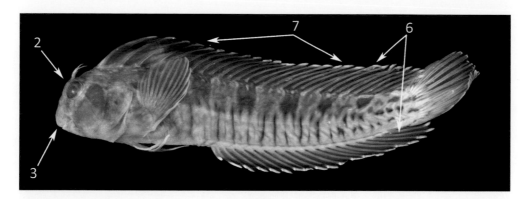

BLENNIID CHARACTERISTICS:

1) body somewhat elongate and laterally compressed
2) forehead bluntly rounded in most species
3) upper jaw not protrusible
4) jaws with comb-like teeth, fixed or freely movable
5) scales absent
6) segmented rays unbranched in all fins except caudal
7) dorsal fin continuous, with more soft rays than spines

ILLUSTRATED SPECIMEN:

Hypsoblennius striatus, SIO 71–253, 54 mm SL

BLENNIIFORMES : LABRISOMIDAE—Labrisomid Blennies

DIVERSITY: 14 genera, 119 species

REPRESENTATIVE GENERA: *Labrisomus, Malacoctenus, Paraclinus, Starksia*

DISTRIBUTION: Western Atlantic and eastern Pacific oceans, with a few species in the central and eastern Atlantic

HABITAT: Marine, coastal; tropical to warm temperate; benthic on rocky and coral reefs

REMARKS: The Labrisomidae is not well defined morphologically, but a recent molecular study (Lin and Hastings, 2013) supported the monophyly of a subset of traditional members including five relatively well-defined lineages (tribes): Labrisomini, Paraclinini, Starksiini, Stathmonotini, and Mnierpini. While most labrisomids are cryptic on subtidal reefs and surrounding areas, the Mnierpini exhibit several unique adaptations to a semi-terrestrial life in and around tidepools (Neider, 2001). Typical of most other blenniiforms, labrisomids are microcarnivores (Kotrschal and Thomson, 1986) and most lay demersal eggs guarded by males (Hastings and Petersen, 2010). Some species of *Starksia* and the single species of *Xenomedea* have internal fertilization, and a few retain eggs within females until hatching (Fishelson et al., 2013).

REFERENCES: Fishelson et al., 2013; Hastings and Petersen, 2010; Hastings and Springer, 2009a; Kotrschal and Thomson, 1986; Lin and Hastings, 2013; Neider, 2001; Williams, in Carpenter, 2003.

(account continued)

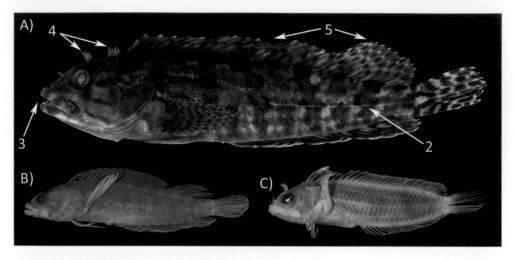

LABRISOMID CHARACTERISTICS:
1) body relatively elongate, laterally compressed in most species
2) scales and trunk lateral-line canal well developed
3) mouth large, terminal in most species
4) head often with well-developed cirri
5) dorsal fin with more spines than soft rays

ILLUSTRATED SPECIMENS:
A) *Labrisomus xanti,* SIO 07–151, 112 mm SL (Labrisomini)
B) *Xenomedea rhodopyga,* SIO 62–212, 42 mm SL, holotype (Starksiini)
C) *Paraclinus tanygnathus,* SIO 62–56, 24 mm SL, holotype (Paraclinini)

BLENNIIFORMES : CHAENOPSIDAE—Tube Blennies

DIVERSITY: 11 genera, 80 species

REPRESENTATIVE GENERA: *Acanthemblemaria, Chaenopsis, Emblemaria, Emblemariopsis*

DISTRIBUTION: Western Atlantic and eastern Pacific oceans

HABITAT: Marine; tropical to warm temperate; coastal, benthic, coral and rocky reefs and surrounding rubble zones

REMARKS: Tube blennies get their common name from their habit of occupying vacant tests of various species of invertebrates. Their systematics and morphology were studied extensively by Stephens (1963), and their phylogenetic relationships have been hypothesized based on both morphological and molecular data (Hastings and Springer, 1994; Lin and Hastings, 2011). Chaenopsids are microcarnivores, feeding on benthic and planktonic invertebrates (Kotrschal and Thomson, 1986). Shelter availability may limit the population density of tube blennies (Hastings and Galland, 2010), forms the basis for interspecific competition (Clarke, 1996), and is critical for male reproductive success (Hastings and Petersen, 2010). Chaenopsids are well known for their dramatic aggressive displays in defense of shelters and the conspicuous courtship displays of males.

REFERENCES: Clarke, 1996; Hastings, in Fischer et al., 1995; Hastings and Galland, 2010;

Hastings and Petersen, 2010; Hastings and Springer, 1994, 2009a; Kotrschal and Thomson, 1986; Lin and Hastings, 2011; Stephens, 1963; Williams, in Carpenter, 2003.

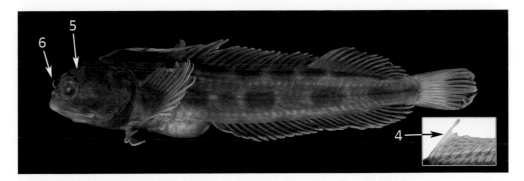

CHAENOPSID CHARACTERISTICS:
1) body somewhat elongate
2) scales absent
3) trunk lateral-line canal absent
4) median fin spines not ossified distally
5) cranial bones often with spines, pits, or ridges
6) supraorbital and nasal cirri present in most species

ILLUSTRATED SPECIMEN:
Acanthemblemaria hastingsi, SIO 65–272, 44 mm SL, holotype.

INSET: Cleared-and-stained anterior dorsal fin of *Tanyemblemaria alleni*, USNM 309963, 54 mm SL

BLENNIIFORM DIVERSITY:
A) CLINIDAE—kelp blennies: *Myxodes ornatus,* SIO 65–678, 54 mm SL, holotype
B) DACTYLOSCOPIDAE—sand stargazers: *Dactyloscopus lunaticus,* SIO 59–218, 42 mm SL

GOBIESOCIFORMES : GOBIESOCIDAE—Clingfishes

DIVERSITY: 1 family, 47 genera, 163 species

REPRESENTATIVE GENERA: *Gobiesox, Rimicola, Sicyases, Tomicodon*

DISTRIBUTION: Atlantic, Indian, and Pacific oceans

HABITAT: Marine to occasionally freshwater; tropical to temperate; intertidal to upper slope, benthic on rocky and coral reefs

REMARKS: The clingfishes are small, benthic fishes that resemble tadpoles, with a broad, rounded, and depressed head, and a narrow, tapering tail region. They share several reductive characteristics: the infraorbital series is represented only by the lacrimal, and they lack scales and the metapterygoid bone. Once thought to be closely related to the dragonets (Callionymidae; Gosline, 1970), current studies show them to be more closely related to the blennies (Hastings and Springer, 2009a; Lin and Hastings, 2013), within the Ovalentaria (Wainwright et al., 2012; Near et al., 2013). Clingfishes are so-named because they are able to attach to the substrate with their modified pelvic fins, a feature that enables them to occupy areas of strong surge, high in the rocky intertidal. Consistent with their cryptoben-

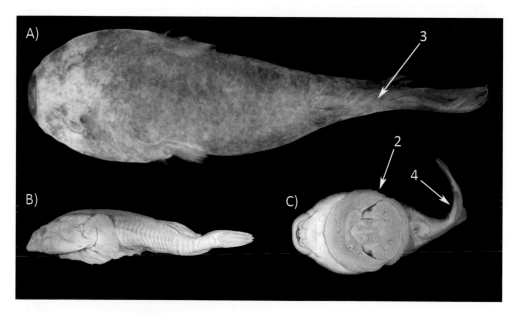

GOBIESOCIFORM CHARACTERISTICS:
1) head and body depressed, tadpole shaped
2) thoracic sucking disk formed by modified pelvic fins
3) single dorsal fin without spines
4) anal fin without spines
5) head and body scaleless
6) gas bladder absent

ILLUSTRATED SPECIMENS:
A) *Pherallodiscus funebris,* SIO 87–182, 55 mm SL (dorsal view)
B) and C) *Sicyases sanguineus,* SIO 87–132, 276 mm SL (lateral and ventral views)

thic habits, clingfishes lack a gas bladder. Most species are plain-colored, but a few reef species are quite strikingly colored. They feed on small invertebrates and algae. Most species are small, usually less than 70 mm, but two Southern Hemisphere species (*Sicyases*) can reach 30 cm or more, and support artisanal fisheries. With a few exceptions (e.g., Williams and Tyler, 2003), the systematics of clingfishes has not been thoroughly reviewed since the landmark study by Briggs (1955).

REFERENCES: Briggs, 1955; Gosline, 1970; Springer and Fraser, 1976; Williams and Tyler, 2003.

GOBIIFORMES—Gobies and Relatives

The gobies and their relatives comprise nine families, approximately 270 genera, and more than 2,200 species. They are characterized by reductive characters including the absence of parietal bones, a gas bladder, and pyloric caecae; two or fewer infraorbital bones; and in most species, united pelvic fins, and a lateral-line system composed of cephalic canals. This large and diverse lineage of generally benthic fishes is found in virtually all coastal habitats as well as freshwaters of tropical and temperate areas. They recently were identified as one of five rapidly radiating lineages of acanthomorph fishes (Near et al., 2013). The size of the group has hampered efforts to resolve their relationships, but in recent years, this daunting task has been taken on by several researchers (Thacker, 2009; reviews in Patzner et al., 2011). Various unique lineages have been recognized as distinct from the main family Gobiidae, including the wormfishes (Microdesmidae), dartfishes (Ptereleotridae), and the paedomorphic infantfishes (Schindleriidae), which are among the world's smallest vertebrates and spend their entire lives in the plankton (Watson and Walker, 2004).

REFERENCES: Near et al., 2013; Patzner et al., 2011; Thacker, 2009; Watson and Walker, 2004; Winterbottom, 1993a.

GOBIIFORMES : GOBIIDAE—Gobies

DIVERSITY: 210 genera, over 1700 species

REPRESENTATIVE GENERA: *Buthygobius, Eviota, Gobionellus, Gobiosoma, Periophthalmus, Trimma*

DISTRIBUTION: Atlantic, Indian, and Pacific oceans and adjacent freshwaters

HABITAT: Marine to freshwater; tropical to temperate; coastal to continental slope, streams, and estuaries, benthic on soft and hard bottoms

REMARKS: The Gobiidae is the most speciose family of largely marine fishes, with over 1,700 valid species and many that remain undescribed. Several lineages within the gobies are recognized, including the Oxudercinae, inhabitants of mangrove and mudflats of the Indo-West Pacific; the Amblyopinae, inhabitants of estuaries around the world; the Sicydiinae, worldwide mostly in freshwaters; the Gobionellinae, found worldwide in estuaries and freshwaters; and the Gobiinae, a diverse worldwide group with over 130 genera found in a wide variety of habitats. Composition and relationships of these lineages have been the focus of much recent research and considerable controversy (see Birdsong et al., 1988; Pezold, 1993; reviews in Patzner et al., 2011; Thacker and Roje, 2011). Gobies characteristically have a proliferation of superficial neuromasts (Asaoka et al., 2012), and are generally microcarnivores (Zander, 2011). Several species act as cleaners of other fishes (Côté and Soares, 2011) or have symbiotic relationships with shrimps and other invertebrates (Karplus and Thompson, 2011). These fishes lay benthic eggs in nests guarded by males, and several species are hermaphroditic (Cole, 2010), including some that exhibit bidirectional sex

GOBIID CHARACTERISTICS:
1) pelvic fins usually united medially, forming rounded disc
2) first dorsal fin usually with two to eight spines, separate from soft dorsal fin
3) caudal fin usually broad and rounded
4) base of anal and second dorsal fins longer than caudal peduncle
5) scales cycloid or ctenoid, absent in some
6) five branchiostegal rays

ILLUSTRATED SPECIMENS:
A) *Bathygobius ramosus,* SIO 67–289, 171 mm SL
B) *Eucyclogobius newberryi,* SIO 90–171, 33 mm SL
C) *Typhlogobius californiensis,* SIO 62–586, 67 mm SL

change (Munday et al., 2010). A few species of gobies enter artisanal fisheries, some in the so-called *tismiche* when larvae of several catadromous gobies, eleotrids, microdesmids, and crustaceans are caught as they migrate through tropical lagoons (Gilbert and Kelso, 1971). Several gobies are widely available in the aquarium trade.

REFERENCES: Asaoka et al., 2012; Birdsong et al., 1988; Cole, 2010; Côté and Soares, 2011; Gilbert and Kelso, 1971; Hoese, in Fischer et al., 1995; Karplus and Thompson, 2011; Larson and Murdy, in Carpenter and Niem, 2001; Munday et al., 2010; Murdy and Hoese, in Carpenter, 2003; Patzner et al., 2011; Pezold, 1993; Thacker, 2009; Thacker and Roje, 2011; Zander, 2011.

GOBIIFORMES : ELEOTRIDAE—Sleepers

DIVERSITY: At least 35 genera, 171 species

REPRESENTATIVE GENERA: *Dormitator, Eleotris, Gobiomorus*

DISTRIBUTION: Atlantic, Indian, and Pacific oceans and adjacent freshwaters

ELEOTRID CHARACTERISTICS:
1) pelvic fins separate but bases may be united
2) two dorsal fins, first with two to eight flexible spines
3) mouth terminal to superior
4) caudal peduncle always equal to or longer than base of anal and second dorsal fins
5) scales cycloid or ctenoid
6) six branchiostegal rays

ILLUSTRATED SPECIMENS:
A) *Dormitator latifrons*, SIO 75–350, 85 mm SL
B) *Gobiomorus maculatus*, SIO 08–156, 223 mm SL

(account continued)

HABITAT: Freshwaters, estuaries, and occasionally coastal marine; tropical to temperate; demersal to benthic on soft and hard bottoms

REMARKS: As their common name implies, many sleepers are relatively inactive, bottom-dwelling fishes. Although similar in appearance to the Gobiidae, sleepers can be distinguished from gobies by their separate pelvic fins, long caudal peduncle, and six branchiostegal rays (Thacker, 2011). They are generally small-bodied and eat a variety of small prey although some species grow as large as 60 cm. Several species of sleepers are diadromous, and as such, are threatened by manipulations to river systems (Nordlie, 2012). At least one species is of commercial importance as a food fish in Southeast Asia, where it is raised in aquaculture facilities. Otherwise, this group is generally not targeted for human consumption other than artisanal fisheries.

REFERENCES: Hoese, in Fischer et al., 1995; Hoese and Gill, 1993; Murdy and Hoese, in Carpenter, 2003; Nordlie, 2012; Pezold and Cage, 2002; Thacker, 2011.

GOBIIFORM DIVERSITY:
A) MICRODESMIDAE—wormfishes: *Cerdale ionthas*, SIO 67–42, 67 mm SL
B) PTERELEOTRIDAE—dartfishes: *Ptereleotris carinata*, SIO 65–295, 87 mm SL

ACANTHURIFORMES—Surgeonfishes and Relatives

Acanthuriform fishes have a compressed body that is often very deep, a narrow caudal peduncle, a small branchial aperture with the gill membranes united across the isthmus, and a small mouth with the upper jaw weakly protrusible to non-protrusible. The limits and relationships of the Acanthuriformes have been studied and debated by a number of researchers (e.g., Tang et al., 1999; Tyler et al., 1989; Winterbottom, 1993b). These studies were reviewed by Holcroft and Wiley (2008), whose molecular analysis did not support the monophyly of the group as construed by earlier workers. As currently defined, the group includes four families: Acanthuridae (surgeonfishes), Siganidae (rabbitfishes), Zanclidae (Moorish Idol), and Luvaridae (Louvar), with nine genera and 119 species (see Wiley and Johnson, 2010). The Acanthuridae, Siganidae, and Zanclidae are generally found in near-shore, tropical reef areas; the Louvar (*Luvarus imperialis*), the only member of the Luvaridae, is a pelagic species reaching lengths of up to 1.8 m.

REFERENCES: Holcroft and Wiley, 2008; Tang et al., 1999; Tyler et al., 1989; Winterbottom, 1993b.

ACANTHURIFORMES : ACANTHURIDAE—Surgeonfishes

DIVERSITY: 6 genera, 85 species

REPRESENTATIVE GENERA: *Acanthurus, Naso, Prionurus, Zebrasoma*

DISTRIBUTION: Atlantic, Indian, and Pacific oceans

HABITAT: Marine; tropical to subtropical; coastal, neritic to demersal, especially prominent on coral reefs

REMARKS: The group's common name is derived from a blade-like spine or spines on the caudal peduncle that probably gives them some protection from predators. Two lineages are recognized, the Nasinae (unicornfishes) and the Acanthurinae (surgeonfishes). Relationships within the group have been studied by a variety of researchers with much disagree-

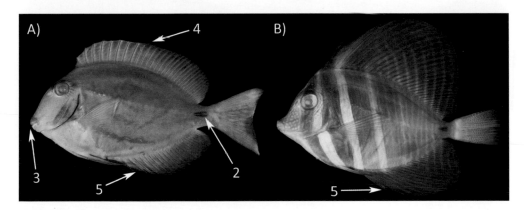

ACANTHURID CHARACTERISTICS:
1) body deeply compressed
2) one or more spines on lateral caudal peduncle
3) mouth small with closely set, incisor-like teeth in both jaws
4) spinous and soft dorsal fins continuous
5) two (Nasinae) or three (Acanthurinae) anal-fin spines
6) four or five branchiostegal rays

ILLUSTRATED SPECIMENS:
A) *Acanthurus bahianus,* SIO 70–175, 121 mm SL
B) *Zebrasoma veliferum,* SIO 64–229, 80 mm SL

ment (e.g., Holcroft and Wiley, 2008; Tang et al., 1999; Tyler et al., 1989; Winterbottom, 1993b). Surgeonfishes have a long-lived larva called an acronurus, accounting for the broad distributions of many species (although cryptic species may be present; see DiBatista et al., 2011). They are important herbivores on coral reefs where some species form large schools that overwhelm damselfishes and other territorial species (Foster, 1985). Juvenile surgeonfishes are valuable in the aquarium trade and adults are eaten by artisanal fishers in many places around the world.

REFERENCES: Borden, 1998; DiBatista et al., 2011; Foster, 1985; Holcroft and Wiley, 2008; Krupp, in Fischer et al., 1995; Randall, 2001b; Randall, in Carpenter and Niem, 2001; Tang et al., 1999; Tyler et al., 1989; Winterbottom, 1993b.

ACANTHURIFORM DIVERSITY:
A) LUVARIDAE—Louvar: *Luvarus imperialis* (juvenile), SIO 82–71, 200 mm SL
B) SIGANIDAE—rabbitfishes: *Siganus canaliculatus,* SIO 73–177, 65 mm SL
C) ZANCLIDAE—Moorish Idol: *Zanclus cornutus,* SIO 60–408, 122 mm SL

XIPHIIFORMES—Billfishes and Swordfishes

DIVERSITY: 2 families, 6 genera, 11 species

REPRESENTATIVE GENERA: *Istiophorus, Kajikia, Tetrapterus, Xiphias*

DISTRIBUTION: Atlantic, Indian, and Pacific oceans

HABITAT: Marine; tropical to temperate; neritic, epipelagic, occasionally mesopelagic

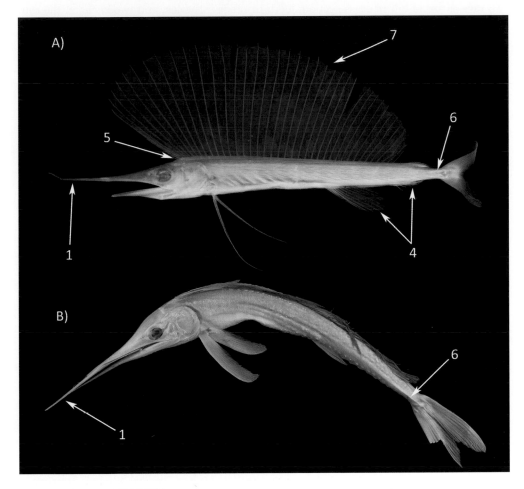

XIPHIIFORM CHARACTERISTICS:
1) elongate bill derived from premaxillae
2) upper jaw nonprotrusible
3) mouth inferior
4) two anal fins
5) origin of dorsal fin over operculum
6) caudal peduncle with one or two keels
7) dorsal fins lacking true spines

ILLUSTRATED SPECIMENS:
A) *Kajikia albida* (juvenile), SIO 96-63, 185 mm SL (Istiophoridae—billfishes)
B) *Xiphias gladius* (juvenile), SIO 69-373, 340 mm SL (Xiphiidae—Swordfish)

(account continued)

REMARKS: Recent studies have shown that the billfishes, long considered members of the Scombriformes, are an independently derived lineage of percomorph fishes whose affinities may lie with the flatfishes, barracudas, and jacks (Little et al., 2010; Miya et al., 2013). A number of characters clearly unite the Xiphiiformes, but there are also significant differences between the two included families, the monotypic Xiphiidae (Swordfish) and the Istiophoridae (billfishes). For example, the billfishes have pelvic fins with a single spine and two rays, a round bill in cross-section, teeth in the jaws, scales, and two keels on the caudal peduncle. The Swordfish lacks pelvic fins and pelvic girdle, teeth, and scales and has a depressed bill and one keel on the caudal peduncle. The billfishes (including the Swordfish) are some of the most highly prized sport-fishes in the world and most are valuable commercial species. The distinctive bill is used to slash and immobilize prey before consumption, and these highly predatory fishes eat other fishes, cephalopods, and crustaceans. The billfishes are some of the fastest swimmers and some of the most highly fecund animals in the world, producing hundreds of millions of eggs at a time. They are also notable for their incredible change in size over time, growing rapidly from a few mm in length and a few thousandths of a gram in weight at hatching to at least 5 m and 900 kg when fully mature.

REFERENCES: Collette et al., 1984; Finnerty and Block, 1995; Johnson, 1986; Little et al., 2010; Miya et al., 2013; Nakamura, 1985; Nakamura, in Carpenter, 2002; Nakamura, in Carpenter and Niem, 2001; Nakamura, in Fischer et al., 1995; Orrell et al., 2006.

SCOMBRIFORMES—Tunas and Relatives

The Scombriformes includes a number of well-known groups of fishes such as the tunas, but also a number of less familiar groups such as the deep-sea gempylids (Nakamura and Parin, 1993). The composition and relationships of the Scombriformes have been controversial (Wiley and Johnson, 2010). Johnson (1986) included seven families (Scombrolabracidae, Gempylidae, Trichiuridae, Scombridae, Sphyraenidae, Istiophoridae, and Xiphiidae), but we follow Helfman and Collette (2011) who excluded the barracudas (Sphyraenidae, considered here under the Perciformes) and the billfishes (Istiophoridae and Xiphiidae, considered under the Xiphiiformes), based on several recent molecular studies (e.g., Betancur et al., 2013; Little et al., 2010; Near et al., 2012; Orrell et al.,2006). This restricted Scombriformes includes four families, 42 genera, and 115 species that are characterized by a non-protrusible upper jaw with the premaxilla firmly bound to skull. Many have a fusiform body, the median fins reduced and finlets present in the dorsal and anal fins. Scombriforms were recently identified as one of five rapidly radiating lineages of acanthomorph fishes (Near et al., 2013).

REFERENCES: Collette et al., 2011; Johnson, 1986; Little et al., 2010; Nakamura and Parin, 1993; Near et al., 2012, 2013; Orrell et al., 2006.

SCOMBRIFORMES : SCOMBRIDAE—Mackerels and Tunas

DIVERSITY: 15 genera, 53 species

REPRESENTATIVE GENERA: *Euthynnus, Sarda, Scomber, Scomberomorus, Thunnus*

DISTRIBUTION: Atlantic, Indian, and Pacific oceans

HABITAT: Marine; tropical to temperate; most species epipelagic or neritic, with a few entering estuaries or rarely freshwaters

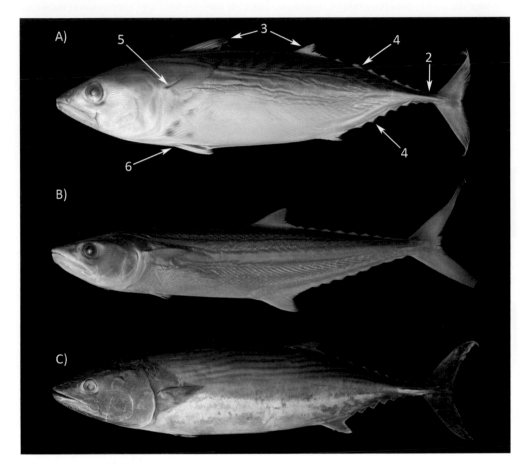

SCOMBRID CHARACTERISTICS:

1) body typically fusiform
2) caudal peduncle slender, with lateral keels
3) two dorsal fins that fold into grooves on dorsal surface
4) series of finlets posterior to second dorsal and anal fins
5) pectoral fins usually positioned high on body
6) pelvic fins thoracic

ILLUSTRATED SPECIMENS:

A) *Euthynnus affinis,* SIO 61–735, 190 mm FL
B) *Scomberomorus sierra,* SIO 60–294, 215 mm FL
C) *Sarda orientalis,* SIO 08–65, 497 mm FL

(account continued)

REMARKS: The tunas are common predators in the epipelagic zone, feeding on fishes and cephalopods. Many species form large schools that search for concentrations of smaller fishes, often herrings, and they, in turn, are joined by foraging birds and marine mammals. Recent tagging efforts have demonstrated the vast extent of migratory behaviors of some species such as the Atlantic Bluefin Tuna (*Thunnus thynnus*), with single individuals repeatedly crossing entire ocean basins (Block et al., 2005). This species is one of the most highly sought-after fishes and can reach over 3 m in length. This and other scombrids are among the world's most valuable marine resources, both for commercial and recreational fisheries (Collette et al., 2011; FAO, 2012). Relationships of the tunas and mackerels have been studied extensively, but remain controversial (Collette, 1983; Johnson, 1986; Orrell et al., 2006). Like the billfishes, the tunas are some of the most highly fecund animals in the world, producing tens of millions of eggs at a time.

REFERENCES: Block et al., 2005; Collette, 1983, 2003b; Collette, in Carpenter, 2003; Collette, in Carpenter and Niem, 2001; Collette, in Fischer et al., 1995; Collette and Chao, 1975; Collette et al., 2001, 2011; Johnson, 1986; Orrell et al., 2006.

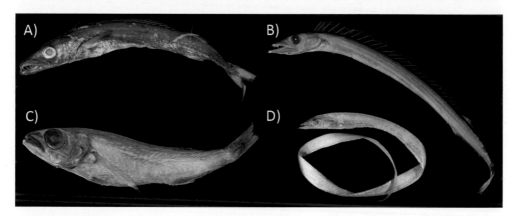

SCOMBRIFORM DIVERSITY:
A) GEMPYLIDAE—snake mackerels: *Promethichthys prometheus,* SIO 74–158, 268 mm SL
B) GEMPYLIDAE: *Gempylus serpens,* SIO 64–666, 174 mm SL
C) SCOMBROLABRACIDAE—Longfin Escolar: *Scombrolabrax heterolepis,* SIO 62–299, 145 mm SL
D) TRICHIURIDAE—cutlassfishes: *Assurger anzac,* SIO 76–292, 1,087 mm SL

STROMATEIFORMES—Butterfishes and Relatives

DIVERSITY: 6 families, 16 genera, 76 species

REPRESENTATIVE GENERA: *Ariomma, Centrolophus, Nomeus, Pampus, Peprilus, Psenes, Stromateus*

DISTRIBUTION: Atlantic, Pacific, and Indian oceans

HABITAT: Marine; tropical to temperate; neritic to epipelagic, often associated with coelenterates, demersal over continental shelf (some centrolophids)

REMARKS: The Stromateiformes is a diverse group that includes small epipelagic species that live in close association with coelenterates (e.g., *Nomeus gronovii* lives within the tentacles of *Physalia,* the Portuguese man-of-war), as well as large epibenthic predators (e.g., *Hyperoglyphe* in the Centrolophidae). The monotypic family Amarsipidae, known only

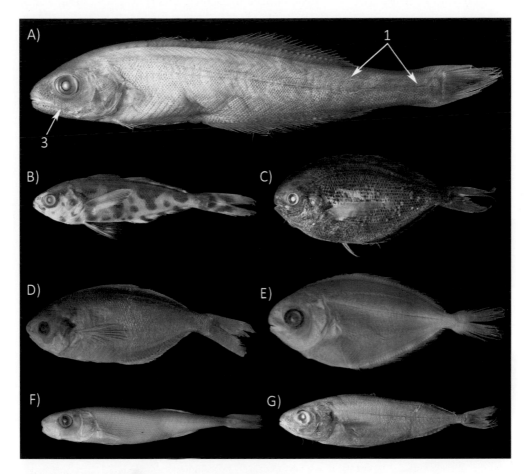

STROMATEIFORM CHARACTERISTICS:
1) small pores scattered over body, communicating to extensive subdermal canal system
2) toothed saccular outgrowths posterior to last gill arch (except for monotypic Amarsipidae)
3) maxilla partially or completely covered by lacrimal
4) scales cycloid or weakly ctenoid
5) five to seven branchiostegal rays

ILLUSTRATED SPECIMENS:
A) *Amarsipus carlsbergi,* SIO 75-125, 120 mm SL (Amarsipidae—Amarsipa)
B) *Nomeus gronovii,* SIO 01-129, 75 mm SL (Nomeidae—driftfishes)
C) *Psenes cyanophrys,* SIO 98-26, 148 mm SL (Nomeidae)
D) *Psenopsis anomala,* SIO 64-259, 164 mm SL (Centrolophidae—medusafishes)
E) *Stromateus stellatus,* SIO 65-667, 43 mm SL (Stromateidae—butterfishes)
F) *Tetragonurus atlanticus,* SIO 77-230, 82 mm SL (Tetragonuridae—squaretails)
G) *Ariomma bondi,* SIO 63-781, 111 mm SL (Ariommatidae—ariommatids)

from a juvenile when first described (Haedrich, 1969), lacks the internal toothed, saccular out-growths typical of all other species, but shares with all others the unique subdermal canal system (Konovalenko and Piotrovskiy, 1989; Wiley and Johnson, 2010). In addition to earlier studies based on morphology (e.g., Haedrich, 1967; Horn, 1984), relationships of stromateiform fishes have been hypothesized by Douichi and colleagues (2004, 2006) based on molecular data. These fishes feed on a wide variety of invertebrates including crustaceans and gelatinous species as well as small fishes.

REFERENCES: Douichi et al., 2004; Douichi and Nakabo, 2006; Haedrich, 1967, 1969; Haedrich, in Carpenter, 2003; Haedrich and Schneider, in Fischer et al., 1995; Horn, 1984; Konovalenko and Piotrovskiy, 1989; Last, in Carpenter and Niem, 2001; Parin and Piotrovsky, 2004.

ICOSTEIFORMES : ICOSTEIDAE—Ragfishes

DIVERSITY: 1 family, 1 genus, 1 species

REPRESENTATIVE GENUS: *Icosteus*

DISTRIBUTION: North Pacific Ocean

HABITAT: Marine; temperate; neritic to bathypelagic

REMARKS: The Ragfish is so named because of its soft, flaccid body, which is easily dam-

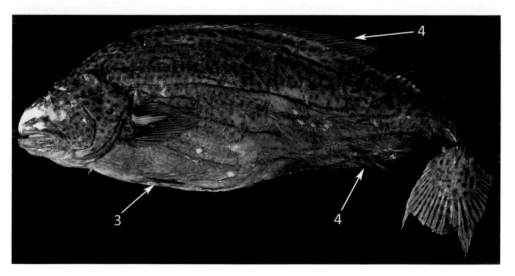

ICOSTEIFORM CHARACTERISTICS:
1) body flaccid, compressed
2) fin spines absent
3) pelvic fins present in juveniles, absent in adults
4) bases of dorsal and anal fins long
5) small prickles on some fin rays
6) skeleton mostly cartilaginous

ILLUSTRATED SPECIMEN:
Icosteus aenigmaticus, SIO 99–95, 302 mm SL

aged during capture. The relationships of this species have long been enigmatic. Although previously considered close to the Stromateiformes, Springer and Johnson (2004) concluded a closer relationship to the Stephanoberyciformes. However, Wiley and Johnson (2010) place this species within the percomorphs (*incertae sedis*), while a recent study (Miya et al., 2013) includes it within the Pelagia, a diverse group including the scombrids and a variety of other epipelagic fishes. Juvenile and adult Ragfish, which reach lengths of up to 2 m, are so different in appearance that they were once described as separate species. The Ragfish is known to eat small fishes and cephalopods.

REFERENCES: Hart, 1973; Mecklenburg et al., 2002; Miya et al., 2013; Springer and Johnson, 2004.

CAPROIFORMES : CAPROIDAE—Boarfishes

DIVERSITY: 1 family, 2 genera, 18 species
REPRESENTATIVE GENERA: *Antigonia, Capros*
DISTRIBUTION: Atlantic, Indian, and Pacific oceans
HABITAT: Marine; tropical to temperate; benthopelagic, over continental shelf and slope

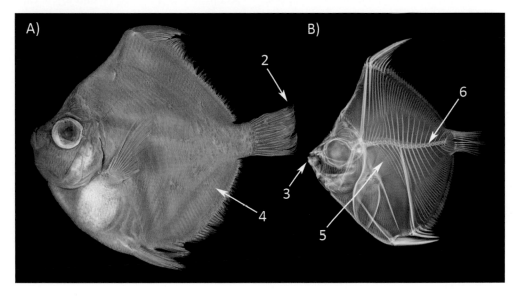

CAPROIFORM CHARACTERISTICS:
1) body relatively deep, compressed
2) caudal fin rounded, with 10–12 branched rays
3) upper jaw protrusible
4) scales small, ctenoid
5) pleural ribs well developed
6) 21–23 vertebrae

ILLUSTRATED SPECIMEN:
A) *Antigonia capros,* SIO 60–397, 143 mm SL
B) *Antigonia capros,* SIO 60–397, radiograph

REMARKS: Traditionally placed within the Zeiformes (Tyler et al., 2003; Zehren, 1987), the boarfishes are considered closer to the percomorphs by recent authors. In addition, the group itself may not be monophyletic. The 17 species of *Antigonia* are characterized by a relatively small mouth, extremely deep, compressed body, and ten branched caudal-fin rays; the single species of *Capros, C. asper,* has a moderately sized mouth, a somewhat deep body, and 12 branched caudal-fin rays. Boarfishes feed on plankton and benthic invertebrates.

REFERENCES: Heemstra, in Carpenter and Niem, 1999; Johnson and Patterson, 1993; Parin, in Carpenter, 2003; Rosen, 1984; Tyler et al., 2003; Zehren, 1987.

ANABANTIFORMES—Gouramies and Snakeheads

DIVERSITY: 4 families, 21 genera, 149 species

REPRESENTATIVE GENERA: *Betta, Channa, Ctenopoma, Helostoma, Osphronemus*

DISTRIBUTION: Africa and Asia

HABITAT: Freshwater to brackish; tropical; lakes, swamps, and slow moving rivers, demersal over soft bottoms

REMARKS: The unique suprabranchial organ of these primarily freshwater fishes permits them to respire in air, allowing many species to occupy low-oxygen waters, and in the case of the snakeheads, to migrate across land (Graham, 1997). Two distinct lineages are recognized, one with the gouramies (Anabantidae), the climbing gouramies (Osphronemidae) and the Kissing Gourami (Helostomatidae), and another with only the snakeheads (Channidae). The 33 species of snakeheads are predatory fishes that feed primarily on other fishes, but also other vertebrates (e.g., frogs, snakes, etc.) and invertebrates (e.g., worms, crustaceans, etc.). Snakeheads are important food fishes, particularly in Southeast Asia, and are targeted for the live food-fish trade, as well as more traditional outlets. The snakeheads'

ANABANTIFORM CHARACTERISTICS:
1) unique suprabranchial organ formed from first epibranchial
2) body variable in shape
3) pelvic fins thoracic, usually with one spine and five rays
4) gill membranes united across isthmus and covered with scales
5) five or six branchiostegal rays

ILLUSTRATED SPECIMENS:
A) *Channa striata,* SIO 64–228, 414 mm SL (Channidae—snakeheads)
B) *Helostoma temminckii,* SIO 64–228, 76 mm SL (Helostomatidae—Kissing Gourami)

adaptability and aggressive behavior have led to their success as invasive species in many areas of the world where they have been introduced. The gouramies include a variety of species important in the aquarium trade, including the Siamese Fighting Fish (*Betta splendens*), one of the most popular aquarium fishes, known for its flowing fins and aggressive behavior. Some gouramies build bubble nests for the incubation of eggs while others are oral brooders. The Kissing Gourami is one of the few species of fishes in the world that can both filter feed for planktonic organisms and scrape algae from hard surfaces. While most bones in its mouth lack teeth, individuals of this species feed on algae and plants using specialized teeth on the lips. This biting behavior and the presence of these modified teeth lead to its common name and have made juveniles very popular in the aquarium trade. Adults of several gouramies are targeted for human consumption.

REFERENCES: Britz, 1994; Courtenay and Williams, 2004; Graham, 1997; Ishimatsu et al., 1979; Liem, 1963; Ruber et al., 2006.

COTTIFORMES—Sculpins, Eelpouts, and Relatives

The Cottiformes as currently construed includes over 20 families, 250 genera, and nearly 1,100 species. It was first recognized by Imamura and Yabe (2002) when, based on morphology, they removed the sculpins and related fishes from the Scorpaeniformes and hypothesized their close relationship with the eelpouts (zoarcoids). This relationship has been supported by several subsequent molecular analyses (e.g., Betancur et al., 2013; Near et al., 2012; Smith and Wheeler, 2004). Wiley and Johnson (2010) list 13 synapomorphies, most of which are internal osteological or soft-tissue features (e.g., six branchiostegal rays, anal-fin spines supported by relatively small pterygiophores, gas bladder absent). Currently, it is not possible to diagnose the group using only external features. Two distinct lineages are evident in the Cottiformes: the Cottoidei, with 11 families, 149 genera, and over 750 species, characterized by the presence of a suborbital stay; and the Zoarcoidei, with nine families, 95 genera, and 340 species, characterized by a single pair of of nostrils (Imamura and Yabe, 2002; Nelson, 2006). We cover four families below and illustrate several additional members of the Cottiformes.

REFERENCES: Imamura and Yabe, 2002; Smith and Wheeler, 2004.

COTTIFORMES : HEXAGRAMMIDAE—Greenlings

DIVERSITY: 5 genera, 12 species

REPRESENTATIVE GENERA: *Hexagrammos, Ophiodon, Oxylebius, Pleurogrammus, Zaniolepis*

DISTRIBUTION: North Pacific Ocean

HABITAT: Marine; temperate; coastal to lower continental shelf, demersal or benthic over soft and hard bottoms

REMARKS: This small but morphologically diverse group of fishes is restricted to the north Pacific Ocean, from northern Mexico to Japan. The composition and relationships of the group are incompletely resolved. The Painted Greenling, *Oxylebius pictus*, often considered a hexagrammid, recently was shown to be closely related to the combfishes (*Zaniolepis*), and together they are hypothesized to be sister group of a large clade including the remaining hexagrammids, cottids, and agonids (Shinohara, 1994; Crow et al., 2004). We follow Eschmeyer and Fong (2013) by including *Oxylebius* and *Zaniolepis* in the Hexagrammidae. Hexagrammids lay demersal eggs in nests guarded by males (Crow et al., 1997). The greenlings

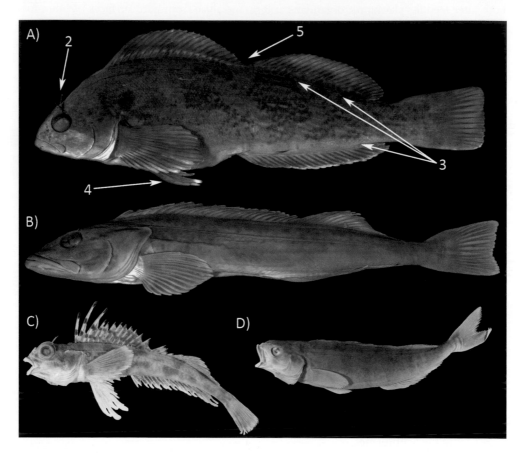

HEXAGRAMMID CHARACTERISTICS:

1) body relatively elongate
2) head bearing cirri, but no spines on cranial bones
3) one or multiple trunk lateral-line canals
4) pelvic fins with one spine and five soft rays
5) dorsal fin long, usually with notch
6) posterior nostril smaller than anterior nostril or sometimes absent

ILLUSTRATED SPECIMENS:

A) *Hexagrammos lagocephalus,* SIO 59–106, 177 mm SL
B) *Ophiodon elongatus,* SIO 63–808, 245 mm SL
C) *Zaniolepis frenata,* SIO 65–106, 153 mm SL
D) *Pleurogrammus monopterygius,* SIO 76–229, 135 mm SL

are generalist predators and include some very large species such as the Lingcod, *Ophiodon elongatus*, that grows to over 1.5 m in length, and is important for recreational and commercial fisheries.

REFERENCES: Crow et al., 1997, 2004; Mecklenburg and Eschmeyer, 2003; Mecklenburg et al., 2002; Quast, 1965; Shinohara, 1994.

COTTIFORMES : COTTIDAE—Sculpins

DIVERSITY: 70 genera, over 250 species

REPRESENTATIVE GENERA: *Clinocottus, Cottus, Hemilepidotus, Icelinus, Ruscarius, Scorpaenichthys*

DISTRIBUTION: Arctic, Atlantic, and Pacific oceans and adjacent freshwaters, most species in Northern Hemisphere

HABITAT: Marine and freshwater; subtropical to polar; coastal to continental slope, benthic to demersal over soft and hard bottoms

REMARKS: Cottids are morphologically and ecologically diverse, occupying habitats from the continental slope, to the intertidal, to freshwaters. Their morphology and systematics were studied in detail by Bolin (1944, 1947) and later by Yabe (1985). Smith and Wheeler (2004) included a number of cottids in their study of scorpaeniform relationships based on molecular data. Relationships of several subgroups of cottids have been hypothesized (e.g., Knope, 2013; Ramon and Knope, 2008; Strauss, 1993), including the freshwater representatives of the genus *Cottus* (Kinziger et al., 2005; Yokoyama and Goto, 2005). Cottids are generally benthic and accordingly, adults lack a gas bladder. Many species are extremely cryptic, and most species are relatively small, but some, such as the Cabezon (*Scorpaenichthys marmoratus*), grow to nearly 1 m and over 11 kg. Cottids are predatory and feed on a wide variety of organisms. They exhibit a number of reproductive strategies (Muñoz, 2010) and parental care varies from male care, to biparental care, to no care (Petersen et al., 2005). Many of the larger sculpins are important in commercial fisheries.

REFERENCES: Bolin, 1944, 1947; Kinziger et al., 2005; Knope, 2013; Mecklenburg et al., 2002; Muñoz, 2010; Petersen et al., 2005; Ramon and Knope, 2008; Smith and Wheeler, 2004; Strauss, 1993; Yabe, 1985; Yokoyama and Goto, 2005.

(account continued)

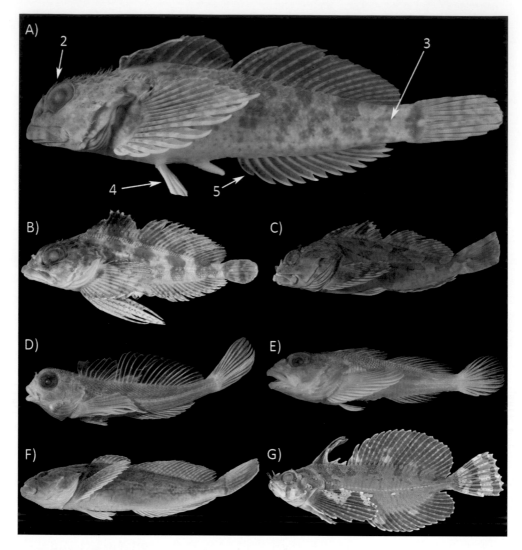

COTTID CHARACTERISTICS:
1) body often naked, some with patches or rows of scales
2) eyes large, dorsally positioned
3) trunk lateral-line canal single
4) pelvic fins with one spine and usually fewer than five soft rays
5) anal fin without spines
6) gas bladder absent in adults

ILLUSTRATED SPECIMENS:
A) *Clinocottus analis,* SIO 46–97, 74 mm SL
B) *Hemilepidotus zapus,* SIO 94–185, 113 mm SL
C) *Scorpaenichthys marmoratus,* SIO 47–93, 150 mm SL
D) *Orthonopias triacis,* SIO 58–487, 48 mm SL
E) *Ruscarius creaseri,* SIO 45–6, 58 mm SL
F) *Cottus asper,* SIO 63–1067, 109 mm SL
G) *Blepsias cirrhosus,* SIO 73–220, 107 mm SL

COTTIFORMES : LIPARIDAE—Snailfishes

DIVERSITY: 29 genera, 415 species

REPRESENTATIVE GENERA: *Careproctus, Crystallichthys, Liparis, Paraliparis, Psednos*

DISTRIBUTION: Arctic, Atlantic, Indian, and Pacific oceans, including Antarctica

HABITAT: Marine; tropical to polar; intertidal to abyssal plain and in deep-sea trenches, benthic or demersal over soft bottoms

REMARKS: Snailfishes occur in an extraordinarily broad array of habitats, from the intertidal to over 7,000 m in depth. This is an amazingly speciose group that has received increasing attention from systematists (e.g., Andriashev, 1986, 2003; Stein, 2012, Stein et al., 2001), with literally hundreds of species described in recent years. The Liparidae also was identified as one of five rapidly radiating lineages of acanthomorph fishes (Near et al., 2013). Knudsen et al. (2007) hypothesized their relationships based on both molecular and morphological characters. Snailfishes lay demersal eggs and have planktonic larvae. They are generalist predators, with some species specializing on invertebrates, while other species eat small fishes.

REFERENCES: Andiashev, 1986, 2003; Chernova et al., 2004; Kido, 1988; Knudsen et al., 2007; Near et al., 2013; Stein, 2012; Stein et al., 2001.

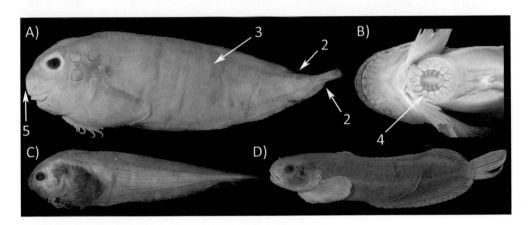

LIPARID CHARACTERISTICS:
1) body somewhat elongate, tapering posteriorly in most species
2) dorsal and anal fins nearly or completely continuous with caudal fin
3) body usually scaleless, skin smooth and soft
4) pelvic fins thoracic and united in a disc, or absent in some species
5) snout bulbous in most species

ILLUSTRATED SPECIMENS:
A) *Crystallichthys cyclospilus,* SIO 94–184, 123 mm SL
B) pelvic disc of *Liparis mucosus,* SIO 74–56, 51 mm SL (ventral view)
C) *Paraliparis rosaceus,* SIO 67–58, 119 mm SL
D) *Liparis mucosus,* SIO 61–404, 41 mm SL

COTTIFORMES : ZOARCIDAE—Eelpouts

DIVERSITY: 46 genera, 303 species

REPRESENTATIVE GENERA: *Bothrocara, Lycenchelys, Lycodes, Pachycara, Thermarces, Zoarces*

DISTRIBUTION: Arctic, Atlantic, Indian, and Pacific oceans, including Antarctica

HABITAT: Marine; tropical to polar; continental shelf to continental slope, mesopelagic, benthic or demersal over soft bottoms

REMARKS: The eelpouts are a diverse group of marine fishes typically associated with soft substrates, but they also occur around hydrothermal vents (e.g., *Thermarces*). Some species

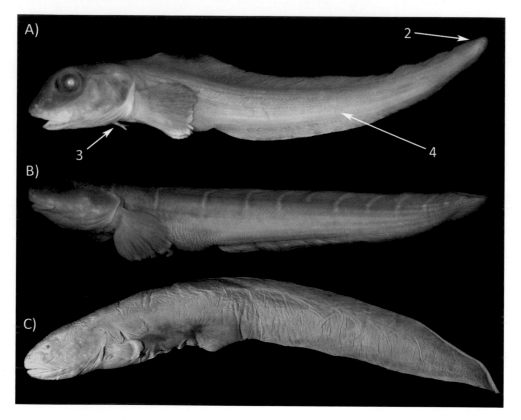

ZOARCID CHARACTERISTICS:
1) body elongate
2) dorsal and anal fins continuous with caudal fin
3) pelvic fins jugular or absent
4) body naked or with small, embedded scales
5) dorsal and anal fins without spines
6) gas bladder absent

ILLUSTRATED SPECIMENS:
A) *Aprodon cortezianus,* SIO 00-157, 151 mm SL
B) *Lycodes brevipes,* SIO 69-106, 178 mm SL
C) *Thermarces cerberus,* SIO 81-155, 250 mm SL, holotype

grow to over 1 m, and most feed on crustaceans or other benthic invertebrates. Most species are oviparous, but those in the genus *Zoarces* have internal fertilization and give birth to well-developed young (Rasmussen et al., 2006). One of these, *Zoarces viviparus,* is considered a bioindicator of environmental conditions in the North and Baltic seas (Hedman et al., 2011).

REFERENCES: Anderson, 1994; Anderson and Fedorov, 2004; Hedman et al., 2011; Rasmussen et al., 2006.

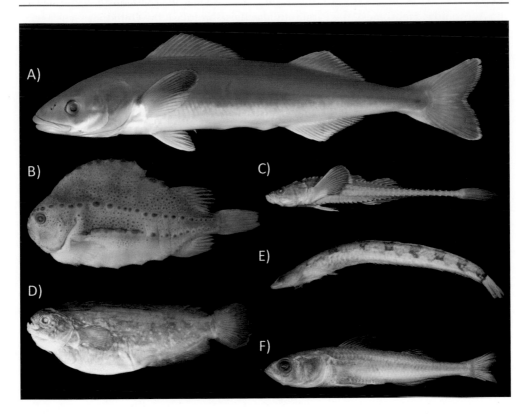

COTTIFORM DIVERSITY:
A) ANOPLOPOMATIDAE—sablefishes: *Anoplopoma fimbria,* SIO 71-145, 173 mm SL
B) CYCLOPTERIDAE—lumpsuckers: *Cyclopterus lumpus,* SIO 82-87, 110 mm SL
C) AGONIDAE—poachers: *Stellerina xyosterna,* SIO 50-237, 74 mm SL
D) ZAPRORIDAE—Prowfish: *Zaprora silenus,* SIO 91-57, 151 mm SL
E) STICHAEIDAE—pricklebacks: *Lumpenopsis clitella,* SIO 02-10, 55 mm SL, holotype
F) NORMANICHTHYIDAE—Mote Sculpin: *Normanichthys crockeri,* SIO 02-134, 95 mm SL

The Ophidiiformes comprises five families, over 100 genera, and approximately 530 species. Cusk-eels and their relatives generally have elongate bodies that taper posteriorly and have long anal- and dorsal-fin bases. These fishes are characterized by jugular or mental pelvic fins with 1–2 soft rays; a pelvic-fin spine is present in some, but pelvic fins are absent in others. The pterygiophores in both the dorsal and anal fins outnumber the adjacent vertebrae (Cohen and Nielsen, 1978). The Ophidiiformes currently includes the Carapidae (pearlfishes, some of which are commensal with sea cucumbers or other invertebrates), the Aphyonidae (aphyonids), the Parabrotulidae (false brotulas), the Ophidiidae (cusk-eels), and the Bythitidae (viviparous brotulas). However, most researchers have struggled to find synapomorphies that successfully unite these five families (Wiley and Johnson, 2010). Long considered members of the Paracanthopterygii (Patterson and Rosen, 1989; Rosen and Patterson, 1969), recent studies place them in the Percomorpha as the sister group of the Batrachoidiformes plus all other percomorphs (Betancur et al., 2013; Near et al., 2012). The two most speciose families are described in more detail below.

REFERENCES: Cohen and Nielsen, 1978; Nielsen et al., 1999, 2006.

OPHIDIIFORMES : OPHIDIIDAE—Cusk-eels

DIVERSITY: 48 genera, 261 species

REPRESENTATIVE GENERA: *Bassogigas, Brotula, Lamprogrammus, Ophidion*

DISTRIBUTION: Atlantic, Indian, and Pacific oceans

HABITAT: Marine; tropical to temperate; coastal to abyssal trenches, benthic or demersal over soft bottoms

REMARKS: The cusk-eels include some of the deepest living fishes in the world. *Abysso-brotula galatheae* has been collected at a depth of 8,370 m in the Puerto Rico trench. Four subfamilies are recognized, the Brotulinae (6 species), the Brotulotaeniinae (4 species), the Ophidiinae (65 species), and the Neobythitinae (185 species). Cusk-eels generally are associated with soft bottoms and feed on a variety of small prey. The otoliths of males and females of some species differ in morphology (Schwarzhans, 1994). Unlike the closely-related Bythitidae, cusk-eels are oviparous. These fishes are typically shorter than 1 m, but one species reaches at least 2 m. Some species are highly regarded as food fishes.

REFERENCES: Cohen and Nielsen, 1978; Lea and Robins, 2003; Nielsen et al., 1999; Schwarzhans, 1994.

OPHIDIID CHARACTERISTICS:
1) body laterally compressed and elongate
2) dorsal and anal fins continuous with caudal fin
3) anal-fin rays equal to or longer than corresponding dorsal-fin rays
4) anterior nostril equally spaced between upper lip and posterior nostril
5) anus generally posterior to end of pectoral fin
6) pelvic fins jugular, rarely absent
7) usually a well-developed spine on opercle
8) scales small, sometimes embedded

ILLUSTRATED SPECIMENS:
A) *Ophidion iris,* SIO 62–42, 116 mm SL (Ophidiinae)
B) *Bassogigas walkeri,* SIO 08–109, 538 mm SL, holotype (Neobythitinae)
C) *Neobythites stelliferoides,* SIO 70–253, 93 mm SL (Neobythitinae)
D) *Thalassobathia nelsoni,* SIO 72–164, 190 mm SL, holotype (Neobythitinae)

OPHIDIIFORMES : BYTHITIDAE—Viviparous Brotulas

DIVERSITY: 37 genera, 209 species

REPRESENTATIVE GENERA: *Bythites, Cataetyx, Ogilbia*

DISTRIBUTION: Atlantic, Indian, and Pacific oceans

HABITAT: Marine, rarely in freshwater; tropical to temperate; coastal to continental slopes, benthic or demersal over soft bottoms and reefs

REMARKS: As the name implies, the viviparous brotulas have internal fertilization and give birth to live young (i.e., they are embryoparous). Most species are relatively small bodied, but some can reach lengths of over 1 m. Bythitids are microcarnivores and eat a variety of small prey. They are mostly a marine lineage, and are found in a variety of habitats ranging from great oceanic depths, to shallow water reefs, to freshwater caves. Their relationships are not well studied, but two subfamilies, the Bythitinae and the Brosmophycinae, are recognized (Nelson, 2006).

REFERENCES: Møller et al., 2004, 2005; Scharzhans et al., 2005.

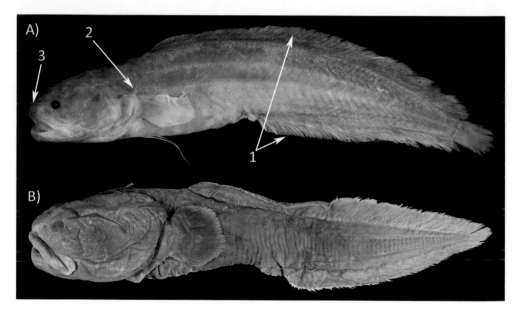

BYTHITID CHARACTERISTICS:

1) anal-fin rays shorter than corresponding dorsal-fin rays
2) strong opercular spine usually present
3) anterior nostril closer to upper lip than to posterior nostril
4) males with intromittent organ
5) scales usually present
6) gas bladder present in all species

ILLUSTRATED SPECIMENS:

A) *Ogilbia boydwalkeri,* SIO 70–165, 54 mm SL, holotype
B) *Thermichthys hollisi,* SIO 88–97, 304 mm SL, holotype

BATRACHOIDIFORMES : BATRACHOIDIDAE—Toadfishes

DIVERSITY: 1 family, 22 genera, 83 species

REPRESENTATIVE GENERA: *Batrachoides, Opsanus, Porichthys, Sanopus, Thalassophryne*

DISTRIBUTION: Atlantic, Indian, and Pacific oceans

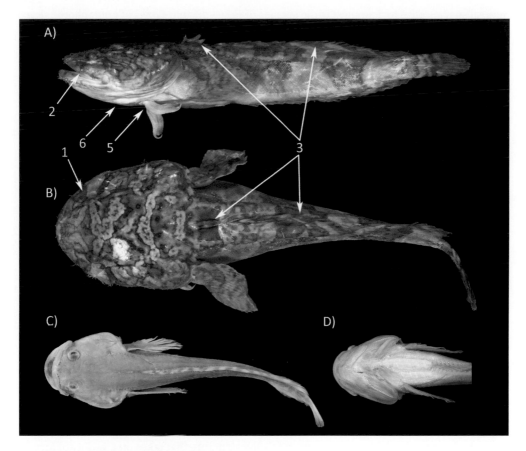

BATRACHOIDIFORM CHARACTERISTICS:

1) head broad, eyes dorsally oriented
2) mouth large, premaxilla and maxilla included in gape
3) small spinous dorsal fin with much longer soft dorsal fin
4) scales absent or, if present, small and cycloid
5) pelvic fins jugular, with one spine and two to three rays
6) gill membranes broadly united with isthmus
7) six branchiostegal rays
8) three gill arches

ILLUSTRATED SPECIMENS:

A) *Batrachoides waltersi,* SIO 73–257, 145 mm SL
B) *Batrachoides waltersi,* SIO 73–257 (dorsal view)
C) *Porichthys myriaster,* SIO 60–452, 118 mm SL (dorsal view)
D) *Porichthys myriaster,* SIO 60–452 (ventral view)

HABITAT: Marine, with a few species entering freshwater; tropical to temperate; coastal, benthic or demersal over soft bottoms and within reefs

REMARKS: Toadfishes are especially interesting fishes with a variety of unusual features. They have a well-developed gas bladder with modified muscles for sound production. They have large demersal eggs with an adhesive disc on the ventral surface of a large yolk sac that persists after hatching, permitting males to guard both eggs and larvae (Knapp et al., 1999). Several species have toxic spines, while species of *Porichthys* have rows of photophores along the ventral portions of the head and body. *Opsanus tau* is commonly used in physiological experiments, including those on sensory biology and muscle physiology, in part because of its robust nature. Some species of toadfishes are highly prized food fishes. Relationships of the Batrachoidiformes remain controversial, as they exhibit a variety of morphological features uniquely shared among disparate lineages of fishes. This complicated history was summarized by Greenfield et al. (2008) and Wiley and Johnson (2010). Though they were once considered members of the Paracanthopterygii, Miya et al. (2005) and Smith and Wheeler (2006) placed them within one of several crown groups of percomorph fishes based on sequence data. Other studies (e.g., Near et al., 2012) place them near the base of the Percomorpha. Within the toadfishes, Greenfield et al. (2008) recognized four subfamilies: Porichthyinae (midshipmen), Thalassophryninae (strongly venomous New World toadfishes), Batrachoidinae (six New World genera including the well-known genus *Opsanus*), and the newly erected Halophryninae (several Old World genera).

REFERENCES: Collette, in Fischer et al., 1995; Collette, in Carpenter, 2003; Collette and Russo, 1981; Greenfield, in Carpenter and Niem, 1999; Greenfield et al., 2008; Knapp et al., 1999; Miya et al., 2005; Smith and Wheeler, 2006; Walker and Rosenblatt, 1988.

LOPHIIFORMES—Anglerfishes

The Lophiiformes includes 18 families, 68 genera, and 325 species of anglerfishes (Miya et al., 2010). The group is characterized by having the first one or two dorsal-fin spines modified to form a structure for attracting prey. This unique feature includes the illicium, an elongate fin spine, and the esca, a terminal fleshy structure that takes on a variety of forms. This structure serves as a visual lure in the shallow-water frogfishes (Antennariidae), where the esca resembles a fish or small invertebrate, and in the deep-sea anglerfishes (Ceratioidei), where it contains bioluminescent bacteria. The same structure in the batfishes (Ogcocephalidae) releases a chemical that is attractive to their invertebrate prey (Nagareda and Shenker, 2008, 2009). The pelvic fins of lophiiforms, when present, are located anterior to the pectoral fins and are composed of one spine and four to five soft rays. The small gill openings are located at or posterior to the pectoral-fin base. Many lophiiforms are benthic, but the 11 families of ceratioid anglerfishes are bathypelagic. The systematics and relationships of lophiiform fishes have been studied extensively. Once considered a member of the Paracanthopterygii, closely related to the toadfishes (Batrachoidiformes), recent studies have placed the Lophiiformes as derived members of the Percomorpha, closely related to the Tetraodontiformes (reviewed in Miya et al., 2010). Some species of goosefishes (Lophiidae) support commercial fisheries and are marketed as "monkfish." Two groups are discussed further below.

REFERENCES: Bradbury, 1967, 2003; Miya et al., 2010; Nagareda and Shenker, 2008, 2009; Pietsch, 1981, 2009; Pietsch and Grobecker, 1987; Shedlock et al., 2004.

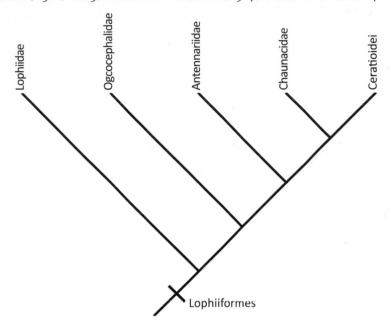

Hypothesized phylogenetic relationships of the Lophiiformes after Miya et al. (2010).

LOPHIIFORMES : ANTENNARIIDAE—Frogfishes

DIVERSITY: 12 genera, 48 species

REPRESENTATIVE GENERA: *Antennarius, Antennatus, Histrio*

DISTRIBUTION: Atlantic, Indian, and Pacific oceans

HABITAT: Marine; tropical to temperate; coastal, benthic, usually on hard bottoms, one species epipelagic

REMARKS: These relatively shallow-water members of the Lophiiformes are cryptic, sit-and-wait predators that use their esca to visually attract small fishes. Their feeding strike is among the most rapid movements recorded in the animal kingdom (Pietsch and Grobecker, 1978). While most frogfishes are benthic, the Sargassumfish (*Histrio histrio*) lives in the epipelagic zone in association with floating *Sargassum,* a pelagic brown alga. The systematics and biology of frogfishes were reviewed in detail by Pietsch and Grobecker (1987), and their phylogenetic relationships were studied by Arnold and Pietsch (2012) based on molecular data. Some frogfishes are targeted for the aquarium trade.

REFERENCES: Arnold and Pietsch, 2012; Miya et al., 2010; Pietsch, 1984; Pietsch and Grobecker, 1978, 1987; Pietsch et al., 2009.

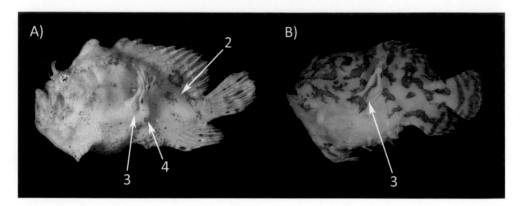

ANTENNARID CHARACTERISTICS:
1) body somewhat laterally compressed compared to other lophiiforms
2) body scaleless or with small denticles, skin loose
3) base of pectoral fins elongate and leg-like
4) gill openings small and pore-like

ILLUSTRATED SPECIMENS:
A) *Antennatus sanguineus,* SIO 65–342, 71 mm SL
B) *Histrio histrio,* SIO 03–34, 62 mm SL

LOPHIIFORMES : CERATIOIDEI—Deep-sea Anglerfishes

DIVERSITY: 11 families, 35 genera, and 167 species

REPRESENTATIVE GENERA: *Ceratias, Himantolophus, Linophryne, Melanocetus, Oneirodes*

DISTRIBUTION: Atlantic, Indian, and Pacific oceans

HABITAT: Marine; tropical to temperate; mesopelagic to bathypelagic

REMARKS: The ceratioid anglerfishes include some of the most remarkable fishes known, exhibiting a variety of "classic" adaptive features associated with their bathypelagic habitat (Pietsch, 2009). They are generally dark in color, they have huge mouths with long needle-like teeth, and most species have symbiotic relationships with bioluminescent bacteria that are harbored in the esca or lure. The illicium (the spine supporting the esca) can be relatively short (approximately head length) or, as in the whipnose anglers (Gigantactinidae), can be much longer than the body. Some species have "parasitic" males (which lack an illicium and esca) and use their modified teeth to attach to female hosts where they remain for life (Pietsch, 2005). Among the 11 families of deep-sea anglerfishes, there are a variety of body types, fin shapes, fin positions, and lure morphologies. Not surprisingly, this group has attracted the attention of several ichthyologists, notably Bertelsen (1951) and Pietsch (2009).

REFERENCES: Bertelsen, 1951; Miya et al., 2010; Pietsch, 2005, 2009; Pietsch and Orr, 2007.

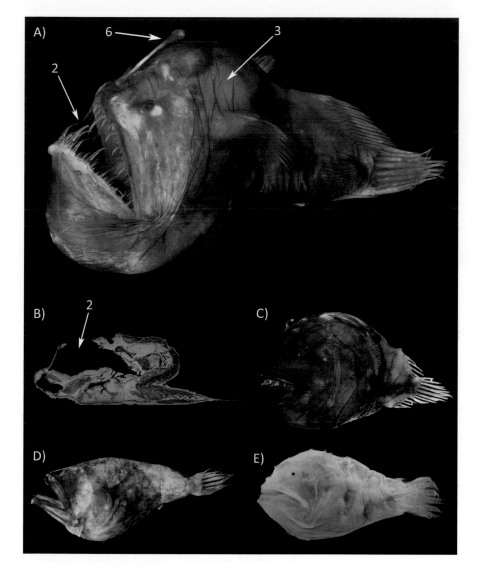

CERATIOID CHARACTERISTICS:

1) body globiform in most species
2) mouth huge, often with long, fang-like teeth
3) scales absent, but some species with small to large dermal spines or plates
4) pelvic fins absent
5) pseudobranch absent
6) females with illicium (absent in males), usually with bioluminescent bacteria in esca

ILLUSTRATED SPECIMENS:

A) *Melanocetus johnsonii,* SIO 69–497, 59 mm SL (Melanocetidae—black seadevils)
B) *Melanocetus johnsonii,* SIO 69–497, 59 mm SL, magnetic resonance image (Melanocetidae from Digital Fish Library, Berquist et al., 2012)
C) *Chaenophryne draco,* SIO 52–409, 50 mm SL (Oneirodidae—dreamers)
D) *Oneirodes rosenblatti,* SIO 69–351, 94 mm SL, holotype (Oneirodidae)
E) *Photocorynus spiniceps,* SIO 70–326, 49 mm SL (Linophrynidae—leftvents)

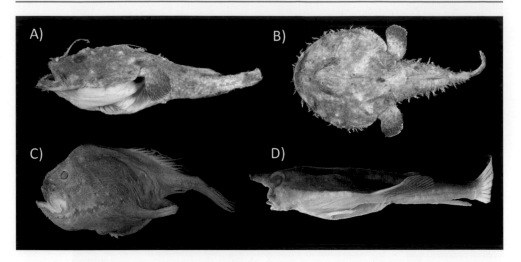

LOPHIIFORM DIVERSITY:
A AND B) LOPHIIDAE—goosefishes: *Lophiodes caulinaris,* SIO 63–514, 86 mm SL
C) CHAUNACIDAE—sea toads: *Chaunacops coloratus,* SIO 93–48, 163 mm SL
D) OGCOCEPHALIDAE—batfishes: *Ogcocephalus darwini,* SIO 51–214, 137 mm SL, holotype

PLEURONECTIFORMES—Flatfishes

Flatfishes are one of the most distinctive groups of teleosts, and are the only vertebrates with eyes on the same side of the head. Their larvae have eyes on both sides of the head as in other fishes, but near the time of settlement, one eye migrates to the opposite side (Brewster, 1987). Fossils document an evolutionary transition to this unique morphology (Friedman, 2008). In addition, flatfishes are characterized by their highly compressed body, anterior placement of the dorsal-fin origin, long dorsal- and anal-fin bases, small body cavity, and the nearly ubiquitous absence of a gas bladder in adults. Additionally, all but one family (Psettodidae—spiny turbots) lack fin spines. The systematics and classification of flatfishes have undergone considerable study and revision in recent decades. The group currently comprises 11 families, 134 genera, and more than 678 species. Recent studies (e.g., Betancur et al., 2013; Little et al., 2010; Near et al., 2012) have hypothesized a relationship of the flatfishes with the billfishes (Xiphiiformes) and the jacks (Carangiformes).

REFERENCES: Brewster, 1987; Chapleau, 1993; Friedman, 2008; Fukuda et al., 2010; Gibson, 2008; Hensley and Ahlstrom, 1984; Hoshino, 2001; Little et al., 2010; Norman, 1934.

PLEURONECTIFORMES : PARALICHTHYIDAE—Sand Flounders

DIVERSITY: 16 genera, 109 species

REPRESENTATIVE GENERA: *Citharichthys, Etropus, Paralichthys, Syacium*

DISTRIBUTION: Atlantic, Indian, and Pacific oceans

HABITAT: Marine to freshwater; tropical to temperate; coastal to upper continental slopes, benthic on soft substrates

REMARKS: The sand flounders represent some of the most valuable flatfish-fisheries in the world (Gibson, 2008). While they generally have eyes on the left side of the body, a few species occasionally have eyes on the right. Phylogenetic evidence places the sand flounders most closely related to the Pleuronectidae (Hoshino, 2001), although some evidence indicates that they are not monophyletic (Betancur, Li, et al., 2013; Chapleau, 1993). They are generalist predators, the smaller species eating benthic invertebrates and the larger species attacking relatively large bony fishes. The sand flounders can rapidly change their

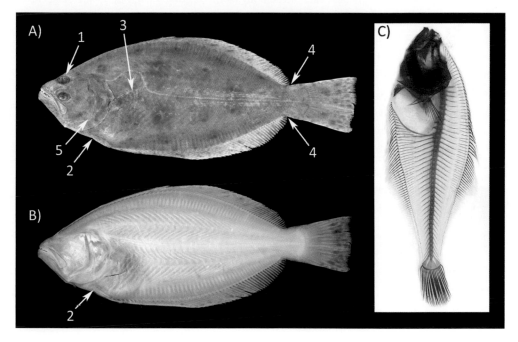

PARALICHTHYID CHARACTERISTICS:
1) eyes usually on left side of body
2) pelvic-fin bases the same length on both sides and usually symmetrically located
3) pectoral-fin rays branched
4) dorsal and anal fins separate from caudal fin
5) margin of preopercle free

ILLUSTRATED SPECIMENS:
A) *Paralichthys californicus*, SIO 64–80, 178 mm SL (eyed side)
B) *Paralichthys californicus*, SIO 64–80 (blind side)
C) *Citharichthys sordidus*, SIO 07–77, 116 mm SL, cleared-and-stained

color patterns in order to match their surroundings, and some species can reach sizes of up to 1.5 m and 30 kg.

REFERENCES: Amaoka and Hensley, in Carpenter and Niem, 2001; Betancur, Li, et al., 2013; Chapleau, 1993; Gibson, 2008; Hensley, in Fischer et al., 1995; Hoshino, 2001; Munroe, in Carpenter, 2003.

PLEURONECTIFORMES : PLEURONECTIDAE—Righteye Flounders

DIVERSITY: 23 genera, 106 species

REPRESENTATIVE GENERA: *Eopsetta, Hippoglossus, Lyopsetta, Microstomus, Pleuronectes*

DISTRIBUTION: Arctic, Atlantic, Indian, and Pacific oceans

HABITAT: Marine to freshwater; tropical to polar; coastal to upper continental slopes, benthic on soft substrates

REMARKS: Although pleuronectids are commonly known as the righteye flounders, a few species may have eyes on the left side of the body (some regularly so), but debate exists as to the nature of this polymorphism (Bergstrom, 2007; Russo et al., 2012). Cooper and Chapleau (1998) presented a morphologically based phylogeny and revised the classification of pleuronectids, recognizing five subfamilies. Righteye flounders are generalist predators, the smaller species relying on benthic invertebrates and the larger species consuming relatively large fishes. The group includes at least 18 commercially exploited species, including the largest flatfish, the Pacific Halibut (*Hippoglossus stenolepis*), which reaches up to 2.7 m and at least 230 kg.

REFERENCES: Bergstrom, 2007; Cooper and Chapleau, 1998; Evseenko, 2004; Hensley, in Carpenter and Niem, 2001; Russo et al., 2012; Sakamoto, 1984.

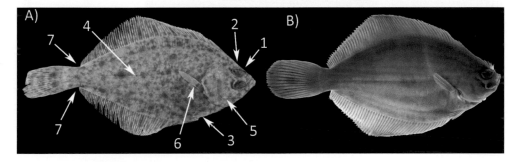

PLEURONECTID CHARACTERISTICS:
1) eyes usually on right side of body
2) dorsal-fin origin over eyes
3) pelvic fins symmetric
4) trunk lateral-line canal well developed on both sides of body
5) margin of preopercle free
6) pectoral fin on eyed side usually larger than on blind side
7) dorsal and anal fins separate from caudal fin

ILLUSTRATED SPECIMENS:
A) *Lepidopsetta bilineata*, SIO 84–145, 86 mm SL
B) *Pleuronichthys coenosus*, SIO 49–37, 175 mm SL

PLEURONECTIFORMES : BOTHIDAE—Lefteye Flounders

DIVERSITY: 23 genera, 166 species

REPRESENTATIVE GENERA: *Bothus, Engyophrys, Monolene, Psettina, Trichopsetta*

DISTRIBUTION: Atlantic, Indian, and Pacific oceans

HABITAT: Marine to freshwater; tropical to temperate; coastal to upper continental slopes, benthic on soft substrates or rubble

REMARKS: The lefteye flounders typically live in shallow, coastal waters, but some species occur down to at least 500 m. They are generalist predators and eat benthic invertebrates and small fishes. Some species have very widely spaced eyes and exhibit conspicuous sexual dimorphism (Kobelkowsky, 2004). The lefteye flounders can rapidly change color patterns in order to match their surroundings. While most species are edible, they are not targeted commercially, though they are undoubtedly captured incidentally in fisheries for other flatfishes.

REFERENCES: Hensley, in Fischer et al., 1995; Hensley and Amaoka, in Carpenter and Niem, 2001; Kobelkowsky, 2004; Munroe, in Carpenter, 2003.

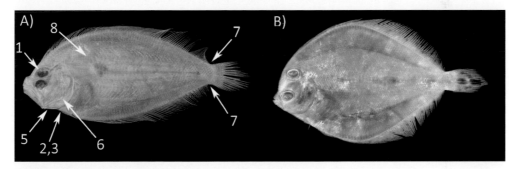

BOTHID CHARACTERISTICS:
1) eyes on left side of body
2) pelvic-fin base longer on eyed side than on blind side
3) pelvic fin of eyed side on midventral line, pelvic fin of blind side above midventral line
4) pectoral- and pelvic-fin rays unbranched
5) branchiostegal membranes connected at isthmus
6) margin of preopercle free
7) dorsal and anal fins separate from caudal fin
8) trunk lateral-line canal arched over pectoral fin

ILLUSTRATED SPECIMENS:
A) *Trichopsetta ventralis,* SIO 64–989, 124 mm SL
B) *Bothus robinsi,* SIO 06–73, 120 mm SL

PLEURONECTIFORMES : ACHIRIDAE—American Soles

DIVERSITY: 7 genera, 36 species

REPRESENTATIVE GENERA: *Achirus, Gymnachirus, Trinectes*

DISTRIBUTION: Atlantic and eastern Pacific oceans

HABITAT: Marine to freshwater; tropical to temperate; coastal to upper continental slope, benthic on soft substrates

REMARKS: The American soles typically live in shallow marine waters but at least one species can penetrate hundreds of miles up rivers. The unusual trunk lateral-line canal occasionally is crossed at right angles by accessory canals, sometimes called achirine lines. They are generalist predators, typically consuming benthic invertebrates and sometimes small fishes. The American soles are not directly targeted commercially but are taken as bycatch in trawl fisheries.

REFERENCES: Krupp, in Fischer et al., 1995; Munroe, in Carpenter, 2003; Ramos, 2003; Ramos et al., 2009; Walker and Bollinger, 2001.

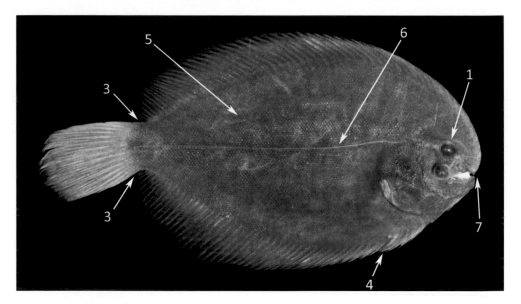

ACHIRID CHARACTERISTICS:
1) eyes on right side of body
2) margin of preopercle usually concealed by skin or represented by naked superficial groove
3) dorsal and anal fins separate from caudal fin
4) right pelvic-fin and anal-fin membranes joined
5) body oval to rounded, covered with fine scales
6) trunk lateral-line canal straight
7) mouth small
8) branchiostegal rays and isthmus covered by skin contiguous with the skin covering the lower jaw

ILLUSTRATED SPECIMEN:
Trinectes xanthurus, SIO 63–292, 75 mm SL, holotype

PLEURONECTIFORMES : CYNOGLOSSIDAE—Tonguefishes

DIVERSITY: 3 genera, 145 species

REPRESENTATIVE GENERA: *Cynoglossus, Paraplagusia, Symphurus*

DISTRIBUTION: Atlantic, Indian, and Pacific oceans

HABITAT: Marine to freshwater; tropical to temperate; coastal to continental slope, benthic on soft substrates

REMARKS: The tonguefishes are relatively elongate flatfishes with extremely small eyes. The lateral-line system is variable in cynoglossids (Fukuda et al., 2010) and some species have numerous superfical neuromasts on the blind side of the head. A number of species have distinct coloration and spotting along the body and median fins. Depths of occurrence range from less than 1 m to over 1,900 m. *Symphurus thermophilus* was recently described from waters adjacent to hydrothermal vents in the western Pacific (Munroe and Hashimoto, 2008). Tonguefishes are generalist predators and feed on small benthic invertebrates. Most species are of little commercial value, but fisheries for the larger-bodied species may develop.

REFERENCES: Chapleau, 1988; Fukuda et al., 2010; Menon, 1977; Munroe, 1992, 1998; Munroe, in Fischer et al., 1995; Munroe, in Carpenter and Niem, 2001; Munroe, in Carpenter, 2003; Munroe and Hashimoto, 2008.

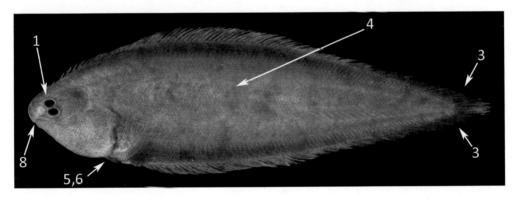

CYNOGLOSSID CHARACTERISTICS:
1) eyes small, closely set, on left side of body
2) margin of preopercle not evident externally, concealed by skin and scales
3) dorsal and anal fins continuous with caudal fin
4) body elongate, tapering posteriorly, covered with fine scales
5) pelvic fin of blind side located along ventral midline, sometimes linked to anal fin
6) pelvic fin and underlying bones on eyed side usually present
7) pectoral fins absent
8) mouth asymmetric
9) branchiostegal rays and isthmus covered by skin contiguous with the skin covering the lower jaw

ILLUSTRATED SPECIMEN:
Symphurus atricaudus, SIO 64–6, 105 mm SL

TETRAODONTIFORMES—Plectognaths

Tetraodontiform fishes are highly variable in body shape, but are characterized by having the hyomandibula and palatine firmly attached to the neurocranium, the maxilla and premaxilla fused, a reduced skull, the gill openings restricted and typically high on the body, and scales usually modified as spines, shields, or plates. This morphologically diverse group includes ten families, over 100 genera, and 433 species. Their evolution, including their extensive fossil record, has been studied by a number of researchers (e.g., Holcroft, 2004, 2005; Matsuura, 1979; Santini and Tyler, 2003, 2004; Tyler, 1980; Winterbottom, 1974b; Yamanoue et al., 2007, 2008). Recent molecular studies have hypothesized a close relationship between the Tetraodontiformes and the Lophiiformes (Betancur et al., 2013; Miya et al., 2010; Near et al., 2012). We cover six families below.

REFERENCES: Holcroft, 2004, 2005; Matsuura, 1979; Miya et al., 2010; Santini and Tyler, 2003, 2004; Tyler, 1980; Winterbottom, 1974b; Yamanoue et al., 2007, 2008).

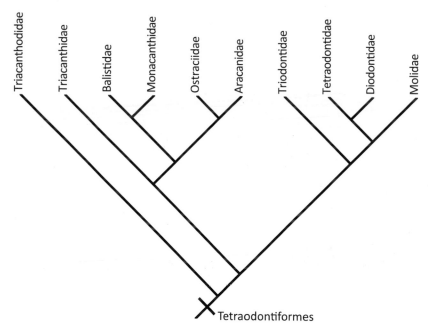

Hypothesized phylogenetic relationships of the Tetraodontiformes after Santini and Tyler (2004).

TETRAODONTIFORMES : BALISTIDAE—Triggerfishes

DIVERSITY: 11 genera, 41 species

REPRESENTATIVE GENERA: *Balistes, Canthidermis, Melichthys, Sufflamen*

DISTRIBUTION: Atlantic, Indian, and Pacific oceans

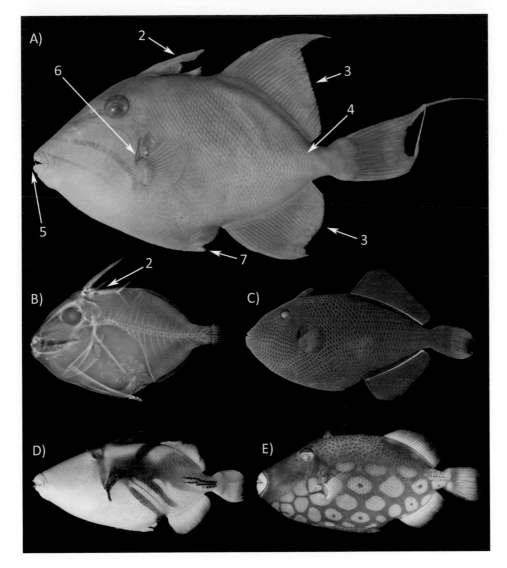

BALISTID CHARACTERISTICS:

1) body compressed
2) two dorsal fins, the first with three spines (third often minute)
3) all soft fins with branched rays
4) scales plate-like, in regular series
5) mouth small, jaws armed with crushing teeth
6) gill slit small, anterior to pectoral-fin base
7) pelvic spine and fins rudimentary or absent

ILLUSTRATED SPECIMENS:

A) *Balistes vetula,* SIO 71–277, 141 mm SL
B) *Balistes polylepis,* SIO 79–212, radiograph
C) *Melichthys niger,* SIO 59–334, 206 mm SL
D) *Rhinecanthus aculeatus,* SIO 92–159, 166 mm SL
E) *Balistoides conspicillum,* SIO 64–229, 131 mm SL

(account continued)

HABITAT: Marine; tropical to temperate; neritic to demersal over soft and hard bottoms, often over or near reefs, juveniles occasionally associated with flotsam far from shore

REMARKS: The common name, triggerfishes, comes from the locking mechanism of the anterior dorsal-fin spines; the large first spine can be held rigidly erect by the second spine. The anatomy and relationships of triggerfishes (and the related filefishes) have been studied by Matsuura (1979), Yamanoue et al. (2008), and Dornburg et al. (2011). The diet of trigger-fishes consists of benthic invertebrates, notably hard-shelled mollusks and crabs, and occasionally zooplankton. Triggerfishes lay demersal eggs in nests that are aggressively guarded by males (Gladstone, 1994). Their flesh is flavorful and many species support artisanal fisheries. Triggerfishes can be pugnacious and are known to harass divers.

REFERENCES: Bussing, in Fischer et al., 1995; Dornburg et al., 2011; Gladstone, 1994; Matsuura, 1979; Matsuura, in Carpenter, 2003; Matsuura, in Carpenter and Niem, 2001; Yamanoue et al., 2008.

TETRAODONTIFORMES : MONACANTHIDAE—Filefishes

DIVERSITY: 32 genera, 109 species

REPRESENTATIVE GENERA: *Aluterus, Cantherhines, Monacanthus, Pervagor*

DISTRIBUTION: Atlantic, Indian, and Pacific oceans

HABITAT: Marine; tropical to temperate; coastal to epipelagic, demersal, neritic, or pelagic, juveniles and adults often epipelagic in association with flotsam

REMARKS: Filefishes are similar to triggerfishs in that the first dorsal-fin spine can be locked erect by the second spine. Sometimes called leatherjackets, monacanthids range in color from drab and cryptic to colorful and conspicuous. The anatomy and relationships of filefishes (and the related triggerfishes) were studied by Matsuura (1979), and the evolution of their unusual pelvic fin morphology by Yamanoue et al. (2008). Filefishes feed on a variety of small invertebrates including, in some species, sponges and corals, as well as plant matter. Some species are known to be monogamous (Barlow, 1984; Whiteman and Côté, 2004). While most species are small, a few grow to a large size and are abundant enough to be exploited by fisheries (Miller and Stewart, 2009). A few others are found in the aquarium industry.

REFERENCES: Barlow, 1984; Bussing and Lavenberg, in Fischer et al., 1995; Hutchins, in Carpenter and Niem, 2001; Matsuura, 1979; Matsuura, in Carpenter, 2003; Miller and Stewart, 2009; Whiteman and Côté, 2004; Yamanoue et al., 2008.

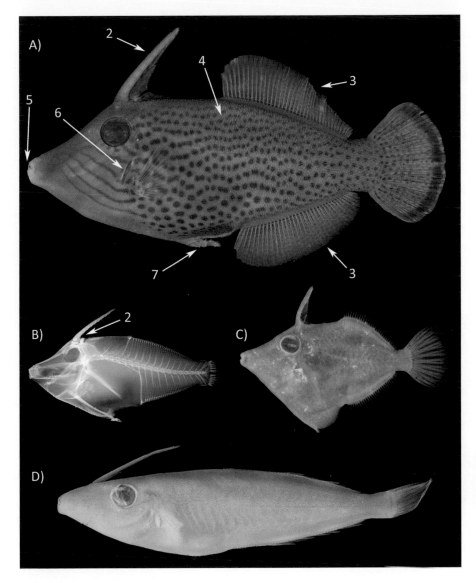

MONACANTHID CHARACTERISTICS:

1) body compressed
2) two dorsal fins, the first with two spines (second spine small or sometimes absent)
3) soft dorsal-, anal-, and pectoral-fin rays unbranched
4) scales small, body prickly or furry to the touch
5) mouth small, jaws with tiny teeth
6) gill slit small, anterior to pectoral-fin base
7) pelvic fins and spines rudimentary or absent

ILLUSTRATED SPECIMENS:

A) *Pervagor spilosoma*, SIO 53–539, 85 mm SL
B) *Pervagor spilosoma*, SIO 53–539, radiograph
C) *Monacanthus ciliatus*, SIO 70–183, 41 mm SL
D) *Pseudalutarius nasicornis*, SIO 73–197, 42 mm SL

TETRAODONTIFORMES : OSTRACIIDAE—Boxfishes

DIVERSITY: 7 genera, 27 species

REPRESENTATIVE GENERA: *Acanthostracion, Lactophrys, Lactoria, Ostracion*

DISTRIBUTION: Atlantic, Indian, and Pacific oceans

HABITAT: Marine; tropical to warm temperate; coastal, from estuaries to reef systems, demersal to neritic, sometimes epipelagic

REMARKS: In addition to their distinctive bony external armor, several boxfish species are known to secrete a skin toxin (ostracitoxin) that is lethal to other fishes, affording them additional protection from predators (Thomson, 1969). This is one reason boxfishes usually are not seen in private, mixed aquariums and are displayed with caution by public aquariums. Ostraciids consume a variety of benthic invertebrates, and a few species support small fisheries. Their relationships were studied by Klassen (1995) based on morphology.

REFERENCES: Klassen, 1995; Matsuura, in Carpenter, 2003; Matsuura, in Carpenter and Niem, 2001; Santini and Tyler, 2003; Thomson, 1969; Winterbottom and Tyler, 1983.

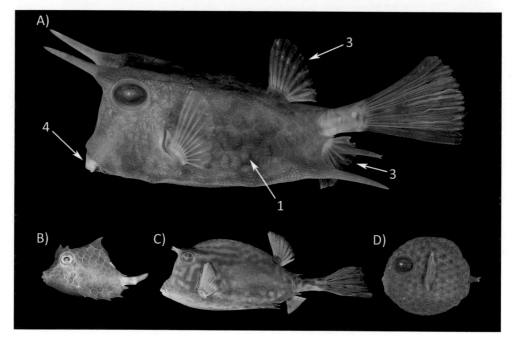

OSTRACIID CHARACTERISTICS:
1) body encased in bony carapace closed behind anal fin
2) pelvic fin and girdle absent
3) dorsal and anal fins lacking spines and located far posteriorly on body
4) upper jaw not protrusible

ILLUSTRATED SPECIMENS:
A) *Lactoria cornuta,* SIO 66–555, 69 mm SL
B) *Tetrasomus gibbosus,* SIO 64–229, 41 mm SL
C) *Acanthostracion quadricornis,* SIO 70–189, 134 mm SL
D) *Lactophrys* sp. (juvenile), SIO 93–213, 7.5 mm SL

DIVERSITY: 19 genera, 190 species

REPRESENTATIVE GENERA: *Arothron, Canthigaster, Lagocephalus, Sphoeroides, Takifugu*

DISTRIBUTION: Atlantic, Indian, and Pacific oceans

HABITAT: Marine; tropical to warm temperate; coastal, some in brackish waters, and a few restricted to freshwaters, demersal to neritic, occasionally epipelagic

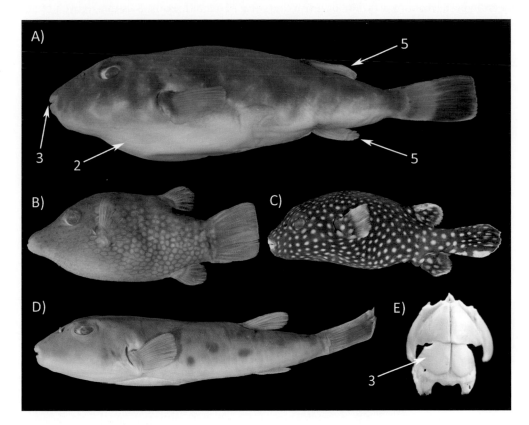

TETRAODONTID CHARACTERISTICS:
1) body inflatable via gulping water into stomach
2) skin naked or with prickles on the belly and rest of body in many species
3) four fused teeth in jaws (two upper, two lower)
4) premaxillaries and dentaries not fused medially
5) dorsal and anal fins located far posteriorly on body
6) pelvic fins absent
7) ribs and epineurals absent

ILLUSTRATED SPECIMENS:
A) *Sphoeroides lispus,* SIO 62–213, 93 mm SL, holotype
B) *Canthigaster punctatissima,* SIO 62–56, 54 mm SL
C) *Arothron meleagris,* SIO 60–147, 104 mm SL
D) *Takifugu vermicularis,* SIO 85–138, 117 mm SL
E) jaws of *Arothron* sp., SIO uncatalogued

(account continued)

REMARKS: Their common name comes from the puffers' ability to inflate their body by ingesting large quantities of water (Wainwright and Turingan, 1997). The organs and flesh of several species are highly toxic, providing adults, their eggs, and their larvae additional protection from predators (Gladstone, 1987). These toxic species serve as the model for the evolution of several mimic species (Randall, 2005). Two subfamilies, the diverse Tetraodontinae and the less diverse Canthigasterinae, are recognized based on both morphological and molecular analyses (Holcroft, 2005). The phylogenetic relationships of the puffers have been hypothesized by a variety of researchers (e.g., Holcroft, 2005; Santini, Nguyen et al., 2013; Tyler, 1980; Winterbottom, 1974). Most puffers are opportunistic carnivores. Puffers have relatively small genomes; thus the famous fugu (genus *Takifugu*), the focus of the renowned, but risky, eating experience, was one of the first fish species to have its genome sequenced (Brenner et al., 1993). Their "compact" genome has served as a comparative tool in vertebrate genomics (Venkatesh and Yap, 2004; www.fugu-sg.org/).

REFERENCES: Brenner et al., 1993; Bussing, in Fischer et al., 1995; Gladstone, 1987; Holcroft, 2005; Matsuura, in Carpenter and Niem, 2001; Randall, 2005; Santini, Nguyen et al., 2013; Shipp, 1974; Shipp, in Carpenter, 2003; Tyler, 1980; Venkatesh and Yap, 2004; Wainwright and Turingan, 1997; Winterbottom, 1974b.

TETRAODONTIFORMES : DIODONTIDAE—Porcupinefishes

DIVERSITY: 6 genera, 18 species

REPRESENTATIVE GENERA: *Chilomycterus, Diodon*

DISTRIBUTION: Atlantic, Indian, and Pacific oceans

HABITAT: Marine; tropical to temperate; coastal, neritic to demersal over reefs and soft bottoms, juveniles often found far offshore in epipelagic waters

REMARKS: Several species of diodontids have broad distributions and are found in tropical waters throughout the world (Leis, 2006). When the stomach is inflated with water (like the puffers), the pungent spines on the body become erect, providing effective protection from predators. However, the epipelagic juveniles are sometimes preyed upon by tunas and billfishes. Some species are thought to mimic opisthobranch mollusks as juveniles (Heck and Weinstein, 1978). The left and right sides of their large, beak-like teeth are fused medially and are used for crushing hard-bodied, benthic invertebrate prey. A few porcupinefishes support artisanal fisheries.

REFERENCES: Heck and Weinstein, 1978; Leis, 1978, 2006; Leis, in Carpenter, 2003; Leis, in Carpenter and Niem, 2001.

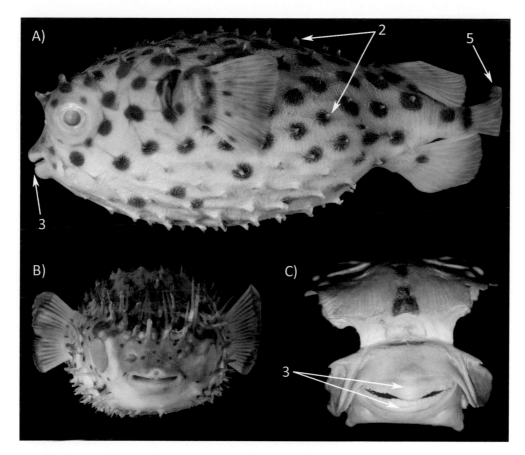

DIODONTID CHARACTERISTICS:

1) body wide, globose, inflatable via gulping water into stomach
2) body covered with well-developed, sharp spines
3) premaxillae and dentaries fused at midline, teeth fused, resembling parrot beak
4) pelvic fins absent
5) caudal fin rounded

ILLUSTRATED SPECIMENS:

A) *Chilomycterus reticulatus,* SIO 60–260, 84 mm SL
B) *Diodon holocanthus,* SIO 65–679, 78 mm SL (frontal view)
C) jaws of *Diodon* sp., SIO uncatalogued

TETRAODONTIFORMES : MOLIDAE—Molas

DIVERSITY: 3 genera, 4 species

REPRESENTATIVE GENERA: *Masturus, Mola, Ranzania*

DISTRIBUTION: Atlantic, Indian, and Pacific oceans

HABITAT: Marine; tropical to temperate; epipelagic, occasionally neritic

REMARKS: This small group of species occurs in the epipelagic zone of the world's oceans, but individuals are occasionally seen in coastal areas. Lacking a caudal fin (Britz and Johnson, 2005), molas are propelled by large dorsal and anal fins. The largest mola, the Ocean Sunfish (*Mola mola*) grows to an enormous size, over 2 m and 2,300 kg, and is the most fecund vertebrate known, with females carrying over 300 million eggs at a time. This spe-

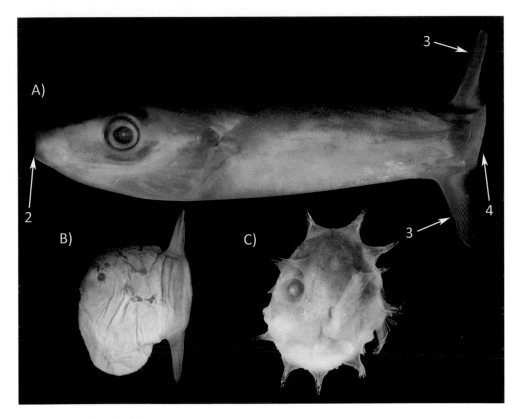

MOLID CHARACTERISTICS:
1) body laterally compressed
2) jaws with fused, beak-like teeth
3) dorsal and anal fins lacking spines, placed far posteriorly on body
4) caudal fin absent

ILLUSTRATED SPECIMENS:
A) *Ranzania laevis*, SIO 76–110, 107 mm TL
B) *Mola mola* (juvenile), SIO 55–71, 54 mm TL
C) *Mola mola* (larva), SIO 73–153, 8 mm TL

cies's common name comes from its "sunning" behavior, orienting horizontally near the surface. Molas feed mostly on pelagic coelenterates, but also take salps, brittlestars, fish larvae, and occasionally larger fishes. Several studies (e.g., Bass et al., 2005; Santini and Tyler, 2002; Yamanoue et al., 2004) have explored their phylogenetic relationships.

REFERENCES: Bass et al., 2005; Britz and Johnson, 2005; Fraser-Brunner, 1951; Hutchins, in Carpenter and Niem, 2001; Parenti, 2003; Santini and Tyler, 2002, 2003; Yamanoue et al., 2004.

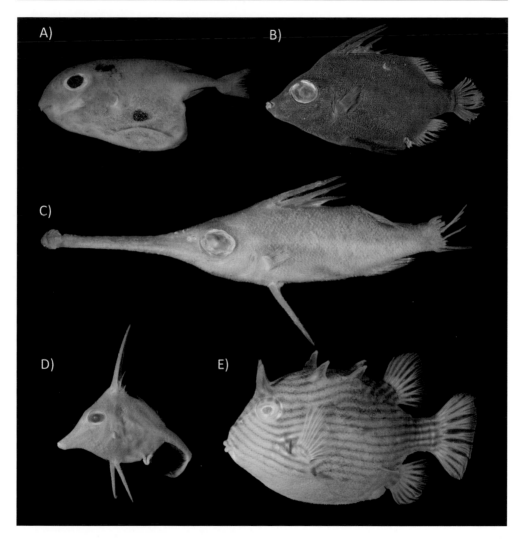

TETRAODONTIFORM DIVERSITY:
A) TRIODONTIDAE—three-toothed puffers: *Triodon macropterus,* SIO 87–76, 84 mm SL
B) TRIACANTHODIDAE—spikefishes: *Triacanthodes anomalus,* SIO 85–138, 83 mm SL
C) TRIACANTHODIDAE: *Macrorhamphosodes uradoi,* SIO 85–3, 106 mm SL
D) TRIACANTHIDAE—triplespines: *Trixiphichthys weberi,* SIO 01–102, 60 mm SL
E) ARACANIDAE—deepwater boxfishes: *Aracana ornata,* SIO 87–72, 69 mm SL

GLOSSARY

ABDOMINAL   Position of the pelvic-fin insertion, well posterior to the pectoral-fin insertion.

ABYSSAL PLAIN   Region of the ocean bottom, generally flat, between approximately 4,000 and 6,000 m depth.

ADIPOSE EYELID   Translucent tissue partially covering the eye in some fishes.

ADIPOSE FIN   Rayless, unpaired fin situated posterior to the main dorsal fin in some fishes.

AMPHISTYLIC   Form of jaw support seen in primitive sharks in which the upper jaw is attached by ligaments to the cranium at two points.

AMPULLAE OF LORENZENI   Electroreceptors in sharks and rays, formed of sensory cells located at the base of small canals open to the surface and filled with conductive jelly.

ANADROMOUS   Species that spawn in freshwater but usually are found in the ocean as adults.

ANTERIOR   Toward the head.

ARTISANAL FISHERY   Small-scale, regional fishery, relying on traditional fishing methods.

AUTOSTYLIC   Form of jaw support seen in tetrapods and lungfishes in which the upper jaw is firmly attached or fused with the cranium.

AXIL   Inside base of the pectoral and pelvic fins.

AXILLARY PROCESS   Modified, usually elongate, scale or scales at the upper or anterior base of the pectoral or pelvic fins.

BAND   Oblique or curved, linear color marking, as distinct from a bar (vertical marking) or a stripe (horizontal marking).

BAR   Elongate, straight-sided, vertical color marking.

BARBEL   Fleshy, flexible, flap-like or finger-like sensory appendage, normally associated with the mouth and often bearing numerous taste buds.

BATHYPELAGIC ZONE   Portion of the ocean water column beyond the continental shelf, extending from approximately 1,000 to 4,000 m depth.

BENTHIC   Occurring in or on the bottom of a body of water.

BIOLUMINESCENCE   Biochemical emission of light by living organisms, occasionally symbiotic in nature.

BONE   Rigid, mineralized skeletal tissue that is either preformed in cartilage (endochondral bone) or not (dermal bone).

BRANCHIOSTEGAL RAYS   Slender, bony elements in the gill membrane, located slightly behind and below the gill cover.

BREEDING TUBERCLE   Keratinized structures on the skin of breeding males of some fishes.

CARTILAGE   Skeletal tissue that is often flexible and not penetrated by blood vessels.

CATADROMOUS   Species usually found in fresh water as adults, but which spawn in the ocean.

CAUDAL PEDUNCLE   Area between the insertion of the dorsal and anal fins and the base of the caudal fin.

CEPHALIC   Referring to the head.

CERATOTRICHIA   Fin-ray elements of chondrichthyans, usually flexible, always unsegmented, and developmentally epidermal.

CHEEK   Area of the head below and posterior to the eye, anterior to the posterior margin of the preopercle.

CIRRUS   Tendril or short, fleshy filament.

COMPRESSED   Laterally (side-to-side) flattened body form.

CRUMENAL ORGAN   Modification of the dorsal portion of the posterior two gill arches that serves to trap food particles in argentiniform fishes.

CTENOID   Scale in which the exposed part and/or posterior margin have small, tooth-like structures (ctenii).

CYCLOID   Scale with the exposed part and posterior margin smooth.

DECIDUOUS SCALES   Scales not firmly attached to the body, easily lost.

DEEP-SCATTERING LAYER   Horizontal, vertically migrating aggregation of mesopelagic organisms that reflects sonar.

DEMERSAL   Occurring near the bottom.

DENTICLE   Tooth-like scale with a hard outer later, found in some elasmobranchs.

DENTIGEROUS   Bearing teeth.

DEPRESSED   Dorso-ventrally (vertically) flattened body form.

DIARTHROSIS   Moveable joint connecting bones.

DISTAL   Near the end; away from the body (see proximal).

DORSAL   Toward the top of the body; upper.

DORSAL-FIN INSERTION   Posterior end of the dorsal-fin base.

EMARGINATE   Caudal-fin shape with a slightly concave posterior margin.

EMBRYOPAROUS   Reproductive mode in which eggs are fertilized and develop internally, and are released from the female after hatching; sometimes called viviparous.

EPIPELAGIC ZONE   Upper portion of the ocean water column beyond the continental shelf, extending from the surface to approximately 200-m depth.

ESCA   Expanded, sometimes highly modified, portion at the end of a modified dorsal-fin spine (the illicium) used by anglerfishes (Lophiiformes) to attract prey.

EXTANT   Still in existence, not extinct.

FINLET   Small fin with one or more rays, usually situated posterior to the dorsal and anal fins.

FLOTSAM   Debris of various kinds usually floating at or near the water's surface.

FORKED   Caudal-fin shape with separate upper and lower lobes joined at a distinct angle.

GAPE   Margin, outline, or rim of the mouth opening.

GAS BLADDER   Gas- or fat-filled organ usually located dorsally in the abdominal cavity; sometimes called a swim bladder.

GILL COVER   Plate-like cover of the gills (operculum).

GILL RAKER   Bony element on the anterior margin of a gill arch opposite the gill filaments.

GULAR PLATE   Flat bone or bones between the left and right sides of the lower jaw.

HERMAPHRODITE   Organism with functional male and female sex organs in the same individual, either simultaneously or sequentially.

HETEROCERCAL   Caudal fin in which the vertebral column is deflected dorsally and extends along the upper, larger caudal-fin lobe.

HOLOSTYLIC   Form of jaw suspension in which the upper jaw is fused to the neurocranium; found in the Holocephali.

HOLOTYPE   Specimen designated by the describer (author) of a species, to which the name of the species is applied.

HYOSTYLIC   Form of jaw suspension in which the upper jaw is connected to the cranium via ligamentous attachments; found in most cartilaginous fishes and all ray-finned fishes.

HYPURALS   Flat bones supporting the principal caudal-fin rays.

ILLICIUM   Modified dorsal-fin spine(s) supporting an esca (or lure), normally near the tip of the snout, used by anglerfishes (Lophiiformes) to attract prey.

INSERTION   Posterior end of the dorsal- or anal-fin base.

ISTHMUS   Area of the throat ventral to the gill openings.

JAKUBOWSKI'S ORGAN   Cluster of sensory cells including the terminal supraorbital neuromasts, located between the nasal and lacrimal bones of beryciform fishes.

JUGULAR   Position of pelvic-fin insertion anterior to pectoral-fin insertion.

KEEL   Lateral ridge on the caudal peduncle.

LATERAL LINE   Sensory system consisting of pores and canals along the head and body for the detection of vibrations and water movement, often associated with perforated scales along the body.

LEPIDOTRICHIA   Fin-ray elements of osteichthyans, segmented or unsegmented, and derived from dermis.

LEPTOCEPHALUS   Larval form of elopomorph fishes, with a small head and a compressed, elongate body.

LIPID HISTOTROPHY   Internal development in myliobatiform fishes in which nutrition to embryos is supplied by yolk, supplemented by lipid secretions from the uterus.

LUNATE   Crescent-shaped caudal fin with deeply emarginate and narrow upper and lower lobes.

MANDIBULAR   Associated with the lower jaw.

MANDIBULAR SYMPHYSIS   Point at which the left and right sides of the lower jaw join.

MAXILLAE   Paired bones of the upper jaw lying posterior to the premaxillae, either included in or excluded from the gape.

MEDIAN   Along the midline of the body.

MEDIAN FINS   Unpaired fins, comprising the dorsal, caudal, and anal fins.

MENTAL   Position of pelvic-fin insertion far anterior, near the mandibular symphysis.

MESOPELAGIC ZONE   Portion of the oceanic water column beyond the continental shelf, extending from approximately 200 to 1,000-m depth.

MOLECULAR PHYLOGENY   Hypothesized relationships of a species or taxonomic group based primarily on DNA sequence data.

MONOPHYLETIC GROUP   Ancestor and all of its descendants.

MORPHOLOGICAL PHYLOGENY   Hypothesized relationships of a species or taxonomic group based on anatomical features.

MUCOID HISTOTROPHY   Internal development in some chondrichthyan fishes in which nutrition to embryos is supplied by yolk, supplemented by mucous secretions from the uterus.

NAKED   Without scales.

NAPE   Dorsal area just posterior to the head.

NERITIC   Occurring in the water column (off the bottom) over the continental shelf; sometimes called coastal pelagic.

NEUROCRANIUM   Skull elements that support and surround the brain.

NICTITATING MEMBRANE   Thin membrane, present in some chondrichthyans, that can be drawn across the eyeball for protection.

OCELLUS   Round or oval eye-like spot with a colored ring around it.

OCCIPUT   Posterior part of the head.

OOPHAGY   Internal development in the lamniform fishes in which nutrition to embryos is supplied by yolk, supplemented by additional eggs produced by the mother.

OPERCULUM   Plate-like structure covering the branchial chamber and consisting of four bones: the opercle, preopercle, subopercle, and interopercle.

ORBITAL   Associated with the eye.

ORIGIN   At the anterior end, in reference to the base of the dorsal and anal fins.

OTOPHYSIC CONNECTION   Diverticula of the gas bladder that penetrates the otic region of the neurocranium, found in a variety of fishes including clupeiforms and holocentrids.

OVIPAROUS   Reproductive mode in which eggs are fertilized externally and development occurs outside the female's body.

PAIRED FINS   Bilateral fins, comprising the pectoral and pelvic fins.

PALATINE   Paired bones in the roof of the mouth, lateral and posterior to the vomer, occasionally bearing teeth.

PARAPHYLETIC GROUP   Ancestor and some, but not all, of its descendants.

PELAGIC   Portion of an ocean, lake, or river not near the margins or bottom.

PERITONEUM   Membrane lining the abdominal cavity.

PHORESY   Attachment of one animal to another for transportation.

PHOTOPHORE   Light-producing organ.

PHYLOGENY   Evolutionary history or relationships of a species or taxonomic group.

PLACENTAL VIVIPARITY   Internal development in which nutrition to embryos is supplied initially by yolk and later by transfer from the uterine lining via a placenta-like extension of the yolk sac.

POSTERIOR   Toward the tail.

PREMAXILLAE   Paired anterior bones of the upper jaw.

PREOPERCLE   Flat bone covering part of the anterior cheek whose margin may be serrated, spinous, or smooth; one of the four bones of the operculum (gill cover).

PRINCIPAL CAUDAL-FIN RAYS   Rays that extend to the posterior margin of the caudal fin, generally supported by the hypural elements, the count usually equal to the number of branched rays plus two (Teleostei).

PROCURRENT CAUDAL-FIN RAYS   Small rays dorsal and ventral to the principal caudal-fin rays.

PROTANDROUS   Maturing initially as a male, then later as a female (see protogynous).

PROTOGYNOUS   Maturing initially as a female, then later as a male (see protandrous).

PROXIMAL   Near the beginning, origin, or base; closer to the body (see distal).

PSEUDOBRANCH   Structure on the underside of the operculum, composed of filaments similar to the gill filaments.

ROSTRAL ORGAN   Collection of sensory receptors in the snout of anchovies and coelacanths.

ROUNDED   Caudal-fin shape with a smoothly convex posterior margin.

SCHRECKSTOFF   Alarm substance found in ostariophysan fishes.

SCUTE   Thickened, hardened scale, or external bony plate.

SEMELPAROUS   Life history condition in which spawning takes place once, just before death, as in anguillids and some salmonids.

SIMPLE FIN RAY   Single element, unbranched for all or most of its length.

SIMULTANEOUS HERMAPHRODITE   Individual with functional male and female sex organs present at the same time.

SNOUT   Area of the head between the tip of the upper jaw and the anterior margin of the orbit.

SOFT-RAY   Rod-like, bony, fin-support element that is bilaterally paired, segmented, and usually flexible.

SPINE   Rod-like, bony, fin-support element that is unpaired, unsegmented, and often pungent.

SPIRACLE   Opening, usually behind the eye, of a duct leading to the gill chamber.

STRIPE   Elongate, straight-sided, horizontal color marking.

SUPERFICIAL NEUROMAST   Mechanosensory organ located on the skin surface.

TAIL   Portion of the body posterior to the anus, including the caudal fin.

TAXONOMY   Branch of science concerned with the identification and classification of organisms.

TETRAPLOID   Having four sets of chromosomes.

THORACIC   Position of the pelvic-fin insertion generally below or slightly posterior to the pectoral-fin insertion.

TRUNCATE   Caudal-fin shape with a vertically straight posterior margin.

TRUNK   Portion of the body between the head and the anus.

VENT   Terminal opening of the alimentary canal (anus).

VENTRAL   Toward the bottom of the body, lower.

VIVIPAROUS   Reproductive mode in which eggs are fertilized and develop internally, and are released from the female after hatching.

VOMER   Typically an unpaired bone in the roof of the mouth, occasionally bearing teeth.

YOLK-SAC VIVIPARITY   Reproductive mode in which eggs are fertilized and develop internally, but all nutrition to the embryo is supplied by yolk.

ZYGOPAROUS   Reproductive mode in which eggs are fertilized internally, but are released shortly thereafter and develop externally.

REFERENCES

Acero, P.A., and R.R. Betancur. 2007. Monophyly, affinities, and subfamilial clades of the sea catfishes (Siluriformes: Ariidae). *Ichthyol Explor. Fresh.* 18:133–43.

Ahlstrom, E.H., W.J. Richards, and S.H. Weitzman. 1984. Families Gonostomatidae, Sternoptychidae, and associated stomiiform groups: Development and relationships, pp. 184–98. In *Ontogeny and systematics of fishes.* H.G. Moser, W.J. Richards, D.M. Cohen, M.P. Fahay, A.W. Kendall, Jr., and S.L. Richardson (eds.). American Society of Ichthyologists and Herpetologists, Spec. Publ. No. 1. Allen Press, Lawrence, Kansas.

Ahnesjö, I., and J.F. Craig, eds. 2011. The biology of Syngnathidae: Pipefishes, seadragons and seahorses. *J. Fish Biol.* 78:1597–602.

Albert, J.S. 2001. Species diversity and phylogenetic systematics of American knifefishes (Gymnotiformes, Teleostei). *Misc. Publ. Mus. Zool. Univ. Mich.* 190:1–127.

Albert, J.S., and W.G.R. Crampton. 2005. Diversity and phylogeny of Neotropical electric fishes, pp. 360–409. In *Electroreception.* T.H. Bullock, C.D. Hopkins, A.N. Popper, and R.R. Fay (eds.). Springer, New York.

Albert, J.S., M.J. Lanoo, and T. Turi. 1998. Testing hypotheses of neural evolution in gymnotiform electric fishes using phylogenetic character data. *Evolution* 52:1760–80.

Albins, M.A., and M.A. Hixon. 2011. Worst case scenario: Potential long-term effects of invasive predatory lionfish (*Pterois volitans*) on Atlantic and Caribbean coral-reef communities. *Env. Biol. Fishes* 96:1151–57.

Alfaro, M.E., F. Santini, C. Brock, H. Alamillo, A. Dornburg, D.L. Rabosky, G. Carnevalef, and L.J. Harmong. 2009. Nine exceptional radiations plus high turnover explain species diversity in jawed vertebrates. *Proc. Natl. Acad. Sci.* 106:13410–14.

Allen, G.R. 1985. *Snappers of the world: An annotated and illustrated catalogue of lutjanid species known to date.* FAO species catalogue. *FAO Fish. Synop.* 125, Vol. 6. FAO, Rome. 208 p.

———. 1991. *Damselfishes of the world.* Melle: Mergus. 271 p.

Allen, G.R., R. Steene, and M. Allen. 1998. *A guide to angelfishes and butterflyfishes.* Odyssey, Perth. 250 p.

Allendorf, F.W., and G.H. Thorgaard. 1984. Tetraploidy and the evolution of salmonid fishes, pp. 1–53. In *Evolutionary genetics of fishes.* B.J. Turner (ed.). Plenum Press, New York.

Alva-Campbell, Y., S.R. Floeter, D.R. Robertson, D.R. Bellwood, and G. Bernardi. 2010. Molecular phylogenetics and evolution of *Holacanthus* angelfishes (Pomacanthidae). *Mol. Phylogen. Evol.* 56:456–61.

Amemiya, C.T., J. Alföldi, A.P. Lee, S. Fan, H. Philippe, I. MacCallum, I. Braasch, et al. 2013. The African coelacanth genome provides insights into tetrapod evolution. *Nature* 496:311–16.

Anderson, M.E. 1994. Systematics and osteology of the Zoarcidae (Teleostei: Perciformes). *Ichthyol. Bull. J. L. B. Smith Inst. Ichthyol.* 60:1–120.

Anderson, M.E., and V.V. Fedorov. 2004. Family Zoarcidae Swainson 1839: Eelpouts. *Calif. Acad. Sci. Annotated Checklists of Fishes* 34. http://research.calacademy.org/ichthyology/checklists.

Andriashev, A.P. 1986. *Review of the snailfish genus Paraliparis (Scorpaeniformes: Liparididae) of the Southern Ocean.* Theses Zoologicae 7. Koeltz, Koenigstein. 204 p.

———. 2003. Liparid fishes (Liparidae, Scorpaeniformes) of the Southern Ocean and adjacent waters. *Biological Results of the Russian Antarctic Expeditions 9. Expl. Fauna Seas* (53) 61:1–476.

Arai, R., and K. Kato. 2003. Gross morphology and evolution of the lateral line system and infraorbital bones in bitterlings (Cyprinidae, Acheilognathinae): With an overview of the lateral line system in the family Cyprinidae. *University Museum, University of Tokyo, Bull.* 40:1–42.

Arnold, R.J., and T.W. Pietsch. 2012. Evolutionary history of frogfishes (Teleostei: Lophiiformes: Antennariidae): A molecular approach. *Mol. Phylogen. Evol.* 62:117–29.

Arratia, G. 1997. Basal teleosts and teleostean phylogeny. *Paleo. Ichthyol.* 7:5–168.

———. 1999. The monophyly of Teleostei and stem-group teleosts. Consensus and disagreements, 265–334. In *Mesozoic fishes 2: Systematics and fossil record.* G. Arratia, and H.-P. Schultze (eds.). Verlag Dr. Friedrich Pfeil, Munchen.

———. 2001. The sister-group of Teleostei: Consensus and disagreements. *J. Vertebr. Paleontol.* 21:767–73.

Asaoka, R., N. Masanori, and K. Sasaki. 2012. The innervation and adaptive significance of extensively distributed neuromasts in *Glossogobius olivaceus* (Perciformes: Gobiidae). *Ichthyol. Res.* 59:143–50.

Aschliman, N. C, M. Nishida, M. Miya, J.G. Inoue, K.M. Rosana, and G.J.P. Naylor. 2012. Body plan convergence in the evolution of skates and rays (Chondrichthyes: Batoidea). *Mol. Phylogen. Evol.* 63:28–42.

Ault, J. (ed.). 2008. *Biology and management of the world tarpon and bonefish fisheries.* Vol. 9. CRC Press, Boca Raton, Florida. 472 p.

Azevedo, M.F.C., C. Oliveira, B.G. Pardo, P. Martínez, and F. Foresti. 2008. Phylogenetic analysis of the order Pleuronectiformes (Teleostei) based on sequences of 12S and 16S mitochondrial genes. *Gen. Mol. Biol.* 31:284–92.

Bailey, R.M., and D.A. Etnier. 1988. Comments on the subgenera of darters (Percidae) with descriptions of two new species of *Etheostoma (Ulocentra)* from southeastern United States. *Misc. Publ. Mus. Zool. Univ. Mich.* 175:1–48.

Baird, R.C. 1971. The systematics, distribution, and zoogeography of the marine hatchetfishes (Family Sternoptychidae). *Bull. Mus. Comp. Zool.* 142:1–128.

Bakke, I., and S.D. Johansen. 2005. Molecular phylogenetics of Gadidae and related Gadiformes based on mitochondrial DNA sequences. *Mar. Biotech.* 7:61–69.

Baldwin, C.C., and G.D. Johnson. 1993. Phylogeny of the Epinephelinae (Teleostei: Serranidae). *Bull. Mar Sci.* 52:240–83.

———. 1996. Interrelationships of Aulopiformes, pp. 355–404. In *Interrelationships of fishes.* M.L.J. Stiassny, L.R. Parenti, and G.D. Johnson (eds.). Academic Press, San Diego, California.

Balushkin, A.V. 2000. Morphology, classification, and evolution of notothenioid fishes of the Southern Ocean (Notothenioidei, Perciformes). *J. Ichthyol.* 40 (Suppl 1):S74–S109.

Balz, D. M. 1983. Life history variation among female surfperches (Perciformes: Embiotocidae). *Env. Biol. Fishes* 10:159–71.

Banford, H. M., and B. B. Collette. 2001. A new species of halfbeak, *Hyporhamphus naos* (Beloniformes: Hemirhamphidae), from the tropical eastern Pacific. *Rev. Biol. Trop.* 49 (Supl. 1):39–49.

Barlow, G. W. 1984. Patterns of monogamy among teleost fishes. *Arch. Fisch. Wiss., Beih.* 1:75–123.

———. 2000. *Cichlid fishes: Nature's grand experiment in evolution.* Perseus Publishing, Cambridge, Massachusetts. 333 p.

Barnett, A., and D. R. Bellwood. 2005. Sexual dimorphism in the buccal cavity of paternal mouth-brooding cardinalfishes (Pisces: Apogonidae). *Mar. Biol.* 148:205–12.

Bass, A. L., H. Dewar, T. Thys, J. T. Streelman, and S. A. Karl. 2005. Evolutionary divergence among lineages of the ocean sunfish family, Molidae (Tetraodontiformes). *Mar. Biol.* 148:405–14.

Bekker, V. E. 1983. *Myctophid fishes of the world ocean.* Science Publisher, Moscow. 248 p.

Bell, M. A. 1974. Reduction and loss of the pelvic girdle in *Gasterosteus* (Pisces): A case of parallel evolution. *Nat. Hist. Mus. Los Angeles Co. Contr. Sci.* 257:1–36.

Bell, M. A., and S. A. Foster, eds. 1994. *The evolutionary biology of the threespine stickleback.* Oxford University Press, Oxford. 571 p.

Bellwood, D. R., L. van Herwerden, and N. Konow. 2004. Evolution and biogeography of marine angelfishes (Pisces: Pomacanthidae). *Mol. Phylogen. Evol.* 33:140–55.

Bemis, W. E., W. W. Burggren, and N. E. Kemp, eds. 1987. *The biology and evolution of lungfishes.* Alan R. Liss, New York. 383 p.

Bemis, W. E., E. K. Findeis, and L. Grande. 1997. An overview of Acipenseriformes. *Env. Biol. Fishes* 48:25–71.

Bergstrom, C. A. 2007. Morphological evidence of correlational selection and ecological segregation between dextral and sinistral forms in a polymorphic flatfish, *Platichthys stellatus*. *J. European Soc. Evol. Biol.* 20:1104–14.

Bernal, D., C. Sepulveda, and J. B. Graham. 2001. Water-tunnel studies of heat balance in swimming mako sharks. *J. Exp. Biol.* 204:4043–54.

Bernardi, G. 1997. Molecular phylogeny of the Fundulidae (Teleostei, Cyprinodontiformes) based on the cytochrome b gene, pp. 189–97. In *Molecular systematics of fishes.* T. D. Kocher and C. A. Stepien (eds.). Academic Press, San Diego, California.

Bernardi, G., and G. Bucciarelli. 1999. Molecular phylogeny and speciation of the surfperches (Embiotocidae, Perciformes). *Mol. Phylogen. Evol.* 13:77–81.

Bernardi, G., and D. A. Powers. 1995. Phylogenetic relationships among nine species from the genus *Fundulus* (Cyprinodontiformes, Fundulidae) inferred from sequences of the cytochrome-b gene. *Copeia* 1995:467–73.

Berquist, R. M., K. M. Gledhill, M. W. Peterson, A. H. Doan, G. T. Baxter, K. E. Yopak, N. Kang, et al. 2012. The digital fish library: Using MRI to digitize, database, and document the morphological diversity of fish. *PLoS ONE* 7(4):e34499.

Berra, T. M. 2001. *Freshwater fish distribution.* Academic Press, San Diego, California. 604 p.

Berra, T. M., and B. J. Pusey. 1997. Threatened fishes of the world: *Lepidogalaxias salamandroides* Mees, 1961 (Lepidogalaxiidae). *Env. Biol. Fishes* 50:201–2.

Bertelsen, E. 1951. The ceratioid fishes: Ontogeny, taxonomy, distribution and biology. *Dana Rept.* 39. 276 p.

Betancur, R. R. 2009. Molecular phylogenetics and evolutionary history of ariid catfishes revisited: A comprehensive sampling. *BMC Evolutionary Biology* 9:175.

Betancur, R. R., R. E. Broughton, E. O. Wiley, K. Carpenter, J. A. López, C. Li, N. I. Holcroft, et al. 2013. The tree of life and a new classification of bony fishes. *PLOS Currents: Tree of Life*. 1st ed. doi: 10.1371/currents.tol.53ba26640dfoccaee75bb165c8c26288.

Betancur, R. R., C. Li, T. A. Munroe, J. A. Ballesteros, and G. Ortí. 2013. Addressing gene tree discordance and non-stationarity to resolve a multi-locus phylogeny of the flatfishes (Teleostei: Pleuronectiformes). *Syst. Biol.* 62:763–85.

Birdsong, R. S., E. O. Murdy, and F. L. Pezold. 1988. A study of the vertebral column and median fin osteology in gobioid fishes with comments on gobioid relationships. *Bull. Mar. Sci.* 42:174–214.

Birkhead, W. S. 1972. Toxicity of stings of ariid and ictalurid catfishes. *Copeia* 1972:790–807.

Birstein, V. J., P. Dloukakis, and R. DeSalle. 2002. Molecular phylogeny of Acipenseridae: Non-monophyly of Scaphirhynchinae. *Copeia* 2002:287–301.

Block, B. A., S. L. H. Teo, A. Walli, et al. 2005. Electronic tagging and population structure of Atlantic bluefin tuna. *Nature* 434:1121–27.

Böhlke, E. B. 1989. Leptocephali, pp. 657–1055. In *Fishes of the western North Atlantic*. E. B. Böhlke (ed.). Part 9. Vol. 2. Sears Foundation for Marine Research, Memoir (Yale University), New Haven, Connecticut.

Böhlke, E. B., and J. E. McCosker. 2001. The moray eels of Australia and New Zealand, with description of two new species (Anguilliformes: Muraenidae). *Rec. Aust. Mus.* 53:71–102.

Böhlke, E. B., J. E. McCosker, and J. E. Böhlke. 1989. Family Muraenidae, pp. 104–206. In *Fishes of the western North Atlantic*. E. B. Böhlke (ed.). Part 9. Vol. 1. Sears Foundation for Marine Research, Memoir (Yale University), New Haven, Connecticut.

Böhlke, J. E., and C. C. G. Chaplin. 1993. *Fishes of the Bahamas and adjacent tropical waters*. 2nd ed. University of Texas Press, Austin, Texas. 771 p.

Bolin, R. L. 1944. A review of the marine cottid fishes of California. *Stanford Ichthyol. Bull.* 3:1–135.
———. 1947. The evolution of the marine Cottidae of California with a discussion of the genus as a systematic category. *Stanford Ichthyol. Bull.* 3:153–68.

Bond, C. E. 1996. *Biology of fishes*. 2nd ed. Harcourt Brace, Orlando, Florida. 750 p.

Bone, Q. and R. Moore. 2008. *Biology of fishes*. 3rd ed. Taylor and Francis Group, New York. 478 p.

Borden, W. C. 1998. Phylogeny of the unicornfishes (*Naso*, Acanthuridae) based on soft anatomy. *Copeia* 1998:104–13.

Bortone, S. A. 1977. Revision of the sea basses of the genus *Diplectrum* (Pisces: Serranidae). *NOAA Tech. Rep. NMFS Circ.* 404:1–49.

Boschung, H. T., Jr., and R. L. Mayden. 2004. *Fishes of Alabama*. Smithsonian Books, Washington, DC. 736 p.

Bowne, P. S. 1994. Systematics and morphology of the Gasterosteiformes, pp. 28–60. In *Evolutionary biology of the threespine stickleback*. M. A. Bell and S. A. Foster (eds.). Oxford University Press, Oxford.

Bradbury, M. G. 1967. The genera of batfishes (family Ogcocephalidae). *Copeia* 1967:399–422.
———. 2003. Family Ogcocephalidae Jordan 1895: Batfishes. *Calif. Acad. Sci. Annotated Checklists of Fishes* 17, http://research.calacademy.org/ichthyology/checklists.

Branch, T. A. 2001. A review of orange roughy *Hoplostethus atlanticus* fisheries, estimation methods, biology and stock structure. *South African J. Mar. Sci.* 23:181–203.

Braun, C. B., and T. Grande. 2008. Evolution of peripheral mechanisms for the enhancement of sound reception. pp. 99–144. In *Fish bioacoustics*. J. F. Webb, R. R. Fay, and A. N. Popper (eds.). Springer Science Business Media, New York.

Brenner, S., G. Elgar, R. Sanford, A. Macrae, B. Venkatesh, and S. Aparicio. 1993. Characterization of the pufferfish (*Fugu*) genome as a compact model vertebrate genome. *Nature* 366:265–68.

Brewster, B. 1987. Eye migration and cranial development during flatfish metamorphosis: A reappraisal (Teleostei: Pleuronectiformes). *J. Fish Biol.* 31:805–33.

Briggs, J. C. 1955. A monograph of the clingfishes (Order Xenopterygii). *Stanford Ichthyol. Bull.* 6:1–224.

Britz, R. 1994. Ontogenetic features of *Luciocephalus* (Perciformes, Anabantoidei) with a revised hypothesis of anabantoid intrarelationships. *Zool. J. Linn. Soc.* 112:491–508.

———. 2004. Egg structure and larval development of *Pantodon buchholzi* (Teleostei: Osteoglossomorpha), with a review of data on reproduction and early life history in other osteoglossomorphs. *Ichthyol. Explor. Fresh.* 15:209–24.

Britz, R., and M. Hoffmann. 2006. Ontogeny and homology of the claustra in Otophysan Ostariophysi (Teleostei). *J. Morph.* 267:909–23.

Britz, R., and G. D. Johnson. 2002. "Paradox Lost": Skeletal ontogeny of *Indostomus paradoxus* and its significance for the phylogenetic relationships of Indostomidae (Teleostei, Gasterosteiformes). *Amer. Mus. Novit.* 3383:1–43.

———. 2003. On the homology of the posteriormost gill arch in polypterids (Cladistia, Actinopterygii). *Zool. J. Linn. Soc.* 138:495–503.

———. 2005. Occipito-vertebral fusion in ocean sunfishes (Teleostei: Tetraodontiformes: Molidae) and its phylogenetic implications. *J. Morph.* 266:74–79.

———. 2012. Ontogeny and homology of the skeletal elements that form the sucking disc of remoras (Teleostei, Echeneoidei, Echeneidae). *J. Morph.* 273:1353–66.

Britz, R., and M. Kottelat. 2003. Descriptive osteology of the family Chaudhuriidae (Teleostei, Synbranchiformes, Mastacembeloidei). *Amer. Mus. Novit.* 3418:1–62.

Bruton, M. N. 1995. Threatened fishes of the world: *Latimeria chalumnae* Smith, 1939 (Latimeriidae). *Env. Biol. Fishes* 43:104.

Bufalino, A. P., and R. L. Mayden. 2010. Phylogenetic relationships of North American phoxinins (Actinopterygii: Cypriniformes: Leuciscidae) as inferred from S7 nuclear DNA sequences. *Mol. Phylogen. Evol.* 55:143–52.

Burgess, W. E. 1974. Evidence for elevation to family status of the angelfishes (Pomacanthidae), previously considered to be a subfamily of the butterflyfishes (Chaetodontidae). *Pac. Sci.* 28:57–71.

———. 1978. *Butterflyfishes of the world.* Tropical Fish Hobbyist, Neptune City, New Jersey. 832 p.

Burkenroad, M. D. 1930. Sound production in the Haemulidae. *Copeia* 1930:17–18.

Burnett, K. G., L. J. Bain, W. S. Baldwin, G. V. Callard, S. Cohen, R. T. Di Giulio, D. H. Evans, et al. 2007. *Fundulus* as the premier teleost model in environmental biology: Opportunities for new insights using genomics. *Comp. Biochem. Physiol.* 2:257–86.

Butler, J. L., M. S. Love, and T. E. Laidig. 2012. *A guide to the rockfishes, thornyheads, and scorpionfishes of the northeast Pacific.* University of California Press, Berkeley, California. 184 p.

Buxton, C. D., and P. A. Garratt. 1990. Alternative reproductive styles in seabreams (Pisces: Sparidae). *Env. Biol. Fishes.* 28:113–24.

Calcagnotto, D., S. A. Schaefer, and R. Desall. 2005. Relationships among characiform fishes inferred from analysis of nuclear and mitochondrial gene sequences. *Mol. Phylogen. Evol.* 36:135–53.

Carpenter, K. E. (ed.). 2003. *The living marine resources of the western central Atlantic.* FAO species identification guide for fishery purposes, and American Society of Ichthyologists and Herpetologists. Spec. Publ. No. 5. Vols. 1–3. FAO, Rome. 2,127 p.

Carpenter, K. E., and G. D. Johnson. 2002. A phylogeny of sparoid fishes (Perciformes, Percoidei) based on morphology. *Ichthyol. Res.* 49:114–27.

Carpenter, K. E., and V. H. Niem, eds. 1998–2001. *The living marine resources of the western central Pacific.* FAO species identification guide for fishery purposes. Vols. 1–6. FAO, Rome. 4,218 p.

Carr, C. E., and L. Maler. 1986. Electroreception in gymnotiform fish: Central anatomy and physiology, pp. 319–73. In *Electroreception*. T. H. Bullock, and W. Heiligenberg (eds.). Wiley, New York.

Carrier, J. C., J. A. Musick, and M. R. Heithaus, eds. 2012. *Biology of sharks and their relatives*. 2nd ed. CRC Press, Boca Raton, Florida. 666 p.

Castle, P. H. J., and J. E. Randall. 1999. Revision of Indo-Pacific garden eels (Congridae: Heterocongrinae), with descriptions of five new species. *Indo-Pacific Fishes* 30:1–52.

Cavender, T. M., and M. M. Coburn. 1992. Phylogenetic relationships of North American Cyprinidae, pp. 293–327. In *Systematics, historical ecology, and North American freshwater fishes*. R. L. Mayden (ed.). Stanford University Press, Stanford, California.

Ceccarelli, D. M., G. P. Jones, and L. J. McCook. 2001. Territorial damselfishes as determinants of the structure of benthic communities on coral reefs. *Oceanogr. Mar. Biol.* 39:355–89.

Chakrabarty, P. 2004. Cichlid biogeography: Comment and review. *Fish and Fish.* 5:97–119.

Chao, L. N. 1978. A basis for classifying western Atlantic Sciaenidae (Teleostei: Perciformes). *NOAA Tech. Rep. NMFS Circ.* 415:1–64.

———. 1986. A synopsis on zoogeography of the Sciaenidae, pp. 570–89. In *Indo-Pacific fish biology: Proc. Second Int'l Conf. Indo-Pacific Fishes, July-August 1985*. T. Uyeno, R. Arai, T. Taniuchi and K. Matsuura (eds.). Ichthyological Society of Japan, Tokyo.

Chapleau, F. 1988. Comparative osteology and intergeneric relationships of the tongue soles (Pisces; Pleuronectiformes; Cynoglossidae). *Can. J. Zool.* 66:1214–32.

———. 1993. Pleuronectiform relationships: A cladistic reassessment. *Bull. Mar. Sci.* 52:516–40.

Chardon, M., E. Parmentier, and P. Vandewalle. 2003. Morphology, development and evolution of the Weberian apparatus in catfish, pp. 71–120. In *Catfishes*. G. Arratia, B. G. Kappor, M. Chardon, and R. Diogo (eds.). Science Publishers, Enfield, New Hampshire.

Chen, J.-N., J. A. López, S. Lavoué, M. Miya, and W.-J. Chen. 2014. Phylogeny of the Elopomorpha (Teleostei): Evidence from six nuclear and mitochondrial markers. *Mol. Phylogen. Evol.* 70:152–61.

Chen, W.-J., C. Bonillo, and G. Lecointre. 2003. Repeatability of clades as a criterion of reliability: A case study for molecular phylogeny of Acanthomorpha (Teleostei) with larger number of taxa. *Mol. Phylogen. Evol.* 26:262–88.

Chen, W.-J., and R. L. Mayden. 2009. Molecular systematics of the Cyprinoidea (Teleostei: Cypriniformes), the world's largest clade of freshwater fishes: Further evidence from six nuclear genes. *Mol. Phylogen. Evol.* 52:544–49.

Chen, W.-J., R. Ruiz-Carus, and G. Orti. 2007. Relationships among four genera of mojarras (Teleostei: Perciformes: Gerreidae) from the western Atlantic and their tentative placement among percomorph fishes. *J. Fish Biol.* 70 (Suppl. B):202–18.

Chen, X. Y., and G. Arratia. 1994. The olfactory organ of Acipenseriformes and comparison with other actinopterygians: Pattern of diversity. *J. Morph.* 222:241–67.

Chernova, N. V., D. L. Stein, and A. P. Andriashev. 2004. Family Liparidae Scopoli 1777: Snailfishes. *Calif. Acad. Sci. Annotated Checklists of Fishes* 31. http://research.calacademy.org/ichthyology/checklists.

Chiba, S. N., Y. Iwatsuki, T. Yoshino, and N. Hanzawa. 2009. Comprehensive phylogeny of the family Sparidae (Perciformes: Teleostei) inferred from mitochondrial gene analyses. *Genes Genet. Syst. Shizuoka* 84:153–70.

Choudhury, A., and T. A. Dick. 1998. The historical biogeography of sturgeons (Osteichthyes: Acipenseridae): A synthesis of phylogenetics, paleontology and palaeogeography. *J. Biogeogr.* 25:623–40.

Claeson, K. M., and M. N. Dean. 2011. Cartilaginous fish skeletal anatomy, pp. 419–27. In *Encyclopedia of fish physiology*. A. P. Farrell (ed.). Vol. 1. Academic Press, San Diego, California.

Clarke, R. D. 1996. Population shifts in two competing fish species on a degrading coral reef. *Mar. Ecol. Prog. Ser.* 137:51–58.

Clements K. D., M. E. Alfaro, J. L. Fessler, and M. W. Westneat. 2004. Relationships of the temperate Australian fish tribe Odacini (Perciformes; Teleostei). *Mol. Phylogen. Evol.* 32:575–87.

Cloutier, R., and P. E. Ahlberg. 1996. Morphology, characters, and interrelationships of basal sarcopterygians, pp. 445–79. In *Interrelationships of fishes*. M. L. J. Stiassny, L. R. Parenti, and G. D. Johnson (eds.). Academic Press, San Diego, California.

Cohen, D. M. (ed.). 1989. Papers on the systematics of gadiform fishes. *Nat. Hist. Mus. Los Angeles Co., Sci. Ser.* 32:1–262.

Cohen, D. M., T. Inada, T. Iwamoto, and N. Scialabba. 1990. *Gadiform fishes of the world (Order Gadiformes)*. FAO species catalogue. *FAO Fish. Synop.* 125, Vol. 10. FAO, Rome. 442 p.

Cohen, D. M., and J. G. Nielsen. 1978. Guide to the identification of genera of the fish order Ophidiiformes with a tentative classification of the order. *NOAA Tech. Rep., NMFS Circ.* 417:1–72.

Cole, K. 2010. Gonad morphology in hermaphroditic gobies, pp. 117–64. In *Reproduction and sexuality in marine fishes: Patterns and processes*. K. S. Cole (ed.). University of California Press, Berkeley, California.

Colin, P. L. 1973. Burrowing behavior of the yellowhead jawfish, *Opistognathus aurifrons. Copeia* 1973: 84–90.

———. 1978. *Serranus incisus,* new species from the Caribbean Sea (Pisces: Serranidae). *Proc. Biol. Soc. Wash.* 91:191–96.

Collette, B. B. 1983. *Scombrids of the world: An annotated and illustrated catalogue of tunas, mackerels, bonitos and related species known to date.* FAO species catalogue. *FAO Fish. Synop.* 125, Vol. 2. FAO, Rome. 137 p.

———. 2003a. Family Belonidae Bonaparte 1832: Needlefishes. *Calif. Acad. Sci. Annotated Checklists of Fishes* 16. http://research.calacademy.org/ichthyology/checklists.

———. 2003b. Family Scombridae Rafinesque 1815: Mackerels, tunas, and bonitos. *Calif. Acad. Sci. Annotated Checklists of Fishes* 19. http://research.calacademy.org/ichthyology/checklists.

———. 2004. Family Hemiramphidae Gill 1859: Halfbeaks. *Calif. Acad. Sci. Annotated Checklists of Fishes* 22. http://research.calacademy.org/ichthyology/checklists.

Collette, B. B., and P. Banarescu. 1977. Systematics and zoogeography of fishes of the family Percidae. *J. Fish. Res. Bd. Can.* 34:1450–63.

Collette, B. B., K. E. Carpenter, B. A. Polidoro, M. J. Juan-Jordá, A. Boustany, D. J. Die, C. Elfes, et al. 2011. High-value and long-lived: Double jeapordy for threatened tunas, mackerels and billfishes. *Science* 333:291–92.

Collette, B. B., and L. N. Chao. 1975. Systematics and morphology of the bonitos (*Sarda*) and their relatives (Scombridae, Sardini). *Fish. Bull.* 73:516–625.

Collette, B. B., T. Potthoff, W. J. Richards, S. Ueyanagi, J. L. Russo, and Y. Nishikawa. 1984. Scombroidei: Development and relationships, pp. 591–620. In *Ontogeny and systematics of fishes.* H. G. Moser, W. J. Richards, D. M. Cohen, M. P. Fahay, A. W. Kendall, Jr., and S. L. Richardson (eds). American Society of Ichthyologists and Herpetologists, Spec. Public. No. 1. Allen Press, Lawrence, Kansas.

Collette, B. B., C. Reeb, and B. A. Block. 2001. Systematics of the tunas and mackerels (Scombridae), pp. 1–33. In *Tuna: Physiology, ecology, and evolution.* B. A. Block and E. D. Stevens (eds.). Fish Physiology 19. Academic Press, San Diego, California.

Collette, B. B., and J. L. Russo. 1981. A revision of the scaly toadfishes, genus *Batrachoides,* with descriptions of two new species. *Bull. Mar. Sci.* 31:197–233.

Compagno, L. J. V. 1984a. *Sharks of the world: An annotated and illustrated catalogue of shark species known to date: Part 1: Hexanchiformes to Lamniformes.* FAO Species Catalogue. *FAO Fish. Synop.* 125, Vol. 4, Part 1:1–249. FAO, Rome.

———. 1984b. *Sharks of the world: An annotated and illustrated catalogue of shark species known to date, Part 2: Carcharhiniformes.* FAO Species Catalogue. *FAO Fish. Synop.* 125, Vol. 4, Part 2:251–655. FAO, Rome.

———. 1988. *Sharks of the order Carcharhiniformes.* Princeton University Press, Princeton, New Jersey. 486 p.

———. 1990. Relationships of the megamouth shark, *Megachasma pelagios* (Lamniformes: Megachasmidae), with comments on its feeding habits. *NOAA Technical Report NMFS* 90:357–79.

———. 2001. *Sharks of the world: An annotated and illustrated catalogue of shark species known to date.* FAO species catalogue for fishery purposes. No. 1, Vol. 2. *Bullhead, mackerel and carpet sharks (Heterodontiformes, Lamniformes and Orectolobiformes).* FAO, Rome. 269 p.

———. 2005. Checklist of Chondrichthyes, pp. 503–47. In *Reproductive biology and phylogeny of Chondrichthyes: Sharks, batoids and chimaeras.* W. C. Hamlett (ed.). Science Publishers, Enfield, New Hampshire.

Compagno, L. J. V., M. Dando, and S. Fowler. 2005. *Sharks of the world.* Princeton University Press, Princeton, New Jersey. 368 p.

Conover, D. O., and B. E. Kynard. 1981. Environmental sex determination: Interaction of temperature and genotype in a fish. *Science* 213:577–79.

Cooke, S., and D. P. Philipp, eds. 2009. *Centrarchid fishes: Diversity, biology and conservation.* Wiley-Blackwell, West Sussex, UK. 560 p.

Cooper, J. A., and F. Chapleau. 1998. Monophyly and interrelationships of the family Pleuronectidae (Pleuronectiformes), with a revised classification. *Fish. Bull.* 69:686–726.

Cooper, W. J. 2006. *The evolution of the damselfishes: Phylogenetics, biomechanics, and development of a diverse coral reef fish family.* Organismal Biology and Anatomy, University of Chicago, Chicago, Illinois. 213 p.

Cooper, W. J., L. L. Smith, and M. W. Westneat. 2009. Exploring the radiation of a diverse reef fish family: Phylogenetics of the damselfishes (Pomacentridae), with new classifications based on molecular analyses of all genera. *Mol. Phylogen. Evol.* 52:1–16.

Corrigan, S., and L. B. Beheregaray. 2009. A recent shark radiation: Molecular phylogeny, biogeography and speciation of wobbegong sharks (family: Orectolobidae). *Mol. Phylogen. Evol.* 52:205–16.

Costa, W. J. E. M. 1998. Phylogeny and classification of the Cyprinodontidae revisited (Teleostei: Cyprinodontiformes): A reappraisal, pp. 527–60. In *Phylogeny and classification of neotropical fishes.* L. R. Malabarba, R. E. Reis, R. P. Vari, Z. M. S. Lucena, and C. A. S. Lucena (eds.). EDIPUCRS, Porto Alegre, Brazil.

———. 2003. Family Cyprinodontidae (pupfishes), pp. 549–54. In *Checklist of the freshwater fishes of South and Central America.* R. E. Reis, S. O. Kullander, and C. J. Ferraris, Jr (eds.). EDIPUCRS, Porto Alegre, Brazil.

Côté, I. M., and M. C Soares. 2011. Gobies as cleaners, pp. 531–58. In *The biology of gobies.* R. A. Patzner, J. L. Van Tassel, M. Kovačić, and B. G. Kapoor (eds.). Science Publishers, Enfield, New Hampshire.

Coulson, M. W., H. D. Marshall, P. Pepin, and S. M. Carr. 2006. Mitochondrial genomics of gadine fishes: Implications for taxonomy and biogeographic origins from whole-genome data sets. *Genome* 49:1115–30.

Courtenay, W. R., Jr., and J. D. Williams. 2004. Snakeheads (Pisces, Channidae): A biological synopsis and risk assessment. *U.S. Geol. Serv.* Circ. 1251:1–143.

Crabtree, R.E., K.J. Sulak, and J. Musick. 1985. Biology and distribution of species of *Polyacantho-notus* (Pisces: Notacanthiformes) in the western North Atlantic. *Bull. Mar. Sci.* 36:235–48.

Craig, J.F., 2000. *Percid fishes: systematics, ecology and exploitation.* Blackwell Scientific Publications, Oxford. 368 p.

Craig, M.T., and P.A. Hastings. 2007. A molecular phylogeny of the groupers of the subfamily Epinephelinae (Serranidae) with a revised classification of the Epinephelini. *Ichthyol. Res.* 54:1–17.

Craig, M.T., Y.J. Sadovy de Mitcheson, and P.C. Heemstra. 2011. *Groupers of the world: A field and market guide.* NISC, Grahamstown, South Africa. 356 p.

Crain, P.K, and P.B. Moyle. 2011. Biology, history, status and conservation of Sacramento perch, *Archoplites interruptus. San Francisco Estu. Water. Sci.* 9:1–37.

Crespi, B.J., and M.J. Fulton. 2004. Molecular systematics of Salmonidae: Combined nuclear data yields a robust phylogeny. *Mol. Phylogen. Evol.* 31:658–79.

Crossman E.J. 1978. Taxonomy and distribution of North American esocids. *Am. Fish. Soc. Spec. Publ.* 11:13–26.

Crossman, E.J., and J.M. Casselman. 1987. *An annotated bibliography of the pike, Esox lucuis (Osteichthys: Salmoniformes).* Royal Ontario Museum, Toronto. 408 p.

Crow, K.D., Z. Kanamoto, and G. Bernardi. 2004. Molecular phylogeny of the hexagrammid fishes using a multi-locus approach. *Mol. Phylogen. Evol.* 32:986–97.

Crow, K.D., D.A. Powers, and G. Bernardi. 1997. Evidence for multiple maternal contributors in nests of kelp greenling (*Hexagrammos decagrammus,* Hexagrammidae). *Copeia* 1997:9–15.

Cui, Z., Y. Liu, C.P. Li, F. You, and K.H. Chu. 2009. The complete mitochondrial genome of the large yellow croaker, *Larimichthys crocea* (Perciformes, Sciaenidae): Unusual features of its control region and the phylogenetic position of the Sciaenidae. *Gene* 432:33–43.

Daget, J., M. Gayet, F.J. Meunier, and J.-Y. Sire. 2001. Major discoveries on the dermal skeleton of fossil and recent polypteriforms: A review. *Fish and Fish.* 2:113–24.

Dasilao, J.C., Jr., and K. Sasaki. 1998. Phylogeny of the flyingfish family Exocoetidae (Teleostei, Beloniformes). *Ichthyol. Res.* 45:347–53.

Davis, M.P., and C. Felitz. 2010. Estimating divergence times of lizardfishes and their allies (Euteleostei: Aulopiformes) and the timing of deep-sea adaptations. *Mol. Phylogen. Evol.* 57:1194–1208.

Davis, W.P., and R.S. Birdsong. 1973. Coral reef fishes which forage in the water column. *Helgo Htnder wiss. Meeresunters.* 24:292–306.

Davison, P.C., D.M. Checkley, Jr., J.A. Koslow, and J. Barlow. 2013. Carbon export mediated by mesopelagic fishes in the northeast Pacific Ocean. *Prog. Oceanog.* 116:14–30.

Dawson, C.E. 1982. Family Syngnathidae, pp. 1–172. In *Fishes of the western North Atlantic.* J.E. Böhlke et al. (eds.). Part 8, Vol. 1. Sears Foundation for Marine Research, Memoir (Yale University), New Haven, Connecticut.

———. 1985. *Indo-Pacific pipefishes (Red Sea to the Americas).* Gulf Coast Research Laboratory, Ocean Springs, Mississippi. 230 p.

Day, J.J. 2002. Phylogenetic relationships of the Sparidae (Teleostei: Percoidei) and implications for convergent trophic evolution. *Biol. J. Linn. Soc.* 76:269–301.

de Carvalho, M.R. 1996. Higher level elasmobranch phylogeny, basal squaleans, and paraphyly, pp. 35–62. In *Interrelationships of fishes.* M.L.J. Stiassny, L.R. Parenti, and G.D. Johnson (eds.). Academic Press, San Diego, California.

———. 2003. Family Pristidae (sawfishes), pp. 17–21. In *Checklist of the freshwater fishes of South and Central America.* R.E. Reis, S.O. Kullander, and C.J. Ferraris, Jr (eds.). EDIPUCRS, Porto Alegre, Brazil.

de Carvalho, M. R., J. G. Maisey, and L. Grande. 2004. Freshwater stingrays of the Green River Formation of Wyoming (early Eocene), with the description of a new genus and species and an analysis of its phylogenetic relationships (Chondrichthyes: Myliobatiformes). *Bull. Amer. Mus. Nat. Hist.* 284:1–136.

Deckert, G. T., and D. W. Greenfield. 1987. A review of the western Atlantic species of the genera *Diapterus* and *Eugerres* (Pisces: Gerreidae). *Copeia* 1987:182–94.

Depczynski, M., and D. R. Bellwood. 2003. The role of cryptobenthic reef fishes in coral reef trophodynamics. *Mar. Ecol. Prog. Ser.* 256:183–91.

de Pinna, M. C. C. 1996. Teleostean monophyly, pp. 147–62. In *Interrelationships of fishes*. M. L. J. Stiassny, L. R. Parenti, and G. D. Johnson (eds.). Academic Press, San Diego, California.

———. 1998. Phylogenetic relationships of Neotropical Siluriformes (Teleostei: Ostariophysi): Historical overview and synthesis of hypotheses, pp. 279–330. In *Phylogeny and classification of neotropical fishes*. L. R. Malabarba, R. E. Reis, R. P. Vari, Z. M. S. Lucena, and C. A. S. Lucena (eds.). EDIPUCRS, Porto Alegre, Brazil.

de Sylva, D. P. 1975. Barracudas (Pisces: Sphyraenidae) of the Indian Ocean and adjacent seas: A preliminary review of their systematics and ecology. *J. Mar. Biol. Assoc. India* 15:74–94.

———. 1984. Sphyraenoidei: Development and relationships, pp. 534–40. In *Ontogeny and systematics of fishes*. H. G. Moser, W. J. Richards, D. M. Cohen, M. P. Fahay, A. W. Kendall, Jr., and S. L. Richardson (eds.). American Society of Ichthyologists and Herpetologists, Spec. Public. No. 1. Allen Press, Lawrence, Kansas.

DiBattista, J. D., C. Wilcox, M. T. Craig, L. A. Rocha, and B. W. Bowen. 2011. Phylogeography of the Pacific blueline surgeonfish, *Acanthurus nigroris*, reveals high genetic connectivity and a cryptic endemic species in the Hawaiian Archipelago. *J. Mar. Biol.* 2011:1–17.

Didier, D. A. 2004. Phylogeny and classification of extant Holocephali, pp. 115–35. In *Biology of sharks and their relatives*. J. C. Carrier, J. A. Musick, and M. R. Heithaus (eds.). CRC Press, Boca Raton, Florida.

Dillman, C. B., D. E. Bergstrom, D. B. Noltie, T. P. Holtsford, and R. L. Mayden. 2011. Regressive progression, progressive regression or neither? Phylogeny and evolution of the Percopsiformes (Teleostei, Paracanthopterygii). *Zool. Scripta* 40:45–60.

Diogo, R., and Z. Peng. 2010. State of the art of siluriform higher-level phylogeny, pp. 465–515. In *Gonorynchiformes and ostariophysan relationships: A comprehensive review*. T. Grande, F. J. Poyato-Ariza, and R. Diogo (eds.). Science Publishers, Enfield, New Hampshire.

Domeier, M. L. (ed.). 2012. *Global perspectives on the biology and life history of the white shark*. CRC Press, Boca Raton, Florida. 567 p.

Donaldson, T. J. 1990. Reproductive behavior and social organization of some Pacific hawkfishes (Cirrhitidae). *Jap. J. Ichthyol.* 36:439–58.

Dornburg, A., B. Sidlauskas, L. Sorenson, F. Santini, T. Near, and M. E. Alfaro. 2011. The influence of an innovative locomotor strategy on the phenotypic diversification of triggerfishes (Family: Balistidae). *Evolution* 65:1912–26.

Douichi, R., and T. Nakabo. 2006. Molecular phylogeny of the stromateoid fishes (Teleostei: Perciformes) inferred from mitochondrial DNA sequences and compared with morphology-based hypotheses. *Mol. Phylogen. Evol.* 39:111–23.

Douichi, R., T. Sato, and T. Nakabo. 2004. Phylogenetic relationships of the stromateoid fishes (Perciformes). *Ichthyol. Res.* 51:202–12.

Dunn, J. R. 1989. A provisional phylogeny of gadid fishes based on adult and early life-history characters, pp. 209–35. In *Papers on the systematics of gadiform fishes*. D. M. Cohen (ed.). *Nat. Hist. Mus. Los Angeles Co., Sci. Ser.* 32.

Dunn, K. A., J. D. McEachran, and R. L. Honeycutt. 2003. Molecular phylogenetics of myliobati-form fishes (Chondrichthyes: Myliobatiformes), with comments on the effects of missing data on parsimony and likelihood. *Mol. Phylogen. Evol.* 27:259–70.

Durand, J.-D., K. N. Shen, W. J. Chen, B. W. Jamandre, H. Blel, K. Diop, M. Nirchio, et al. 2012. Systematics of the grey mullets (Teleostei: Mugiliformes: Mugilidae): Molecular phylogenetic evidence challenges two centuries of morphology-based taxonomy. *Mol. Phylogen. Evol.* 64:73–92.

Dyer, B. S. 1997. Phylogenetic revision of Atherinopsinae (Teleostei, Atherinopsidae), with com-ments on the systematics of the South American freshwater fish genus *Basilichthys* Girard. *Misc. Publ. Mus. Zool. Univ. Mich.* 185:1–64.

———. 1998. Phylogenetic systematics and historical biogeography of the Neotropical silverside family Atherinopsidae (Teleostei: Atherinformes), pp. 519–36. In *Phylogeny and classification of neotropical fishes*. L. R. Malabarba, R. E. Reis, R. P. Vari, Z. M. S. Lucena, and C. A. S Lucena (eds.). EDIPUCRS, Porto Alegre, Brazil.

Dyer, B. S., and B. Chernoff. 1996. Phylogenetic relationships among atheriniform fishes (Tele-ostei: Atherinomorpha). *Zool. J. Linn. Soc.* 117:1–69.

Eakin, R. R. 1981. Biology of the Antarctic Seas IX: Osteology and relationships of the fishes of the Antarctic family Harpagiferidae (Pisces, Notothenioidei). *Antarc. Res. Ser.* 31:81–147.

Eastman, J. T. 2005. The nature of the diversity of Antarctic fishes. *Polar. Biol.* 28:93–107.

Eastman, J. T., and R. R. Eakin. 2000. An updated species list for notothenioid fish (Perciformes; Notothenioidei), with comments on Antarctic species. *Arch. Fish. Mar. Res.* 48:11–20.

Ebeling, A. W., and W. H. Weed, III. 1973. Order Xenoberyces (Stephanoberyciformes), pp. 397–478. In *Fishes of the western North Atlantic*. D. M. Cohen et al. (eds.). Part 6. Vol. 1. Sears Foundation for Marine Research, Memoir (Yale University), New Haven, Connecticut.

Ebert, D. A. 2003. *Sharks, rays, and chimaeras of California*. University of California Press, Berke-ley, California. 284 p.

Ebert, D. A., and L. J. V. Compagno. 2007. Biodiversity and systematics of skates (Chondrichthyes: Rajiformes: Rajoidei). *Env. Biol. Fishes* 80:111–24.

Ebert, D. A., and J. A. Sulikowski (eds). 2008. *Biology of skates.* Springer, Netherlands. 243 p.

Echelle, A. A., E. W. Carson, A. F. Echelle, R. A. Van Den Bussche, T. E. Dowling, and A. Meyer. 2005. Historical biogeography of the new world pupfish genus *Cyprinodon* (Teleostei: Cyprin-odontidae). *Copeia* 2005:320–39.

Echelle, A. A., A. F. Echelle, and C. D. Crozier. 1983. Evolution of an all-female fish, *Menidia clarkhubbsi* (Atherinidae). *Evolution* 37:772–84.

Endo, H. 2002. Phylogeny of the order Gadiformes (Teleostei, Paracanthopterygii). *Mem. Grad. School Fish. Sci. Hokkaido Univ.* 49:75–149.

Erisman, B. E., M. C. Craig, and P. A. Hastings. 2009. A phylogenetic test of the size-advantage model: Evolutionary changes in mating behavior influence the loss of sex change in a reef fish lineage. *Amer. Nat.* 174:E83–99.

Erisman, B. E., and P. A. Hastings. 2011. Evolutionary transitions in the sexual patterns of fishes: Insights from a phylogenetic analysis of the seabasses (Teleostei: Serranidae). *Copeia* 2011:257–64.

Eschmeyer, W. N. 1969. A systematic review of the scorpionfishes of the Atlantic Ocean (Pisces: Scorpaenidae). *Occas. Pap. Cal. Acad. Sci.* 79:1–143.

———. 1997. A new species of Dactylopteridae (Pisces) from the Philippines and Australia, with a brief synopsis of the family. *Bull. Mar. Sci.* 60:727–38.

Eschmeyer, W. N. (ed.). 2013. Catalog of fishes: Genera, species, references. http://research.calacademy.org/research/ichthyology/catalog/fishcatmain.asp. Accessed on 7 December 2013.

Eschmeyer, W. N., and J. D. Fong. 2013. Species by family/subfamily. http://research.calacademy. org/research/ichthyology/catalog/SpeciesByFamily.asp. Accessed on 7 December 2013.

Eschmeyer, W. N., and E. S. Herald. 1983. *A field guide to Pacific coast fishes of North America.* Houghton Mifflin, Boston, Massachusetts. 336 p.

Evans, J. P., A. Pilastro, and I. Schlupp (eds). 2011. *Ecology and evolution of poeciliid fishes.* University of Chicago Press, Chicago, Illinois. 424 p.

Evseenko, S. A. 2004. Family Pleuronectidae Cuvier 1816: Righteye flounders. *Calif. Acad. Sci. Annotated Checklists of Fishes* 37. http://research.calacademy.org/ichthyology/checklists.

Faircloth, B. C., L. Sorenson, F. Santini, and M. E. Alfaro. 2013. A phylogenomic perspective on the radiation of ray-finned fishes based upon targeted sequencing of ultraconserved elements (UCEs). *PLoS ONE* 8:e65923.

FAO Fisheries and Aquaculture Department. 2012. *The state of world fisheries and aquaculture.* FAO, Rome. 209 p.

Farrell, A. P. (ed.). 2011. *Encyclopedia of fish physiology: From genome to environment.* Vol. 1. *The senses, supporting tissues, reproduction, and behavior.* Academic Press, San Diego, California. 789 p.

Fernholm, B. 1998. Hagfish systematics, pp. 33–44. In *The biology of hagfishes.* J. M. Jorgensen, J. P. Lomholt, R. E. Weber, and H. Malte (eds.). Chapman and Hall, London.

Ferraris, C. J., Jr. 2007. Checklist of catfishes, recent and fossil (Osteichthyes: Siluriformes), and catalogue of siluriform primary types. *Zootaxa* 1418:1–628.

Ferraris, C. J., Jr., and M. C. C. de Pinna. 1999. Higher-level names for catfishes (Actinopterygii: Ostariophysi: Siluriformes). *Proc. Cal. Acad. Sci.* 51:1–17.

Ferris, S. D., and G. S. Whitt. 1978. Phylogeny of tetraploid catostomid fishes based on the loss of duplicate gene expression. *Syst. Zool.* 27:189–206.

Fessler, J. L., and M. W. Westneat. 2007. Molecular phylogenetics of the butterflyfishes (Chaetodontidae): Taxonomy and biogeography of a global coral reef fish family. *Mol. Phylogen. Evol.* 45:50–68.

Fink, S. V., and W. L. Fink. 1981. Interrelationships of the ostariophysan fishes (Pisces, Teleostei). *Zool. J. Linn. Soc.* 72:297–353.

———. 1996. Interrelationships of ostariophysan fishes (Teleostei), pp. 405–26. In *Interrelationships of fishes.* M. L. J. Stiassny, L. R. Parenti, and G. D. Johnson (eds.). Academic Press, San Diego, California.

Fink, W. L. 1985. Phylogenetic interrelationships of the stomiid fishes (Teleostei: Stomiiformes). *Misc. Publ. Mus. Zool. Univ. Mich.* 171:1–127.

Fink, W. L., and S. H. Weitzman. 1982. Relationships of the stomiiform fishes (Teleostei), with a redescription of *Diplophos. Bull. Mus. Comp. Zool.* 150:31–93.

Finnerty, J. R., and B. A. Block. 1995. Evolution of cytochrome b in the Scombroidei (Teleostei): Molecular insights into billfish (Istiophoridae and Xiphiidae) relationships. *Fish. Bull.* 93:78–96.

Fishelson, L., C. C. Baldwin, and P. A. Hastings. 2013. Gonad morphology, gametogenesis, and reproductive modes in fishes of the tribe Starksiini (Teleostei, Blenniiformes). *J. Morph.* 274:496–511.

Fischer, W., F. Krupp, W. Schneider, C. Sommer, K. E. Carpenter, and V. H. Niem (eds.). 1995. *Guia FAO para la identificacion de especies para los fines de la pesca: Pacifico central-oriental.* Vols. 1–3. FAO, Rome. 1,813 p.

Forey, P. L. 1973. A revision of the elopiform fishes, fossil and recent. *Bull. Brit. Mus. Nat. Hist. (Geol.) Suppl.* 10:1–222.

———. 1980. *Latimeria:* A paradoxical fish. *Proc. Royal Soc. London* B 208:369–84.

————. 1991. *Latimeria* and its pedigree. *Env. Biol. Fishes* 32:75–97.

————. 1995. Agnathans recent and fossil, and the origin of jawed vertebrates. *Rev. Fish Biol. Fish.* 5:267–303.

————. 1998. *History of the coelacanth fishes.* London: Chapman and Hall. 419 p.

Forey, P. L., D. T. J. Littlewood, P. Ritchie, and A. Meyer. 1996. Interrelationships of elopomorph fishes, pp. 175–91. In *Interrelationships of fishes.* M. L. J. Stiassny, L. R. Parenti, and G. D. Johnson (eds.). Academic Press, San Diego, California.

Foster, S. A. 1985. Size-dependent territory defense by a damselfish, *Stegastes dorsopunicans,* a determinant of resource use by group-foraging surgeonfishes. *Oecologia* 67:499–505.

Foster, S. J., and A. C. J. Vincent. 2004. Life history and ecology of seahorses: Implications for conservation and management. *J. Fish Biol.* 65:1–61.

Fouts, W. R., and D. R. Nelson. 1999. Prey capture by the Pacific angel shark, *Squatina californica:* Visually mediated strikes and ambush-site characteristics. *Copeia* 1999:304–12.

Fraser, T. H. 1972. Comparative osteology of the shallow water cardinal fishes (Perciformes: Apogonidae) with reference to the systematics and evolution of the family. *Ichthyol. Bull. J. L. B. Smith Inst. Ichthyol.* 34:1–105.

————. 2005. A review of the species in the *Apogon fasciatus* group with a description of a new species of cardinalfish from the Indo-West Pacific (Perciformes: Apogonidae). *Zootaxa* 924:1–30.

————. 2013. A new genus of cardinalfish (Apogonidae: Percomorpha), redescription of *Archamia* and resemblances and relationships with *Kurtis* (Kurtidae: Percomorpha). *Zootaxa* 3714:1–63.

Fraser-Brunner, A. 1951. The ocean sunfishes (family Molidae). *Bull. Brit. Mus. Nat. Hist. (Zool.)* 1:89–121.

Freihofer, W. C. 1978. Cranial nerves of the percoid fish *Polycentrus schomburgkii* (family Nandidae), a contribution to the morphology and classification of the order Perciformes. *Occas. Pap. Cal. Acad. Sci.* 128:1–78.

Fricke, R. 2009. Systematics of the Tripterygiidae (triplefins), pp. 31–67. In *The biology of blennies.* R. A. Patzner, E. J. Gonçalves, P. A. Hastings, and B. G. Kapoor (eds.). Science Publishers, Enfield, New Hampshire.

Friedman, M. 2008. The evolution of flatfish asymmetry. *Nature* 454:209–12.

Friedman, M., Z. Johanson, R. C. Harrington, T. J. Near, and M. R. Graham. 2013. An early fossil remora (Echeneoidea) reveals the evolutionary assembly of the adhesion disc. *Proc. Royal Soc. B* 280:20131200.

Fritzsche, R. A. 1980. Revision of the eastern Pacific Syngnathidae (Pisces: Syngnathiformes), including both recent and fossil forms. *Proc. Cal. Acad. Sci.* 42:181–227.

Froese, R., and D. Pauly. 2000. *FishBase 2000: Concepts, design and data sources.* ICLARM, Los Baños, Laguna, Philippines.

Fryer, G., and T. D. Iles. 1972. *The cichlid fishes of the Great Lakes of Africa.* Oliver and Boyd, Edinburgh. 641 p.

Fujita, K. 1990. *The caudal skeleton of teleostean fishes.* Tokai University Press, Tokyo. 897 p.

Fukuda, E., M. Nakae, R. Asaoka, K. Sasaki. 2010. Branching patterns of trunk lateral line nerves in Pleuronectiformes: Uniformity and diversity. *Ichthyol. Res.* 57:148–60.

Fuller, R. C., K. E. Mcghee, and M. Schrader. 2007. Speciation in killifish and the role of salt tolerance. *J. Evol. Biol.* 20:1962–75.

Gans, C. 1987. The neural crest: A spectacular invention, pp. 261–379. In *Developmental and evolutionary aspects of the neural crest.* P. F. A. Mederson (ed.). John Wiley and Sons, New York.

Gans, C., and R. G. Northcutt. 1983. Neural crest and the origin of vertebrates: A new head. *Science* 220:268–74.

Garrick, J. A. F. 1982. Sharks in the genus *Carcharhinus*. *NOAA Tech. Rep. NMFS Circ.* 445:1–194.

Gayet, M., F. J. Meunier, and C. Werner. 2002. Diversification in Polypteriformes and special comparison with the Lepisosteiformes. *Palaeont.* 45:361–76.

Genner, M. J., O. Seehausen, D. H. Lunt, D. A. Joyce, P. W. Shaw, G. R. Carvalho, and G. F. Turner. 2007. Age of cichlids: New dates for ancient lake fish radiations. *Mol. Biol. Evol.* 24:1269–82.

Genner, M. J., and G. F. Turner. 2005. The Mbuna cichlids of Lake Malawi: A model for rapid speciation and adaptive radiation. *Fish Fish.* 6:1–34.

Géry, J. 1977. *Characoids of the world.* T. F. H. Publications, Neptune City, New Jersey. 672 p.

Ghedotti, M. J. 2000. Phylogenetic analysis and taxonomy of the poecilioid fishes (Teleostei: Cyprinodontiformes). *Zool. J. Linn. Soc.* 130:1–53.

Ghedotti, M. J., and M. P. Davis. 2013. Phylogeny, classification, and evolution of salinity tolerance of the North American topminnows and killifishes, family Fundulidae (Teleostei: Cyprinodontiformes). *Fieldiana Life and Earth Sciences* 1564(7):1–65.

Gibson, R. N. 2008. *Flatfishes: Biology and exploitation.* Wiley, Hoboken, New Jersey. 416 p.

Gilbert, C. R. 1967. A revision of the hammerhead sharks (family Sphyrnidae). *Proc. U.S. Natl. Mus.* 119:1–88.

Gilbert, C. R., and D. P. Kelso. 1971. Fishes of the Tortuguero area, Caribbean Costa Rica. *Bull. Florida State Mus.: Biol. Sci.* 16:1–54.

Gill, H. S., C. B. Renaud, F. Chapleau, R. L. Mayden, and I. C. Potter. 2003. Phylogeny of living parasitic lampreys (Petromyzonitiformes) based on morphological data. *Copeia* 2003:687–703.

Gladstone, W. 1987. The eggs and larvae of the sharpnose puffer fish *Canthigaster valentini* are unpalatable to other reef fishes. *Copeia* 1987:227–30.

———. 1994. Lek-like spawning, parental care and mating periodicity of the triggerfish *Pseudobalistes flavimarginatus* (Balistidae). *Env. Biol. Fishes* 39:249–57.

Gold, J. R., G. Voelker, and M. A. Renshaw. 2011. Phylogenetic relationships of tropical western Atlantic snappers in subfamily Lutjaninae (Lutjanidae: Perciformes) inferred from mitochondrial DNA sequences. *Biol. J. Linn. Soc.* 102:915–929.

Gomon, M. F., C. D. Struthers, and A. L. Stewart. 2013. A new genus and two new species of the family Aulopidae (Aulopiformes), commonly referred to as aulopus, flagfins, sergeant bakers or threadsails, in Australasian waters. *Spec. Divers.* 18:141–61.

Gon, O., and P. C. Heemstra (eds.). 1990. *Fishes of the southern ocean.* J. L. B. Smith Institute of Ichthyology, Grahamstown, South Africa. 462 p.

Gosline, W. A. 1968. The suborders of perciform fishes. *Proc. U.S. Natl. Mus.* 124:1–78.

———. 1970. A reinterpretation of the teleostean fish order Gobiesociformes. *Proc. Cal. Acad. Sci.* 37:363–82.

———. 1971. *Functional morphology and classification of teleostean fishes.* University Press of Hawaii, Honolulu. 208 p.

———. 1983. The relationships of the mastacembelid and synbranchid fishes. *Jap. J. Ichthyol.* 29:323–28.

———. 1984. Structure, function, and ecology of the goatfishes (family Mullidae). *Pac. Sci.* 38:312–23.

Gosse, J.-P. 1984. Polypteridae, pp. 18–29. In *Checklist of the freshwater fishes of Africa.* J. Daget, J. P. Gosse, G. G. Teugels, and D. F. E. Thys van den Audenaerde (eds.). Vol. 1. MRAC, Tervuren; ORSTOM, Paris.

———. 1988. Revision systematique de deux species du genre *Polypterus* (Pisces, Polypteridae). *Cybium* 12:239–45.

Goto, T. 2001. Comparative anatomy, phylogeny, and cladistic classification of the order Orectolobiformes (Chondrichthyes: Elasmobranchii). *Mem. Grad. Sch. Fish Sci. Hokkaido Univ.* 48:1–100.

Graham, J. B. 1997. *Air-breathing fishes.* Academic Press, San Diego, California. 299 p.

Grande, L. 1985. Recent and fossil clupeomorph fishes with material for revision of the subgroups of clupeoids. *Bull. Amer. Mus. Nat. Hist.* 181:231–372.

———. 2010. *An empirical synthetic pattern study of gars (Lepisosteiformes) and closely related species, based mostly on skeletal anatomy: The resurrection of Holostei.* American Society of Ichthyologists and Herpetologists, Spec. Publ. 6:1–871; *Copeia* 2010 (Suppl. 2A).

Grande, L., and W. E. Bemis. 1991. Osteology and phylogenetic relationshships of fossil and recent paddlefishes (Polyodontidae) with comments on the interrelationships of Acipenseriformes. *J. Vertebr. Paleontol.* 11(1):1–121.

———. 1996. Interrelationships of Acipenseriformes with comments on "Chondrostei," pp. 85–115. In *Interrelationships of fishes.* M. L. J. Stiassny, L. R. Parenti, and G. D. Johnson (eds.). Academic Press, San Diego, California.

———. 1998. A comprehensive phylogenetic study of amiid fishes (Amiidae) based on comparative skeletal anatomy. *J. Vert. Paleontol.,* Spec. Memoir 4, Suppl. Vol. 18: 1–690.

———. 1999. Historical biogeography and historical paleoecology of Amiidae and other halecomorph fishes, pp. 413–24. In *Mesozoic fishes 2: Systematics and fossil record.* G. Arratia and H.-P. Schultze (eds.). Verlag Dr. Friedrich Pfeil, Munchen.

Grande, T., H. Laten, and J. A. Lopez. 2004. Phylogenetic relationships of extant esocid species (Teleostei: Salmoniformes) based on morphological and molecular characters. *Copeia* 2004:743–57.

Grande, T., and F. J. Poyato-Ariza. 1999. Phylogenetic relationships of fossil and recent gonorynchiform fishes (Teleostei: Ostariophysi). *Zool. J. Linn. Soc.* 125:197–238.

Grande, T., F. J. Poyato-Ariza, and R. Diogo (eds.). 2010. *Gonorynchiformes and ostariophysan relationships: A comprehensive review.* Science Publishers, Enfield, New Hampshire. 592 p.

Gray, K. N., J. R. McDowell, B. B. Collette, and J. E. Graves. 2009. A molecular phylogeny of remoras and their relatives. *Bull. Mar. Sci.* 84:183–98.

Greenfield, D., W., R. Winterbottom, and B. B. Collette. 2008. Review of the toadfish genera (Teleostei: Batrachoididae). *Proc. Cal. Acad. Sci.* 59:665–710.

Greenwood, P. H. 1970a. On the genus *Lycoptera* and its relationships with the family Hiodontidae (Pisces, Osteoglossomorpha). *Bull. Br. Mus. Nat. Hist. (Zool.)* 19:257–85.

———. 1970b. Skull and swimbladder connections in fishes of the family Megalopidae. *Bull. Brit. Mus. Nat. Hist. (Zool.)* 19:121–35.

———. 1976. A review of the family Centropomidae (Pisces, Perciformes). *Bull. Brit. Mus. Nat. Hist. (Zool.)* 29:1–81.

Greenwood, P. H., D. E. Rosen, S. H. Weitzman, and G. S. Myers. 1966. Phyletic studies of teleostean fishes, with a provisional classification of living forms. *Bull. Am. Mus. Nat. Hist.* 131:339–456.

Gregory, W. K. 1933 (repr. 2002). *Fish skulls.* Krieger: Malabar, Florida. 481 p. Orig.: *Amer. Phil. Soc.* Vol. 23, Part 2.

Grey, M., A.-M. Blais, B. Hunt, and A. C. J. Vincent. 2006. The USA's international trade in fish leather, from a conservation perspective. *Environ. Conserv.* 33:100–108.

Grogan, E. D., and R. Lund. 2004. The origin and relationship of early Chondrichthyes, pp. 3–31 In *Biology of sharks and their relatives.* J. C. Carrier, J. A. Musick, and M. R. Heithaus (eds.). CRC Press, Boca Raton, Florida.

Grogan, E. D., R. Lund, and D. Didier. 1999. Description of the chimaera jaw and its phylogenetic origins. *J. Morph.* 239:45–59.

Gross, J.B. 2012. The complex origin of Astyanax cavefish. *BMC Evol. Biol.* 12:105.

Gushiken, S. 1988. Phylogenetic relationships of the perciform genera of the family Carangidae. *Japan. J. Ichthyol.* 34:443–61.

Haedrich, R.L. 1967. The stromateoid fishes: Systematics and a classification. *Bull. Mus. Comp. Zool.* 135:31–139.

―――. 1969. A new family of aberrant stromateoid fishes from the equatorial Indo-Pacific. *Dana Rept.* 76. 14 p.

Hamlett, W.C. (ed.). 2005. *Reproductive biology and phylogeny of Chondrichthyes: Sharks, batoids and chimaeras.* Science Publishers, Enfield, New Hampshire. 562 p.

Hamlett, W.C., G. Komanik, M. Storrie, B. Stevens, and T.I. Walker. 2005. Chondrichthyan parity, lecithotrophy and matrotrophy, pp. 395–434. In *Reproductive biology and phylogeny of Chondrichthyes.* W.C. Hamlett (ed.). Science Publishers, Enfield, New Hampshire.

Hardisty, M.W. 1979. *Biology of the cyclostomes.* Chapman and Hall, London. 428 p.

Hardisty, M.W., and I.C. Potter. 1971. Paired species, pp. 249–77. In *The biology of lampreys.* M.W. Hardisty and I.C. Potter (eds.). Vol. 1. Academic Press, London.

Harold, A.S. 1993. Phylogenetic relationships of the sternoptychid genus *Argyropelecus* (Teleostei: Stomiiformes). *Copeia* 1993:123–33.

―――. 1994. A taxonomic revision of the sternoptychid genus *Polyipnus* (Teleostei: Stomiiformes) with an analysis of phylogenetic relationships. *Bull. Mar. Sci.* 54:428–534.

―――. 1998. Phylogenetic relationships of the Gonostomatidae (Teleostei: Stomiiformes). *Bull. Mar. Sci.* 62:715–41.

Harold, A.S., and S.H. Weitzman. 1996. Interrelationships of stomiiform fishes, pp. 333–53. In *Interrelationships of fishes.* M.L.J. Stiassny, L.R. Parenti, and G.D. Johnson (eds.). Academic Press, San Diego, California.

Harris, P.M., and R.L. Mayden. 2001. Phylogenetic relationships of major clades of Catostomidae (Teleostei: Cypriniformes) as inferred from mitochondrial SSU and LSU rDNA sequences. *Mol. Phylogen. Evol.* 20:225–37.

Hart, J.L. 1973. Pacific fishes of Canada. *Bull. Fish. Res. Bd. Canada* 180. 740 p.

Hastings, P.A. 1993. Relationships of fishes of the perciform suborder Notothenioidei, pp. 99–107. In *History and atlas of the fishes of the Antarctic Ocean.* R.G. Miller (ed.). Foresta Institute, Carson City, Nevada.

―――. 2011. Complementary approaches to systematic ichthyology. *Zootaxa* 2011:57–59.

Hastings, P.A., and G.R. Galland. 2010. Ontogeny of microhabitat use and two-step recruitment in a specialist reef fish, the Browncheek Blenny (Chaenopsidae). *Coral Reefs* 29:155–64.

Hastings, P.A., and C.W. Petersen. 2010. Parental care, oviposition sites and mating systems of blennioid fishes. pp. 91–116. In *Reproduction in marine fishes: Evolutionary patterns and innovations.* K.S. Cole (ed.). University of California Press, Berkeley, California.

Hastings, P.A., and V.G. Springer. 1994. Review of *Stathmonotus,* with redefinition and phylogenetic analysis of the Chaenopsidae (Teleostei: Blennioidei). *Smithson. Contr. Zool.* 558:1–48.

―――. 2009a. Systematics of the Blennioidei and the included families Chaenopsidae, Clinidae, Labrisomidae and Dactyloscopidae, pp. 3–30. In *The biology of blennies.* R.A. Patzner, E.J. Gonçlaves, P.A. Hastings, and B.G. Kapoor (eds.). Science Publishers, Enfield, New Hampshire.

―――. 2009b. Systematics of the Blenniidae (Blennioidei), pp. 69–91. In *The biology of blennies.* R.A. Patzner, E.J. Gonçlaves, P.A. Hastings, and B.G. Kapoor (eds.). Science Publishers, Enfield, New Hampshire.

Heck, K.L., Jr., and M.P. Weinstein. 1978. Mimetic relationships between tropical burrfishes and opisthobranchs. *Biotropica* 10:78–79.

Hedman, J. E., H. Rüdel, J. Gercken, S. Bergek, J. Strand, M. Quack, M. Appelberg, et al. 2011. Eelpout (*Zoarces viviparus*) in marine environmental monitoring. *Mar. Poll. Bull.* 62:2015–29.

Heemstra, P. C., and J. E. Randall. 1993. *Groupers of the world (family Serranidae, subfamily Epinephelinae) an annotated and illustrated catalogue of the grouper, rockcod, hind, coral grouper and lyretail species known to date.* FAO species catalogue. *FAO Fish. Synop.* 125, Vol. 16. 382 p.

Heimberg, A. M., R. Cowper-Sal·lari, M. Sémon, P. C. J. Donoghue, and K. J. Peterson. 2010. MicroRNAs reveal the interrelationships of hagfish, lampreys, and gnathostomes and the nature of the ancestral vertebrate. *Proc. Nat. Acad. Sci.* 107:19379–83.

Helfman, G., and B. Collette. 2011. *Fishes: The animal answer guide.* Johns Hopkins Press, Baltimore, Maryland. 178 p.

Helfman, G. S., B. B. Collette, D. E. Facey and B. W. Bowen. 2009. *The diversity of fishes. Biology, evolution, and ecology.* 2nd ed. Wiley-Blackwell. Hoboken, New Jersey. 720 p.

Helfman, G., J. L. Meyer, and W. N. McFarland. 1982. The ontogeny of twilight migration patterns in grunts (Pisces: Haemulidae). *Anim. Behav.* 30:317–26.

Hensley, D. A., and E. H. Ahlstrom. 1984. Pleuronectiformes: Relationships, pp. 670–87. In *Ontogeny and systematics of fishes.* H. G. Moser, W. J. Richards, D. M. Cohen, M. P. Fahay, A. W. Kendall, Jr., and S. L. Richardson (eds.). American Society of Ichthyologists and Herpetologists, Spec. Public. No. 1. Allen Press, Lawrence, Kansas.

Hertwig, S. H. 2008. Phylogeny of the Cyprinodontiformes (Teleostei, Atherinomorpha): The contribution of cranial soft tissue characters. *Zool. Scripta* 37:141–74.

Hickey, A. J. R., S. D. Lavery, D. A. Hannan, C. S. Baker, and K. D. Clements. 2009. New Zealand triplefin fishes (family Tripterygiidae): Contrasting population structure and mtDNA diversity within a marine species flock. *Mol. Ecol.* 18:680–96.

Hidaka, K., Y. Iwatsuki, and J. E. Randall. 2008. A review of the Indo-Pacific bonefishes of the *Albula argentea* complex, with a description of a new species. *Ichthyol. Res.* 55:53–64.

Hilton, E. J. 2003. Comparative osteology and phylogenetic systematics of fossil and living bony-tongue fishes (Actinopterygii, Teleostei, Osteoglossomorpha). *Zool. J. Linn. Soc.* 137:1–100.

———. 2011. Bony fish skeleton, pp. 434–48. In *Encyclopedia of fish physiology.* A. P. Farrell (ed.). Vol. 1. Academic Press, San Diego, California.

Hilton, E. J., L. Grande, and W. E. Bemis. 2011. Skeletal anatomy of the shortnose sturgeon, *Acipenser brevirostrum* Lesueur 1818, and the systematics of sturgeons (Acipenseriformes, Acipenseridae). *Fieldiana (Life and Earth Sciences)* 3:1–168.

Hilton, E. J., and G. D. Johnson. 2007. When two equals three: Developmental osteology of the caudal skeleton in carangid fishes (Perciformes: Carangidae). *Evol. Dev.* 9:178–89.

Hilton, E. J., G. D. Johnson, and W. F. Smith-Vaniz. 2010. Osteology and systematics of *Parastromateus niger* (Perciformes: Carangidae), with comments on the carangid dorsal gill-arch skeleton. *Copeia* 2010:312–33.

Hochleithner, M., and J. Gessner. 2001. *The sturgeons and paddlefishes (Acipenseriformes) of the world: Biology and aquaculture.* Aqua Tech Publications, Kitzbuehel. 248 p.

Hoegg, S., H. Brinkmann, J. S. Taylor, and A. Meyer. 2004. Phylogenetic timing of the fish-specific genome duplication correlates with the diversification of teleost fish. *J. Mol. Evol.* 59:190–203.

Hoese, D. F., and A. C. Gill. 1993. Phylogenetic relationships of eleotridid fishes (Perciformes: Gobioidei). *Bull. Mar. Sci.* 52:415–40.

Holcroft, N. I. 2004. A molecular test of alternative hypotheses of tetraodontiform (Acanthomorpha: Tetraodontiformes) sistergroup relationships using data from the RAG1 gene. *Mol. Phylogen. Evol.* 32:749–60.

———. 2005. A molecular analysis of the interrelationships of tetraodontiform fishes (Acanthomorpha: Tetraodontiformes). *Mol. Phylogen. Evol.* 34:525–44.

Holcroft, N.I., and E.O. Wiley. 2008. Acanthuroid relationships revisited: A new nuclear gene-based analysis that incorporates tetraodontiform representatives. *Ichthyol. Res.* 55:274–83.

Holder, M.T., M.V. Erdmann, T.P. Wilcox, R.L. Caldwell, and D.M. Hillis. 1999. Two Living Species of Coelacanths? *Proc. Nat. Acad. Sci.* 96:12616–20.

Horn, M.H., 1984. Stromateoidei: Development and relationships, pp. 620–28. In *Ontogeny and systematics of fishes.* H.G. Moser, W.J. Richards, D.M. Cohen, M.P. Fahay, A.W. Kendall, Jr., and S.L. Richardson (eds.). American Society of Ichthyologists and Herpetologists, Spec. Public. No. 1. Allen Press, Lawrence, Kansas.

Hoshino, K. 2001. Monophyly of the Citharidae (Pleuronectoidei: Pleuronectiformes: Teleostei) with considerations of pleuronectoid phylogeny. *Ichthyol. Res.* 48:391–404.

Howes, G.J. 1989. Phylogenetic relationships of macrouroid and gadoid fishes based on cranial morphology. *Sci. Ser. Nat. Hist. Mus. Los Angeles Co.* 32:113–28.

———. 1991. Biogeography of gadoid fishes. *J. Biogeogr.* 18:595–622.

Hrbek, T., and A. Meyer. 2003. Closing of the Tethys Sea and the phylogeny of Eurasian killifishes (Cyprinodontiformes: Cyprinodontidae). *J. Evol. Biol.* 16:17–36.

Hrbek, T., J. Seckinger, and A. Meyer. 2007. A phylogenetic and biogeographic perspective on the evolution of poeciliid fishes. *Mol. Phylogen. Evol.* 43:986–98.

Hubbs, C.L., and K.F. Lagler. 1958. *Fishes of the Great Lakes region.* University of Michigan Press, Ann Arbor, Michigan. 213 p.

Hubbs, C.L., and I.C. Potter. 1971. Distribution, phylogeny, and taxonomy, pp. 1–65. In *The biology of lampreys.* M.W. Hardisty and I.C. Potter (eds.). Academic, London.

Hundt, P.J., S.P. Iglésias, A.S. Hoey, A.M. Simons. 2013. A multilocus molecular phylogeny of combtooth blennies (Percomorpha: Blennioidei: Blenniidae): Multiple invasions of intertidal habitats. *Mol. Phylogen. Evol.* 70:47–56.

Hurley, I.A., R.L. Mueller, K.A. Dunn, K.J. Schmidt, M. Friedman, R.K. Ho, V.E. Prince, et al. 2007. A new time-scale for ray-finned fish evolution. *Proc. R. Soc. B* 274:489–98.

Hyde, J.R., and R.D. Vetter. 2007. The origin, evolution, and diversification of rockfishes of the genus *Sebastes* (Cuvier). *Mol. Phylogen. Evol.* 44:790–811.

Imamura, H. 1996. Phylogeny of the family Platycephalidae and related taxa (Pisces: Scorpaeniformes). *Spec. Divers.* 1:123–33.

———. 2000. An alternative hypothesis on the phylogenetic position of the family Dactylopteridae (Pisces: Teleostei), with a proposed new classification. *Ichthyol. Res.* 47:203–22.

———. 2004. Phylogenetic relationships and new classification of the superfamily Scorpaenoidea (Actinopterygii: Perciformes). *Spec. Divers.* 9:1–36.

Imamura, H., S.M. Shirai, and M. Yabe. 2005. Phylogenetic position of the family Trichodontidae (Teleostei: Perciformes), with a revised classification of the perciform suborder Cottoidei. *Ichthyol. Res.* 52:264–74.

Imamura, I., and G. Shinohara. 1998. Scorpaeniform fish phylogeny: An overview. *Bull. Natl. Sci. Mus. Tokyo, Ser. A* 24:185–212.

Imamura, I., and M. Yabe. 2002. Demise of the Scorpaeniformes (Actinopterygii: Percomorpha): An alternative phylogenetic hypothesis. *Bull. Fish. Sci. Hokkaido Univ.* 53:107–28.

Inoue, J.G., M. Miya, M.J. Miller, T. Sado, R. Hanel, K. Hatooka, J. Aoyama, et al. 2010. Deep-ocean origin of the freshwater eels. *Biol. Lett.* 6:363–66.

Inoue, J.G., M. Miya, K. Tsukamoto, and M. Nishida. 2001. A mitogenomic perspective on the basal teleostean phylogeny: Resolving higher-level relationships with longer DNA sequences. *Mol. Phylogen. Evol.* 20:275–85.

———. 2003. Basal actinopterygian relationships: A mitogenomic perspective on the phylogeny of the "ancient fish." *Mol. Phylogen. Evol.* 26:110–20.

————. 2004. Mitogenomic evidence for the monophyly of elopomorph fishes (Teleostei) and the evolutionary origin of the leptocephalus larva. *Mol. Phylogen. Evol.* 32:274–86.

Ishida, M. 1994. Phylogeny of the suborder Scorpaenoidei (Pisces: Scorpaeniformes). *Bull. Nansei Natl. Fish. Res. Inst.* 27:1–112.

Ishiguro, N. B., M. Miya, and M. Nishida. 2003. Basal euteleostean relationships: A mitogenomic perspective on the phylogenetic reality of the "Protacanthopterygii." *Mol. Phylogen. Evol.* 27:476–88.

Ishimatsu, A., Y. Itazawa, and T. Takeda. 1979. On the circulatory systems of the snakeheads *Channa maculata* and *Channa argus* with reference to bimodal breathing. *Jap. J. Ichthyol.* 26:167–80.

IUCN. 2013. *The IUCN Red List of threatened species.* Version 2013.2. www.iucnredlist.org. Accessed on 21 November 2013.

Iwamoto, T. 1989. Phylogeny of grenadiers (suborder Macrouroidei): Another interpretation, pp. 159–73. In *Papers on the systematics of gadiform fishes.* D. M. Cohen (ed.). *Nat. Hist. Mus. Los Angeles Co., Sci. Ser.* 32.

————. 2008. A brief taxonomic history of grenadiers, pp. 3–13. In *Grenadiers of the world oceans: Biology, stock assessment, and fisheries.* A. M. Orlov, and T. Iwamoto (eds.). Symposium 63. American Fisheries Society, Bethesda, Maryland.

Jacobsen, I. P., and M. B. Bennett. 2009. A taxonomic review of the Australian butterfly ray *Gymnura australis* (Ramsay and Ogilby, 1886) and other members of the family Gymnuridae (Order Rajiformes) from the Indo-West Pacific. *Zootaxa* 2228:1–28.

Jakubowski, M. 1974. Structure of the lateral-line canal system and related bones in the berycoid fish *Hoplostethus mediterraneus* Cuv. et Val. (Trachichthyidae, Pisces). *Acta Anat.* 87:261–74.

Janvier, P. 1996. The dawn of the vertebrates: Characters versus common ascent in the rise of current vertebrate phylogenies. *Palaeontology* 39:259–87.

Johns, G. C., and J. C. Avise. 1998. Tests for ancient species flocks based on molecular phylogenetic appraisals of *Sebastes* rockfishes and other marine fishes. *Evolution* 52:1135–46.

Johnson, G. D. 1980. The limits and relationships of the Lutjanidae and associated families. *Bull. Scripps Inst. Oceanog.* 24. 114 p.

————. 1983. *Niphon spinosus*: A primitive epinepheline serranid, with comments on the monophyly and interrelationships of the Serranidae. *Copeia* 1983:777–87.

————. 1984. Percoidei: Development and relationships, pp. 464–98. In *Ontogeny and systematics of fishes.* H. G. Moser, W. J. Richards, D. M. Cohen, M. P. Fahay, A. W. Kendall, Jr., and S. L. Richardson (eds.). American Society of Ichthyologists and Herpetologists, Spec. Public. No. 1. Allen Press, Lawrence, Kansas.

————. 1986. Scombroid phylogeny: An alternative hypothesis. *Bull. Mar. Sci.* 39:1–41.

————. 1992. Monophyly of the euteleostean clades Neoteleostei, Eurypterygii and Ctenosquamata. *Copeia* 1992:8–25.

————. 1993. Percomorph phylogeny: Progress and problems. *Bull. Mar. Sci.* 52:3–28.

Johnson, G. D., C. C. Baldwin, M. Okiyama, and Y. Tominaga. 1996. Osteology and relationships of *Pseudotrichonotus altivelis* (Teleostei: Aulopiformes: Pseudotrichonotidae). *Ichthyol. Res.* 43:17–45.

Johnson, G. D., and R. Britz. 2010. Occipito-vertebral fusion in actinopterygians: Conjecture, myth and reality. Part 2: Teleosts, pp. 95–110. In *Origin and phylogenetic interrelationships of teleosts.* J. S. Nelson, H.-P. Schultze , and M. V. H. Wilson (eds.). Verlag Dr. Friedrich Pfeil, Munchen.

Johnson, G. D., and R. A. Fritzsche. 1989. *Graus nigra* Philippi, an omnivorous girellid, with comments on relationships of the Girellidae (Pisces: Perciformes). *Proc. Acad. Natur. Sci. Philadelphia* 141:1–27.

Johnson, G.D., H. Ida, J. Sakaue, T. Sado, T. Asahida, and M. Miya. 2012. A "living fossil" eel (Anguilliformes: Protanguillidae, fam. nov.) from an undersea cave in Palau. *Proc. Royal Society B: Biol. Sci.* 279:934–43.

Johnson, G.D., and C. Patterson. 1993. Percomorph phylogeny: A survey of acanthomorphs and a new proposal. *Bull. Mar. Sci.* 52:554–626.

———. 1996. Relationships of lower eutelostean fishes, pp. 251–332. In *Interrelationships of fishes*. M.L.J. Stiassny, L.R. Parenti, and G.D. Johnson (eds.). Academic Press, San Diego, California.

Johnson, G.D., J.R. Paxton, T.T. Sutton, T.P. Satoh, T. Sado, M. Nishida, and M. Miya. 2009. Deep-sea mystery solved: Astonishing larval transformations and extreme sexual dimorphism unite three fish families. *Biol. Lett.* 5:235–39.

Johnson, G.D., and R.H. Rosenblatt. 1988. Mechanisms of light organ occlusion in flashlight fishes, family Anomalopidae (Teleostei: Beryciformes), and the evolution of the group. *Zool. J. Linn. Soc.* 94:65–96.

Johnson, R.K. 1982. Fishes of the families Evermannellidae and Scopelarchidae: Systematics, morphology, interrelationship and zoogeography. *Fieldiana Zool. (N. Ser.)* 12:1–252.

Jones, F.C., M.G. Grabherr, Y.F. Chan, P. Russell, E. Mauceli, J. Johnson, R. Swofford, et al. 2012. The genomic basis of adaptive evolution in threespine sticklebacks. *Nature* 484:55–61.

Jorgensen, J.M., J.P. Lomholt, R.E. Weber, and H. Malte (eds.). 1998. *The biology of hagfishes*. Chapman and Hall, London. 578 p.

Joyce, D.A., D.H. Lunt, M.J. Genner, G.F. Turner, R. Bills, and O. Seehausen. 2011. Repeated colonization and hybridization in Lake Malawi cichlids. *Curr. Biol.* 21(3):R108–9.

Kailola, P.J. 2004. A phylogenetic exploration of the catfish family Ariidae. *The Beagle, Rec. Mus. Art Galleries N. Terr.* 20:87–166.

Karplus, I., and A.R, Thompson. 2011. The partnership between gobiid fishes and burrowing alpheid shrimps, pp. 559–608. In *The biology of gobies*. R.A. Patzner, J.L. Van Tassel, M. Kovačić, and B.G. Kapoor (eds.). Science Publishers, Enfield, New Hampshire

Kassler, T.W., J.B. Koppelman, T.J. Near, C.B. Dillman, J.M. Levengood, D.L. Swofford, J.L. VanOrman, et al. 2002. Molecular and morphological analyses of the black basses (*Micropterus*): Implications for taxonomy and conservation. *Am. Fish. Soc. Symp.* 31:291–322.

Kaufman, L.S., and K.F. Liem. 1982. Fishes of the suborder Labroidei (Pisces: Perciformes): Phylogeny, ecology and evolutionary significance. *Breviora* 472:1–19.

Kawahara, R., M. Miya, K. Mabuchi, S. Lavoué, J.G. Inoue, T.P. Satoh, A. Kawaguchi, and M. Nishida. 2008. Interrelationships of the eleven gasterosteiform families (sticklebacks, pipefishes, and their relatives): A new perspective based on whole mitogenome sequences from 75 higher teleosts. *Mol. Phylogen. Evol.* 46:224–36.

Kazancioğlu, E., and S.H. Alonzo. 2010. A comparative analysis of sex change in Labridae supports the size advantage hypothesis. *Evolution* 64:2254–64.

Keenleyside, M.H.A. (ed.). 1991. *Cichlid fishes: Behaviour, ecology, and evolution*. Chapman and Hall, London. 378 p.

Keivany, Y., and J.S. Nelson. 2000. Taxonomic review of the genus *Pungitius*, ninespine sticklebacks (Gasterosteidae). *Cybium* 24:107–22.

———. 2006. Interrelationships of Gasterosteiformes (Actinopterygii, Percomorpha). *J. Ichthyol.* 46 (Suppl. 1):S84-S96.

Kelly, N.B., T.J. Near, and S.H. Alonzo. 2012. Diversification of egg-deposition behaviours and the evolution of male parental care in darters (Teleostei: Percidae: Etheostomatinae). *J. Evol. Biol.* 25:836–46.

Kenaley, C.P. 2007. Revision of the stoplight loosejaw genus *Malacosteus* (Teleostei: Stomiidae: Malacosteinae), with description of a new species from the temperate Southern Hemisphere and Indian Ocean. *Copeia* 2007:886–900.

————. 2009. Revision of Indo-Pacific species of the loosejaw dragonfish genus *Photostomias* (Teleostei: Stomiidae: Malacosteinae). *Copeia* 2009:175–89.

————. 2010. Comparative innervation of cephalic photophores of the loosejaw dragonfishes (Teleostei: Stomiiformes: Stomiidae): Evidence for parallel evolution of long-wave bioluminescence. *J. Morph.* 271:418–37.

Kido, K. 1988. Phylogeny of the family Liparididae, with the taxonomy of the species found around Japan. *Mem. Fac. Fish. Hokkaido Univ.* 35:125–256.

Kim, B.-J. 2002. Comparative anatomy and phylogeny of the family Mullidae (Teleostei: Perciformes). *Mem. Grad. Sch. Fish. Sci. Hokkaido Univ.* 49:1–74.

Kim, B.-J., M. Yabe, and K. Nakaya. 2001. Barbels and related muscles in Mullidae (Perciformes) and Polymixiidae (Polymixiiformes). *Ichthyol. Res.* 48:409–13.

Kim, S. H., K. Shimada, and C. K. Rigsby. 2013. Anatomy and evolution of heterocercal tail in lamniform sharks. *Anat. Rec.* 296:433–42.

Kinziger, A. P., R. M. Wood, and D. A. Neely. 2005. Molecular systematics and classification of *Cottus* and the Baikalian cottoids (Scorpaeniformes). *Copeia* 2005:303–11.

Klassen, G. 1995. Phylogeny and biogeography of Ostraciinae (Tetraodontiformes: Ostraciidae). *Bull. Mar. Sci.* 57:393–441.

Klepadlo, C. 2011. Three new species of the genus *Photonectes* (Teleostei: Stomiiformes: Stomiidae: Melanostomiinae) from the Pacific Ocean. *Copeia* 2011:201–10.

Klimley, A. P. 2013. *The biology of sharks and rays.* University of Chicago Press, Chicago, Illinois. 512 p.

Knapp, R., J. C. Wingfield and A. H. Bass. 1999. Steroid hormones and paternal care in the midshipman fish (*Porichthys notatus*). *Horm. Beh.* 35:81–89.

Knope, M. L. 2013. Phylogenetics of the marine sculpins (Teleostei: Cottidae) of the North American Pacific coast. *Mol. Phylogen. Evol.* 66:341–49.

Knudsen, S. W., P. R. Miller, and P. Gravlund. 2007. Phylogeny of the snailfishes (Teleostei: Liparidae) based on molecular and morphological data. *Mol. Phylogen. Evol.* 44:649–66.

Kobelkowsky, A. 2004. Sexual dimorphism of the flounder *Bothus robinsi* (Pisces: Bothidae). *J. Morphol.* 260:165–71.

Kobyliansky, S. G. 1990. Taxonomic status of microstomatid fishes and problems of classification of suborder Argentinoidei (Salmoniformes, Teleostei). *Trudy Inst. Okeanol. Akad. Nauk SSSR* 125:148–77.

Kocher, T. D. 2004. Adaptive evolution and explosive speciation: The cichlid fish model. *Nature Rev. Genet.* 5:288–98.

Konovalenko, I. I., and A. S. Piotrovskiy. 1989. First description of a sexually mature Amarsipa, *Amarsipus carlsbergi. J. Ichthyol.* 28:86–89.

Kotlyar, A. N. 1987. Classification and distribution of fishes of the family Anoplogastridae (Beryciformes). *J. Ichthyol.* 27:133–53.

————. 1996. *Beryciform fishes of the world.* VNIRO publishing, Moscow. 368 p.

————. 2003. Family Anoplogastridae Gill 1893: Fangtooths. *Calif. Acad. Sci. Annotated Checklists of Fishes* 20. http://research.calacademy.org/ichthyology/checklists.

Kotrschal, K., and D. A. Thomson. 1986. Feeding patterns in eastern tropical Pacific blennioid fishes (Teleostei: Tripterygiidae, Labrisomidae, Chaenopsidae, Blenniidae). *Oecologia* 70:367–78.

Kramer, B. 1996. *Electroreception and communication in fishes.* Georg Fischer Verlag, Stuttgart. 119 p.

Kuehne, R. A., and R. W. Barbour. 1983. *The American darters.* University Press of Kentucky, Lexington, Kentucky. 177 p.

Kuiter, R. H. 2001. Revision of the Australian seahorses of the genus *Hippocampus* (Syngnathiformes: Syngnathidae) with descriptions of nine new species. *Rec. Aust. Mus.* 53:292–340.

Kullander, S. O. 1998. A phylogeny and classification of the South American Cichlidae (Teleostei: Perciformes), pp. 461–98. In *Phylogeny and classification of neotropical fishes*. L. R. Malabarba, R. E. Reis, R. P. Vari, Z. M. S. Lucena, and C. A. S. Lucena (eds.). EDIPUCRS, Porto Alegre, Brazil.

———. 2003. Family Cichlidae (cichlids), pp. 605–54. In *Checklist of the freshwater fishes of South and Central America*. R. E. Reis, S. O. Kullander, and C. J. Ferrais, Jr. (eds.). EDIPUCRS, Porto Alegre, Brazil.

Kuo, C. H., S. Huang, and S. C. Lee. 2003. Phylogeny of hagfish based on the mitochondrial 16S rRNA gene. *Mol. Phylogen. Evol.* 28:448–57.

Kuraku, S., and S. Kuratani. 2006. Time scale for cyclostome evolution inferred with a phylogenetic diagnosis of hagfish and lamprey cDNA sequences. *Zool. Sci.* 23:1053–64.

Kurlansky, M. 1997. *Cod: Biography of the fish that changed the world*. Walker, New York. 294 p.

Lampert, K. P., K. Blassmann, K. Hissmann, J. Schauer, P. Shunula, Z. el Kharousy, B. P. Ngatunga, et al. 2013. Single-male paternity in coelacanths. *Nat. Commun.* 4:2488.

Lauder, G. V., and K. F. Liem. 1983. The evolution and interrelationships of the actinopterygian fishes. *Bull. Mus. Comp. Zool.* 150:95–197.

Lavoué, S., M. Miya, J. G. Inoue, K. Saitoh, N. B. Ishiguro, M. Nishida. 2005. Molecular systematics of the gonorynchiform fishes (Teleostei) based on whole mitogenome sequences: Implications for higher-level relationships within the Otocephala. *Mol. Phylogen. Evol.* 37:165–77.

Lea, R. N., and C. R. Robins. 2003. Four new species of the genus *Ophidion* (Pisces: Ophidiidae) from the western Atlantic Ocean. *Univ. Kansas Mus. Nat. Hist., Sci. Pap.* 31:1–9.

Lecointre, G. 2010. Gonorynchiformes in the teleostean phylogeny: Molecules and morphology used to investigate interrelationships of the Ostariophysi, pp. 356–400. In *Gonorynchiformes and ostariophysan relationships: A comprehensive review*. T. Grande, F. J. Poyato-Ariza, and R. Diogo (eds.). Science Publishers, Enfield, New Hampshire.

Lecointre, G., and G. Nelson. 1996. Clupeomorpha, sister-group of Ostariophysi, pp. 193–207. In *Interrelationships of fishes*. M. L. J. Stiassny, L. R. Parenti, and G. D. Johnson (eds.). Academic Press, San Diego, California.

Le Comber, S. C., and C. Smith. 2004. Polyploidy in fishes: Patterns and processes. *Biol. J. Linn. Soc.* 82:431–42.

Lee, D. S., C. Gilbert, C. Hocutt, R. Jenkins, D. E. McAllister, and J. R. Stauffer. 1980. *Atlas of North American freshwater fishes*. North Carolina Biol. Survey. 854 p.

Leis, J. M. 1978. Systematics and zoogeography of the porcupinefishes (*Diodon*, Diodontidae, Tetraodotiformes) with comments on egg and larval development. *Fish. Bull.* 76:535–67.

———. 2006. Nomenclature and distribution of the species of the porcupinefish family Diodontidae (Pisces: Teleostei). *Mem. Mus. Vict.* 63:77–90.

Lewallen, E. A., R. L. Pitman, S. L. Kjartson, and N. R. Lovejoy. 2011. Molecular systematics of flyingfishes (Teleostei: Exocoetidae): Evolution in the epipelagic zone. *Biol. J. Linn. Soc.* 102:161–74.

Li, B., A. Dettaï, C. Cruaud, A. Couloux, M. Desoutter-Meniger, and G. Lecointre. 2009. RNF213, a new nuclear marker for acanthomorph phylogeny. *Mol. Phylogen. Evol.* 50:345–63.

Li, C. H., B. R. Ricardo, W. L. Smith, and G. Orti. 2011. Monophyly and interrelationships of snook and barramundi (Centropomidae *sensu* Greenwood) and five new markers for fish phylogenetics. *Mol. Phylogen. Evol.* 60:463–71.

Li, G.-Q., and M. V. H. Wilson. 1994. An Eocene species of *Hiodon* from Montana, its phylogenetic relationships, and the evolution of the postcranial skeleton in the Hiodontidae (Teleostei). *J. Vert. Paleontol.* 14:153–67.

————. 1996. Phylogeny of Osteoglossomorpha, pp. 163–74. In *Interrelationships of fishes*. M. L. J. Stiassny, L. R. Parenti, and G. D. Johnson (eds.). Academic Press, San Diego, California.

Li, G.-Q., M. V. H. Wilson, and L. Grande. 1997. Review of *Eohiodon* (Teleostei: Osteoglossomorpha) from western North America, with a phylogenetic reassessment of Hiodontidae. *J. Paleontol*. 71:1109–24.

Li, S.-Z. 1981. On the origin, phylogeny and geographical distribution of the flatfishes (Pleuronectiformes). *Trans. Chinese Ichthyol. Soc*. 1981:11–20.

Liem, K. F. 1963. The comparative osteology and phylogeny of the Anabantoidei (Teleostei, Pisces). *Illinois Biol. Monogr*. 30:1–149.

————. 1968. Geographical and taxonomic variation in the pattern of natural sex reversal in the teleost fish order Synbranchiformes. *J. Zool. London* 156:225–38.

————. 1973. Evolutionary strategies and morphological innovations: Cichlid pharyngeal jaws. *Syst. Zool*. 22:425–41.

————. 1986. The pharyngeal jaw apparatus of the Embiotocidae (Teleostei): A functional and evolutionary perspective. *Copeia* 1986:311–23.

Lim, D. D., P. J. Motta, K. Mara, and A. P. Martin. 2010. Phylogeny of hammerhead sharks (Family Sphyrnidae) inferred from mitochondrial and nuclear genes. *Mol. Phylogen. Evol*. 55:572–79.

Lin, H. C., and P. A. Hastings. 2011. Evolution of a neotropical marine fish lineage (Subfamily Chaenopsinae, Suborder Blennioidei) based on phylogenetic analysis of combined molecular and morphological data. *Mol. Phylogen. Evol*. 60:236–48.

————. 2013. Phylogeny and biogeography of a shallow water fish clade (Teleostei: Blenniiformes). *BMC Evol. Biol*. 13:210.

Little, A. G., S. C. Lougheed, and C. D. Moyes. 2010. Evolutionary affinity of billfishes (Xiphiidae and Istiophoridae) and flatfishes (Plueronectiformes): Independent and trans-subordinal origins of endothermy in teleost fishes. *Mol. Phylogen. Evol*. 56:897–904.

Littlewood, D. T. J., S. M. Mcdonald, and A. C. Gill. 2004. Molecular phylogenetics of *Chaetodon* and the Chaetodontidae (Teleostei: Perciformes) with reference to morphology. *Zootaxa* 779:1–20.

Lloris, D., J. Matallanas, and P. Oliver. 2005. *Hakes of the world (Family Merlucciidae): An annotated and illustrated catalogue of hake species known to date*. FAO Species Catalogue. *FAO Species Catalogue for Fishery Purposes*. No. 2. FAO, Rome. 57 p.

López, J. A., W. J. Chen, and G. Orti. 2004. Esociform phylogeny. *Copeia* 2004:449–64.

López, J. A., J. A. Ryburn, O. Fedrigo, G. J. P. Naylor. 2006. Phylogeny of sharks of the family Triakidae (Carcharhiniformes) and its implications for the evolution of carcharhiniform placental viviparity. *Mol. Phylogen. Evol*. 40:50–60.

Love, M. S., M. Yoklavich, and L. Thorsteinson. 2002. The rockfishes of the northeast Pacific. University of California Press, Berkeley. 404 p.

Lovejoy, N. R. 1996. Systematics of myliobatoid elasmobranchs: With emphasis on the phylogeny and historical biogeography of neotropical freshwater stingrays (Potamotrygonidae: Rajiformes). *Zool. J. Linn. Soc*. 117:207–57.

————. 2000. Reinterpreting recapitulation: Systematics of needlefishes and their allies. (Teleostei: Beloniformes). *Evolution* 54:1349–62.

Lovejoy, N. R., and B. B. Collette. 2001. Phylogenetic relationships of New World needlefishes (Teloeosteia: Belonidae) and the biogeography of transitions between marine and freshwater habitats. *Copeia* 2001:324–38.

Lovejoy, N. R., M. Iranpour, and B. B. Collette. 2004. Phylogeny and jaw ontogeny of Beloniform fishes. *Integr. Comp. Biol*. 44:366–77.

Lucinda, P. H. F., and R. E. Reis. 2005. Systematics of the subfamily Poeciliinae Bonaparte (Cyprinodontiformes: Poeciliidae), with an emphasis on the tribe Cnesterodontini Hubbs. *Neotrop. Ichthyol*. 3:1–60.

Lund, R., and E. D. Grogan. 1997. Relationships of the Chimaeriformes and the basal radiation of the Chondrichthyes. *Rev. Fish Biol. Fish.* 7:65–123.

Lundberg, J. G. 1992. The phylogeny of ictalurid catfishes: A synthesis of recent work, pp. 392–420. In *Systematics, historical ecology, and North American freshwater fishes.* R. L. Mayden (ed.). Stanford University Press, Stanford, California.

Lundberg, J. G., and J. N. Baskin. 1969. The caudal skeleton of the catfishes, order Siluriformes. *Amer. Mus. Novit.* 2398:1–49.

Lundberg, J. G., and E. Marsh. 1976. Evolution and functional anatomy of the pectoral fin rays in cyprinoid fishes, with emphasis on the suckers (family Catostomidae). *Amer. Midl. Nat.* 96:332–49.

Mabee, P. M. 1993. Phylogenetic interpretation of ontogenetic change: Sorting out the actual and artefactual in an empirical case study of centrarchid fishes. *Zool. J. Linn. Soc.* 107:175–291.

Mabee, P. M., E. A. Grey, G. Arratia, N. Bogutskaya, A. Boron, M. M. Coburn, K. W. Conway, et al. 2011. Gill arch and hyoid diversity and cypriniform phylogeny: Distributed integration of morphology and web-based tools. *Zootaxa* 2877:1–40.

Mabuchi, K., M. Miya, Y. Azuma, and M. Nishida. 2007. Independent evolution of the specialized pharyngeal jaw apparatus in cichlid and labrid fishes. *BMC Evol. Biol.* 7:10. doi:10.1186/1471–2148–7-10.

Maisey, J. G. 2012. What is an "elasmobranch"? The impact of paleontology in understanding elasmobranch phylogeny and evolution. *J. Fish Biol.* 80:918–51.

Maisey, J. G., G.J.P. Naylor, and D.J. Ward. 2004. Mesozoic elasmobranchs, neoselachian phylogeny and the rise of modern elasmobranch diversity, pp. 17–56. In *Mesozoic Fishes 3.* G. Arratia and A. Tintori (eds.). Verlag F. Pfeil, Munich.

Marceniuk, A. P., and C. J. Ferraris, Jr. 2003. Family Ariidae (sea catfishes), pp. 447–55. In *Checklist of the freshwater fishes of South and Central America.* R. E. Reis, S. O. Kullander, and C. J. Ferraris, Jr. (eds.). EDIPUCRS, Porto Alegre, Brazil.

Marceniuk, A. P., and N. A. Menezes. 2007. Systematics of the family Ariidae (Ostariophysi, Siluriformes), with a redefinition of the genera. *Zootaxa* 1416:1–126.

Markle, D. F. 1989. Aspects of character homology and phylogeny of the Gadiformes. *Nat. Hist. Mus. Los Angeles Co. Sci. Ser.* 32:59–88.

Matheson, R. E., Jr., and J. D. McEachran. 1984. Taxonomic studies of the *Eucinostomus argenteus* complex (Pisces: Gerreidae): Preliminary studies of external morphology. *Copeia* 1984:893–902.

Matsuura, K. 1979. Phylogeny of the superfamily Balistoidea (Pisces: Tetraodontiformes). *Mem. Fac. Fish. Hokkaido Univ.* 26:49–169.

Mattern, M. Y. 2004. Molecular phylogeny of the Gasterosteidae: The importance of using multiple genes. *Mol. Phylogen. Evol.* 30:366–77.

Mattern, M. Y., and D. A. McLennan. 2004. Total evidence phylogeny of Gasterosteidae: Combining molecular, morphological and behavioral data. *Cladistics* 20:14–22.

Mayden, R. L. (ed.). 1992. *Systematics, historical ecology, and North American freshwater fishes.* Stanford University Press. Stanford, California. 970 p.

Mayden, R. L., W.-J. Chen, H. L. Bart, M. H. Doosey, A. M. Simons, K. L. Tang, R. M. Wood, et al. 2009. Reconstructing the phylogenetic relationships of the earth's most diverse clade of freshwater fishes: Order Cypriniformes (Actinopterygii: Ostariophysi): A case study using multiple nuclear loci and the mitochondrial genome. *Mol. Phylogen. Evol.* 51:500–14.

Mayden, R. L., K. L. Tang, R. M. Wood, W.-J. Chen, M. K. Agnew, K. W. Conway, L. Yang, et al. 2008. Inferring the tree of life of the order Cypriniformes, the earth's most diverse clade of freshwater fishes: Implications of varied taxon and character sampling. *J. System. Evol.* 46:424–38.

McAllister, D. E. 1963. A revision of the smelt family, Osmeridae. *Bull. Nat. Mus. Can.* 191:1–53.

———. 1968. Evolution of branchiostegals and classification of teleostome fishes. *Bull. Nat. Mus. Can.* 221:1–239.

McBride, R. S., C. R. Rocha, R. Ruiz-Carus, and B. W. Bowen. 2010. A new species of ladyfish, of the genus *Elops* (Elopiformes: Elopidae), from the western Atlantic Ocean. *Zootaxa* 2346:29–41.

McCormick, M. I., and M. J. Milicich. 1993. Late pelagic-stage goatfishes: Distribution patterns and inferences on schooling behaviour. *J. Exp. Mar. Biol. Ecol.* 174:15–42.

McCosker, J. E. 1977. The osteology, classification, and relationships of the eel family Ophichthidae. *Proc. Cal. Acad. Sci.* 41:1–123.

———. 2010. Deepwater Indo-Pacific species of the snake-eel genus *Ophichthus* (Anguilliformes: Ophichthidae), with the description of nine new species. *Zootaxa* 2505:1–39.

McCosker, J. E., E. B. Böhlke, and J. E. Böhlke. 1989. Family Ophichthidae, pp. 254–412. In *Fishes of the western North Atlantic*. E. B. Böhlke (ed.). Part 9. Vol. 1. Orders Anguilliformes and Saccopharyngiformes. Sears Foundation for Marine Research, Memoir (Yale University), New Haven, Connecticut.

McCosker, J. E., and M. D. Lagios (eds.). 1979. The biology and physiology of the living coelacanth. *Occ. Pap. Cal. Acad. Sci.* 134:1–175.

McCosker, J. E., and R. H. Rosenblatt. 1998. A revision of the Eastern Pacific snake-eel genus *Ophichthus* (Anguilliformes: Ophichthidae) with the description of six new species. *Proc. Cal. Acad. Sci.* 50:397–432.

McEachran, J. D., and N. Aschliman. 2004. Phylogeny of Batoidea, pp. 79–113. In *Biology of sharks and their relatives*. J. C. Carrier, J. A. Musick, and M. R. Heithaus (eds.). CRC Press, Boca Raton, Florida.

McEachran, J. D., and K. A. Dunn. 1998. Phylogenetic analysis of skates, a morphologically conservative clade of elasmobranches (Chondrichthyes: Rajidae). *Copeia* 1998:271–90.

McEachran, J. D., K. A. Dunn, and T. Miyake. 1998. Interrelationships of the batoid fishes (Chondrichthyes: Batoidea), pp. 63–84. In *Interrelationships of fishes*. M. L. J. Stiassny, L. R. Parenti, and G. D. Johnson (eds.). Academic Press, San Diego, California.

McEachran, J. D., and J. D. Fechhelm. 1998. *Fishes of the Gulf of Mexico. Volume 1: Myxiniformes to Gasterosteiformes*. University of Texas Press, Austin. 1,112 p.

McEachran, J. D., and J. D. Fechhelm. 2005. *Fishes of the Gulf of Mexico. Volume 2: Scorpaeniformes to Tetraodontiformes*. University of Texas Press, Austin. 1,004 p.

McKinnon, J. S., and H. D. Rundle. 2002. Speciation in nature: The threespine stickleback model systems. *Trends Ecol. Evol.* 17:480–88.

Mecklenburg, C. W., and W. N. Eschmeyer. 2003. Family Hexagrammidae Gill 1889: Greenlings. *Calif. Acad. Sci. Annotated Checklists of Fishes*. No. 3. http://research.calacademy.org/ichthyology/checklists.

Mecklenburg, C. W., T. A. Mecklenburg, and L. K. Thorsteinson. 2002. *Fishes of Alaska*. American Fisheries Society, Bethesda, Maryland. 1,037 p.

Meffe, G. K., and F. F. Snelson, Jr. (eds.). 1989. *Ecology and evolution of livebearing fishes (Poeciliidae)*. Prentice Hall, Inc., Englewood Cliffs, New Jersey. 453 p.

Mehta, R. S., and P. C. Wainwright. 2008. Functional morphology of the pharyngeal jaw apparatus in moray eels. *J. Morph.* 269:604–19.

Menon, A. G. K. 1977. A systematic monograph of the tongue soles of the genus *Cynoglossus* Hamilton-Buchanan (Pisces: Cynoglossidae). *Smithson. Contr. Zool.* 238:1–129.

Meyer, A., and Y. Van de Peer. 2005. From 2R to 3R: Evidence for a fish-specific genome duplication (FSGD). *Bioessays* 27:937–45.

Meyer, A., and R. Zardoya. 2003. Recent advances in the (molecular) phylogeny of vertebrates. *Annu. Rev. Ecol. Evol. Syst.* 34:311–38.

Miller, D. J., and R. N. Lea. 1972. *Guide to the coastal marine fishes of California.* Cal. Dept. Fish Game, Fish Bull. 157. 249 p.

Miller, M. E., and J. Stewart. 2009. The commercial fishery for ocean leatherjackets (*Nelusetta ayraudi*, Monacanthidae) in New South Wales, Australia. *Asian Fish Sci.* 22:257–64.

Miller, R. G. 1993. *History and atlas of the fishes of the Antarctic Ocean.* Foresta Institute for Ocean and Mountain Studies, Carson City, Nevada. 792 p.

Miller, R. R., W. L. Minckley, and S. M. Norris. 2005. *Freshwater fishes of Mexico.* University of Chicago Press, Chicago, Illinois. 652 p.

Miller, T. L., and T. H. Cribb. 2007. Phylogenetic relationships of some common Indo-Pacific snappers (Perciformes: Lutjanidae) based on mitochondrial DNA sequences, with comments on the taxonomic position of the Caesioninae. *Mol. Phylogen. Evol.* 44:450–60.

Minckley, W. L., and J. E. Deacon. 1991. *Battle against extinction.* University of Arizona Press, Tucson, Arizona. 517 p.

Miya, M., M. Friedman, T. P. Satoh, H. Takeshima, T. Sado, W. Iwasaki, Y. Yamanoue, et al. 2013. Evolutionary origin of the Scombridae (tunas and mackerels): Members of a Paleogene adaptive radiation with fourteen other pelagic fish families. *PLoS One* 8 (9):e73535.

Miya, M., N. I. Holcroft, T. P. Satoh, M. Yamaguchi, M. Nishida, and E. O. Wiley. 2007. Mitochondrial genome and a nuclear gene indicate a novel phylogenetic position of deep-sea tube-eye fish (Stylephoridae). *Ichthyol. Res.* 54:323–32.

Miya, M., and M. Nishida. 1996. Molecular phylogenetic perspective on the evolution of the deep-sea fish genus *Cyclothone* (Stomiiformes: Gonostomatidae). *Ichthyol. Res.* 43:375–98.

———. 1997. Speciation in the open ocean. *Nature* 389:803–4.

———. 1998. Molecular phylogeny and evolution of the deep-sea fish genus *Sternoptyx*. *Mol. Phylogen. Evol.* 10:11–22.

———. 2000. Molecular systematics of the deep-sea fish genus *Gonostoma* (Stomiiformes: Gonostomatidae): Two paraphyletic clades and resurrection of *Sigmops*. *Copeia* 2000:378–89.

Miya, M., T. W. Pietsch, J. W. Orr, R. J. Arnold, T. P. Satoh, A. M. Shedlock, H. C. Ho, M. Shimazaki, M. Yabe, and M. Nishida. 2010. Evolutionary history of anglerfishes (Teleostei: Lophiiformes): A mitogenomic perspective. *BMC Evol. Biology* 10:58–85.

Miya, M., T. P. Satoh, and M. Nishida. 2005. The phylogenetic position of toadfishes (order Batrachoidiformes) in the higher ray-finned fish as inferred from partitioned Bayesian analysis of 102 whole mitochondrial genome sequences. *Biol. J. Linn. Soc.* 85:289–306.

Miya, M., H. Takeshima, H. Endo, N. B. Ishiguro, J. G. Inoue, T. Mukai, T. P. Satoh, et al. 2003. Major patterns of higher teleostean phylogenies: A new perspective based on 100 complete mitochondrial DNA sequences. *Mol. Phylogen. Evol.* 26:121–38.

Møller, P. R., W. Schwarzhans, and J. G. Nielsen. 2004. Review of the American Dinematichthyini (Teleostei, Bythitidae). Part I: *Dinematichthys, Gunterichthys, Typhliasina* and two new genera. *Aqua, J. Ichthyol. Aquat. Biol.* 8:141–92.

Møller, P. R., W. Schwarzhans, and J. G. Nielsen. 2005. Review of the American Dinematichthyini (Teleostei, Bythitidae). Part II: *Ogilbia. Aqua, J. Ichthyol. Aquat. Biol.* 10:133–205.

Montgomery, J., and K. Clements. 2000. Disaptation and recovery in the evolution of Antarctic fishes. *Trends Ecol. Evol.* 15:267–71.

Mooi, R. D., and A. C. Gill. 1995. Association of epaxial musculature with dorsal-fin pterygiophores in acanthomorph fishes, and its phylogenetic significance. *Bull. Nat. Hist. Mus. London (Zool.)* 61:121–37.

Mooi, R. D., and G. D. Johnson. 1997. Dismantling the Trachinoidei: Evidence of a scorpaenoid relationship for the Champsodontidae. *Ichthyol. Res.* 44:143–76.

Moore, J. A. 1993. The phylogeny of the Trachichthyiformes (Teleostei: Percomorpha). *Bull. Mar. Sci.* 52:114–36.

Morita, T. 1999. Molecular phylogenetic relationships of the deep-sea fish genus *Coryphaenoides* (Gadiformes: Macrouridae) based on mitochondrial DNA. *Mol. Phylogen. Evol.* 13:447–54.

Moser, H. G., and E. H. Ahlstrom. 1978. Larvae and pelagic juveniles of blackgill rockfish, *Sebastes melanostomus*, taken in midwater trawls off southern California. *J. Fish. Res. Bd. Can.* 35:981–96.

Motomura, H. 2002. Revision of the Indo-Pacific threadfin genus *Polydactylus* (Perciformes: Polynemidae) with a key to the species. *Bull. Nat. Sci. Mus., Tokyo, Series A (Zoology)*, 28:171–94.

———. 2004. *Threadfins of the world (family Polynemidae): An annotated and illustrated catalogue of polynemid species known to date. FAO Species Catalogue for Fishery Purposes.* No. 3. FAO, Rome. 117 p.

Moyle, P. B. and J. J. Cech, Jr. 2004. *Fishes: An introduction to ichthyology.* 5th ed. Prentice Hall, Upper Saddle River, New Jersey. 726 p.

Munday, P. L., T. Kuwamura, and F. J. Kroon. 2010. Bi-directional sex change in marine fishes, pp. 241–71. In *Reproduction and Sexuality in Marine Fishes: Patterns and Processes.* K. S. Cole (ed.). University of California Press, Berkeley, California.

Muñoz, M. 2010. Reproduction in Scorpaeniformes, pp. 65–89. In *Reproduction and Sexuality in Marine Fishes: Patterns and Processes.* K. S. Cole (ed.). University of California Press, Berkeley, California.

Munroe, T. A. 1992. Interdigitation pattern of dorsal-fin pterygiophores and neural spines, an important diagnostic character for symphurine tonguefishes (*Symphurus:* Cynoglossidae: Pleuronectiformes). *Bull. Mar. Sci.* 50:357–403.

———. 1998. Systematics and ecology of the tonguefishes of the genus *Symphurus* (Cynoglossidae: Pleuronectiformes) from the western Atlantic Ocean. *Fish Bull.* 96:1–182.

Munroe, T. A., and J. Hashimoto. 2008. A new western Pacific tonguefish (Pleuronectiformes: Cynoglossidae): The first pleuronectiform discovered at active hydrothermal vents. *Zootaxa* 1839:43–59.

Murray, A. M., and M. V. H. Wilson. 1999. Contributions of fossils to the phylogenetic relationships of the percopsiform fishes (Teleostei: Paracanthopterygii): Order restored, pp. 397–411. In *Mesozoic Fishes 2: Systematics and Fossil Record.* G. Arratia and H.-P. Schultze (eds.). Die Deutsche Bibliothek, Munchen.

Musick, J. A. 2011. Chondrichthyan reproduction, pp. 3–19. In *Reproduction in marine fishes: Evolutionary patterns and innovations.* K. S. Cole (ed.). University of California Press, Berkeley, California.

Musick, J. A., M. N. Bruton, and E. K. Balon. 1991. *The biology of Latimeria chalumnae and the evolution of coelacanths.* Kluwer Acad. Publ., Netherlands. 438 p.

Musick, J. A., and J. K. Ellis. 2005. Reproductive evolution of chondrichthyans, pp. 45–79. In *Reproductive biology and phylogeny of Chondrichthyes.* W. C. Hamlett (ed.). Science Press, Enfield, New Hampshire.

Myrberg, A. A., Jr. 1972. Ethology of the bicolor damselfish *Eupomacentrus partitus* (Pisces: Pomacentridae): A comparative analysis of laboratory and field behavior. *Anim. Beh. Monog.* 5:197–283.

Nafpaktitis, B. G. 1977. Family Neoscopelidae, pp. 1–12. In *Fishes of the western North Atlantic.* R. H. Gibbs, Jr., et al. (eds.). Part 7. Vol. 1. Sears Foundation for Marine Research, Memoir (Yale University), New Haven, Connecticut.

———. 1978. Systematics and distribution of lanternfishes of the genera *Lobianchia* and *Diaphus* (Myctophidae) in the Indian Ocean. *Sci. Bull. Nat. Hist. Mus. Los Angeles Co.* 30:1–92.

Nafpaktitis, B.G., et. al. 1977. Family Myctophidae, pp. 13–265. In *Fishes of the western North Atlantic*. R.H. Gibbs, Jr., et al. (eds.). Part 7. Vol. 1. Sears Foundation for Marine Research, Memoir (Yale University), New Haven, Connecticut.

Nagareda, B.H., and J.M. Shenker. 2008. Dietary analysis of batfishes (Lophiiformes: Ogcocephalidae) in the Gulf of Mexico. *Gulf Mex. Sci.* 26: 28–35.

———. 2009. Evidence for chemical luring in the polka-dot batfish *Ogcocephalus cubifrons* (Teleostei: Lophiiformes: Ogcocephalidae). *Florida Sci.* 72:11–17.

Nakamura, I. 1985. *Billfishes of the world: An annotated and illustrated catalogue of marlins, sailfishes, spearfishes and swordfishes known to date*. FAO species catalogue. *FAO Fish. Synop.* 125, Vol. 5. FAO, Rome. 65 p.

Nakamura, I., and N.V. Parin. 1993. *Snake mackerels and cutlassfishes of the world (families Gempylidae and Trichiuridae)*. FAO Species Catalogue. *FAO Fish. Synop.* 125, Vol. 15. FAO, Rome. 136 p.

Nakatani, M., M. Miya, K. Mabuchi, K. Saitoh, and M. Nishida. 2011. Evolutionary history of Otophysi (Teleostei), a major clade of the modern freshwater fishes: Pangaean origin and Mesozoic radiation. *BMC Evol. Biol.* 11:177.

Naylor, G.J.P. 1992. The phylogenetic relationships among requiem and hammerhead sharks: Inferring phylogeny when thousands of equally parsimonious trees result. *Cladistics* 8:295–318.

Naylor, G.J.P., J.N. Caira, K. Jensen, K.A.M. Rosana, W.T. White, and P.R. Last. 2012. A DNA sequence–based approach to the identification of shark and ray species and its implications for global elasmobranch diversity and parasitology. *Bull. Amer. Mus. Nat. Hist.* 367:1–262.

Naylor, G.J.P., A.P. Martin, E.G. Mattison, and W.M. Brown. 1997. Interrelationships of lamniform sharks: Testing phylogenetic hypotheses with sequence data, pp. 199–218. In *Molecular Systematics of Fishes*. T.D. Kocher and C.A. Stepien (eds.). Academic Press, San Diego, California.

Naylor, G.J.P., J.A. Ryburn, O. Fedrigo, and J.A. López. 2005. Phylogenetic relationships among the major lineages of modern elasmobranchs, pp. 1–26. In *Reproductive biology and phylogeny of Chondrichthyes*. W.C. Hamlett (ed.). Science Press, Enfield, New Hampshire.

Near, T.J., D.I. Bolnick, and P.C. Wainwright. 2004. Investigating phylogenetic relationships of sunfishes and black basses (Actinopterygii: Centrarchidae) using DNA sequences from mitochondrial and nuclear genes. *Mol. Phylogen. Evol.* 32:344–57.

Near, T.J., C.M. Bossu, G.S. Bradburd, R.L. Carlson, R.C. Harrington, P.R. Hollingsworth, B.P. Keck, and D.A. Etnier. 2011. Phylogeny and temporal diversification of darters (Percidae: Etheostomatinae). *Syst. Biol.* 60 (5):565–95.

Near, T.J., A. Dornburg, R.I. Eytan, B.P. Keck, W.L. Smith, K.L. Kuhn, J.A. Moore, et al. 2013. Phylogeny and tempo of diversification in the superradiation of spiny-rayed fishes. *Proc. Nat. Acad. Sci.* 110:12738–43.

Near, T.J., R.I. Eytan, A. Dornburg, K.L. Kuhn, J.A. Moore, M.P. Davis, P.C. Wainwright, M. Friedman, and W.L. Smith. 2012. Resolution of ray-finned fish phylogeny and timing of diversification. *Proc. Nat. Acad. Sci.* 109:13698–703.

Near, T.J., T.W. Kassler, J.B. Koppelman, C.B. Dillman, and D.P. Philipp. 2003. Speciation in North American black basses, *Micropterus*. *Evolution* 57:1610–21.

Near, T.J., J.J. Pesavento, and C.-H.C. Cheng. 2003. Mitochondrial DNA, morphology, and the phylogenetic relationships of Antarctic icefishes (Notothenioidei: Channichthyidae). *Mol. Phylogen. Evol.* 28:87–98.

———. 2004. Phylogenetic investigations of Antarctic notothenioid fishes (Perciformes: Notothenioidei) using complete gene sequences of the mitochondrial encoded 16S rRNA. *Mol. Phylogen. Evol.* 32:881–91.

Near, T. J., M. Sandel, K. L. Kuhn, P. J . Unmack, P. C. Wainwright, and W. L. Smith. 2012. Nuclear gene-inferred phylogenies resolve the relationships of the enigmatic pygmy sunfishes, *Elassoma* (Teleostei: Percomorpha). *Mol. Phylogen. Evol.* 63:388–95.

Nelson, G. J. 1972. Cephalic sensory canals, pitlines, and the classification of esocoid fishes, with notes on galaxiids and other teleosts. *Amer. Mus. Novit.* 2492:1–49.

———. 1984. Notes on the rostral organ of anchovies (Family Engraulidae). *Jap. J. Ichthyol.* 31:86–87.

Nelson, J. S. 2006. (and previous editions) *Fishes of the world.* John Wiley and Sons, Hoboken, New Jersey. 601 p.

Nelson, J. S., H.-P. Schultze, and M. V. H. Wilson (eds.). 2010. *Origin and phylogenetic interrelationships of teleosts.* Verlag Dr. Friedrich Pfeil, Munchen. 480 p.

Nieder, J. 2001. Amphibious behaviour and feeding ecology of the four-eyed blenny (*Dialommus fuscus*, Labrisomidae) in the intertidal zone of the island of Santa Cruz (Galapagos, Ecuador). *J. Fish Biol.* 58:755–67.

Nielsen, J. G., E. Bertelsen, and A. Jespersen. 1989. The biology of *Eurypharynx pelecanoides* (Pisces, Eurypharyngidae). *Acta Zool. (Stokh.)* 70:187–97.

Nielsen, J. G., D. M. Cohen, D. F. Markle, and C. R. Robins. 1999. *Ophidiiform fishes of the world (Order Ophidiiformes): An annotated and illustrated catalogue of pearlfishes, cusk-eels, brotulas and other ophidiiform fishes known to date.* FAO Species Catalogue. *FAO Fish. Synop.* 125, Vol. 18. FAO, Rome. 178 p.

Nielsen, J. G., P. R. Møller, and M. Segonzac. 2006. *Ventichthys biospeedoi* n. gen. et sp. (Teleostei, Ophidiidae) from a hydrothermal vent in the South East Pacific. *Zootaxa* 1247:13–24.

Nolf, D., and J. C. Tyler. 2006. Otolith evidence concerning interrelationships of caproid, zeiform and tetraodontiform fishes. *Bull. L'Inst. Royal Sci. Nat. Belgique, Biol.* 76:147–80.

Nordlie, F. G. 2012. Life-history characteristics of eleotrid fishes of the western hemisphere, and perils of life in a vanishing cnvironment. *Rev. Fish Biol. Fish.* 22:189–224.

Norman, J. R. 1934. *A systematic monograph of the flatfishes (Heterosomata).* Vol. I. *Psettodidae, Bothidae, Pleuronectidae.* British Museum (Natural History), London. 459 p.

Notarbartolo-Di-Sciara, G. 1987. A revisionary study of the genus *Mobula* Rafinesque, 1810 (Chondrichthyes: Mobulidae) with the description of a new species. *Zool. J. Linn. Soc.* 91:1–91.

Obermiller, L. E., and E. Pfeiler. 2003. Phylogenetic relationships of elopomorph fishes inferred from mitochondrial ribosomal DNA sequences. *Mol. Phylogen. Evol.* 26:202–14.

Odum, W. E. 1970. Utilization of the direct grazing and plant detritus food chains by the striped mullet *Mugil cephalus*, pp. 222–40. In *Marine food chains.* J. H. Steele (ed.). University of California Press, Berkeley, California.

Oliveira, C., G. Avelino, K. Abe, T. Mariguela, R. Benine, G. Ortí, R. Vari, et al. 2011. Phylogenetic relationships within the speciose family Characidae (Teleostei: Ostariophysi: Characiformes) based on multilocus analysis and extensive ingroup sampling. *BMC Evol. Biol.* 11:275.

Olney, J. E., G. D. Johnson, and C. C. Baldwin. 1993. Phylogeny of lampridiform fishes. *Bull. Mar. Sci.* 52:137–69.

Orlov, A. M., and T. Iwamoto (eds.). 2008. *Grenadiers of the world oceans: Biology, stock assessment, and fisheries.* Symposium 63. American Fisheries Society, Bethesda, Maryland. 484 p.

Orrell, T. M., K. E. Carpenter, J. A. Musick, and J. E. Graves. 2002. Phylogenetic and biogeographic analysis of the Sparidae (Perciformes: Percoidei) from cytochrome b sequences. *Copeia* 2002:618–31.

Orrell, T. M., B. B. Collette, and G. D. Johnson. 2006. Molecular data support separate scombroid and xiphioid clades. *Bull. Mar. Sci.* 79:505–19.

Ortí, G., and A. Meyer. 1997. The radiation of characiform fishes and the limits of resolution of mitochondrial ribosomal DNA sequences. *Syst. Biol.* 46:75–100.

Otero, O. 2004. Anatomy, systematics, and phylogeny of both recent and fossil latid fishes (Teleostei, Perciformes, Latidae). *Zool. J. Linn. Soc.* 141:81–133.

O'Toole, B. 2002. Phylogeny of the species of the superfamily Echeneoidea (Perciformes: Carangoidei: Echeneidae, Rachycentridae, and Coryphaenidae), with an interpretation of the echeneid hitchhiking behaviour. *Can. J. Zool.* 80:596–623.

Page, L. M. 1981. The genera and subgenera of darters (Percidae, Etheostomatini). *Occas. Pap. Mus. Nat. Hist. Univ. Kansas* 90:1–69.

———. 1983. *Handbook of darters.* T.F.H. Publications, Neptune City, New Jersey. 271 p.

Page, L. M., and B. M. Burr. 2011. *A field guide to freshwater fishes of North America north of Mexico.* 2nd ed. Peterson Field Guides. Houghton Mifflin Harcourt, Boston, Massachusetts. 688 p.

Page, L. M., and D. L. Swofford. 1984. Morphological correlates of ecological specialization in darters. *Env. Biol. Fishes.* 11:139–59.

Parenti, L. R. 1981. A phylogenetic and biogeographic analysis of the cyprinodontiform fishes (Teleostei, Atherinomorpha). *Bull. Amer. Mus. Nat. Hist.* 168:335–557.

———. 1993. Relationships of atherinomorph fishes (Teleostei). *Bull. Mar. Sci.* 52:170–96.

———. 2005. The phylogeny of atherinomorphs: Evolution of a novel fish reproductive system, pp. 13–30. In *Viviparous fishes.* M. C. Uribe and H. J. Grier (eds.). New Life Publications, Homestead, Florida.

———. 2008. A phylogenetic analysis and taxonomic revision of ricefishes, *Oryzias* and relatives (Beloniformes, Adrianichthyidae). *Zool. J. Linn. Soc.* 154:494–610.

Parenti, L. R., and M. Rauchenberger. 1989. Systematic overview of the poeciliins, pp. 3–12. In *Ecology and evolution of livebearing fishes (Poeciliidae).* G. K. Meffe and F. F. Snelson, Jr. (eds.). Prentice Hall, Englewood Cliffs, New Jersey.

Parenti, L. R., and J. Song. 1996. Phylogenetic significance of the pectoral-pelvic fin association in acanthomorph fishes: A reassessment using comparative neuroanatomy, pp. 427–44. In *Interrelationships of fishes.* M. L. J. Stiassny, L. R. Parenti, and G. D. Johnson (eds.). Academic Press, San Diego, California.

Parenti, P. 2003. Family Molidae Bonaparte 1832: Molas or ocean sunfishes. *Calif. Acad. Sci. Annotated Checklists of Fishes* 18. http://research.calacademy.org/ichthyology/checklists.

Parenti, P., and J. E. Randall. 2000. An annotated checklist of the species of the labroid fish families Labridae and Scaridae. *Ichthyol. Bull. J. L. B. Smith Inst. Ichthyol.* 68:1–97.

———. 2010. Checklist of the species of the families Labridae and Scaridae: An update. *Smithiana Bull.* 13:29–44.

Parin, N. V., and A. S. Piotrovsky. 2004. Stromateoid fishes (suborder Stromateoidei) of the Indian Ocean (species composition, distribution, biology, and fisheries). *J. Ichthyol.* 44 (Suppl. 1):S33-S62.

Partridge, J. C., and R. H. Douglas. 1995. Far-red sensitivity of dragon fish. *Nature* 375:21–22.

Patterson, C. 1965. The phylogeny of the chimaeroids. *Phil. Trans. Royal Soc. London* 249:101–219.

———. 1973. Interrelationships of holosteans, pp. 233–305. In *Interrelationships of Fishes.* P. H. Greenwood, R. S. Miles, and C. Patterson (eds.). *Zool. J. Linn. Soc.* 53, Suppl. 1. Academic Press, London.

———. 1982. Morphology and interrelationships of primitive actinopterygian fishes. *Amer. Zool.* 22:241–59.

Patterson, C., and G. D. Johnson. 1995. The intermuscular bones and ligaments of teleostean fishes. *Smithson. Contr. Zool.* 559:1–83.

———. 1997. Comments on Begle's "Monophyly and relationships of argentinoid fishes." *Copeia* 1997: 401–9.

Patterson, C., and D. E. Rosen. 1989. The Paracanthopterygii revisited. *Sci. Ser. Nat. Hist. Mus. Los Angeles Co.* 32:5–36.

Patzner, R.A., E.J. Gonçlaves, P.A. Hastings, and B.G. Kapoor (eds.). 2009. *The biology of blennies.* Science Publishers, Enfield, New Hampshire. 482 p.

Patzner, R.A., J.L. Van Tassel, M. Kovačić, and B.G. Kapoor (eds). 2011. *The biology of gobies.* Science Publishers, Enfield, New Hampshire. 685 p.

Paxton, J.R. 1972. Osteology and relationships of the lanternfishes (family Myctophidae). *Bull. Nat. Hist. Mus. Los Angeles Co.* 13:1–81.

———. 1979. Nominal genera and species of lanternfishes (family Myctophidae). *Nat. Hist. Mus. Los Angeles Co. Contr. Sci.* 322:1–28.

———. 1989. Synopsis of the whalefishes (family Cetomimidae) with descriptions of four new genera. *Rec. Aust. Mus.* 41:135–206.

Paxton, J.R., G.D. Johnson, and T. Trnski. 2001. Larvae and juveniles of the deepsea "whalefishes" *Barbourisia* and *Rondeletia* (Stephanoberyciformes: Barbourisiidae, Rondeletiidae), with comments on family relationships. *Rec. Aust. Mus.* 53:407–25.

Peng, Z., S. He, J. Wang, W. Wang, and R. Diogo. 2006. Mitochondrial molecular clocks and the origin of the major Otocephalan clades (Pisces: Teleostei): A new insight. *Gene* 370:113–24.

Petersen, C.W. 1995. Male mating success and female choice in permanently territorial damselfishes. *Bull. Mar. Sci.* 57:690–704.

Petersen, C.W., C. Mazzoldi, K.A. Zarrella, and R.E. Hale. 2005. Fertilization mode, sperm characteristics, mate choice and parental care patterns in *Artedius* spp. (Cottidae). *J. Fish Biol.* 67:239–59.

Pezold, F. 1993. Evidence for a monophyletic Gobiinae. *Copeia* 1993:634–43.

Pezold, F., and B. Cage. 2002. A review of the spinycheek sleepers, genus *Eleotris* (Teleostei: Eleotridae), of the Western Hemisphere, with comparisons to the west African species. *Tulane Stud. Zool. Bot.* 31:19–63.

Pfeiler, E., B.G. Bitler, R. Ulloa, A.M. van der Heiden, and P.A. Hastings. 2008. Molecular identification of the bonefish *Albula esuncula* (Albuliformes: Albulidae) from the tropical eastern Pacific, with comments on distribution and morphology. *Copeia* 2008:763–70.

Pietsch, T.W. 1978. Evolutionary relationships of the sea moths (Teleostei: Pegasidae) with a classification of gasterosteiform families. *Copeia* 1978:517–29.

———. 1981. The osteology and relationships of the anglerfish genus *Tetrabrachium,* with comments on lophiiform classification. *Fish. Bull.* 79:387–419.

———. 1984. The genera of frogfishes (Family Antennariidae). *Copeia* 1984:27–44.

———. 1989. Phylogenetic relationships of trachinoid fishes of the family Uranoscopidae. *Copeia* 1989:253–303.

———. 2005. Dimorphism, parasitism, and sex revisited: Modes of reproduction among deep-sea ceratioid anglerfishes (Teleostei: Lophiiformes). *Ichthyol. Res.* 52:207–36.

———. 2009. *Oceanic anglerfishes: Extraordinary diversity in the deep sea.* University of California Press, Berkeley, California. 557 p.

Pietsch, T.W., R.J. Arnold, and D.J. Hall. 2009. A bizarre new species of frogfish of the genus *Histiophryne* (Lophiiformes: Antennariidae) from Ambon and Bali, Indonesia. *Copeia* 2009:37–45.

Pietsch, T.W., and D.B. Grobecker. 1978. The compleat angler: Aggressive mimicry in an antennariid anglerfish. *Science* 201:369–70.

———. 1987. *Frogfishes of the world.* Stanford University Press, Stanford, California. 420 p.

Pietsch, T.W., and J.W. Orr. 2007. Phylogenetic relationships of deep-sea anglerfishes of the suborder Ceratioidei (Teleostei: Lophiiformes) based on morphology. *Copeia* 2007:1–34.

Pietsch, T.W., and C.P. Zabetian. 1990. Osteology and interrelationships of the sand lances (Teleostei: Ammodytidae). *Copeia* 1990:78–100.

Poly, W. J., and J. E. Wetzel. 2003. Transbranchial spawning: Novel reproductive strategy observed for the pirate perch, *Aphredoderus sayanus* (Aphredoderidae). *Ichthyol. Explor. Fresh.* 14:151–58.

Potter, I. C. 1980. The Petromyzontiformes with particular reference to paired species. *Can. J. Fish. Aquat. Sci.* 37:1595–1615.

Potts, G. W. 1985. The nest structure of the corkwing wrasse, *Crenilabrus melops* (Labridae: Teleostei). *J. Mar. Biol. Assoc. UK* 65:531–46.

Poulsen, J. Y., P. R. Møller, S. Lavoué, S. W. Knudsen, M. Nishida, and M. Miya. 2009. Higher and lower-level relationships of the deep-sea fish order Alepocephaliformes (Teleostei: Otocephala) inferred from whole mitogenome sequences. *Biol. J. Linn. Soc. Lond.* 98:923–36.

Poulson, T. L. 1963. Cave adaptation in amblyopsid fishes. *Amer. Midl. Nat.* 70:257–90.

Pouyaud, L., S. Wirjoatmodjo, I. Rachmatika, A. Tjakrawidjaja, R. Hadiaty, and W. Hadie. 1999. Une nouvelle espece de coelacanthe: Preuves genetiques et morphologiques. *C. R. Acad. Sci. Paris, Sci. Vie* 322:261–67.

Poyato-Ariza, F. J. 1996. A revision of the ostariophysan fish family Chanidae, with special reference to the Mesozoic forms. *Palaeo. Ichthyol.* 6:1–52.

Pyle, R. L., and J. E. Randall. 1994. A review of hybridization in marine angelfishes (Perciformes: Pomacanthidae). *Env. Biol. Fishes* 41:127–45.

Quast, J. C. 1965. Osteological characteristics and affinities of the hexagrammid fishes, with a synopsis. *Proc. Cal. Acad. Sci. Ser.* 4:563–600.

Quéro, J.-C., J.-C. Hureau, C. Karrer, A. Post, and L. Saldanha (eds.). 1990. *Check-list of the fishes of the eastern tropical Atlantic.* CLOFETA. Vols. 1–3. Junta Nac. Invest. Cien. Tech., Lisbon; UNESCO, Paris. 1,492 p.

Ramon, M. L., and M. L. Knope. 2008. Molecular support for marine sculpin (Cottidae: Oligocottinae) diversification during the transition from the subtidal to intertidal habitat in the Northeastern Pacific Ocean. *Mol. Phylogen. Evol.* 46:475–83.

Ramos, R. T. C. 2003. Systematic review of *Apionichthys* (Pleuronectiforms: Achiridae), with description of four new species. *Ichthyol. Explor. Fresh.* 14:97–126.

Ramos, R. T. C., T. P. A. Ramos, and P. R. D. Lopes. 2009. New species of *Achirus* (Pleuronectiformes: Achiridae) from Northeastern Brazil. *Zootaxa* 2113:55–62.

Ramsden, S. D., H. Brinkmann, C. W. Hawryshyn, and J. S. Taylor. 2003. Mitogenomics and the sister of Salmonidae. *Trends Ecol. Evol.* 18:607–10.

Randall, J. E. 1958. A review of ciguatera tropical fish poisoning with a tentative explanation of its cause. *Bull. Mar. Sci.* 8:236–67.

———. 1963. Review of the hawkfishes (family Cirrhitidae). *Proc. U.S. Nat. Mus.* 114:389–451.

———. 1980. Revision of the fish genus *Plectranthias* (Serranidae: Anthiinae) with descriptions of 13 new species. *Micronesica* 16:101–87.

———. 1998. Revision of the Indo-Pacific squirrelfishes (Beryciformes: Holocentridae: Holocentrinae) of the genus *Sargocentron*, with descriptions of four new species. *Indo-Pacific Fishes* 27:1–105.

———. 2001a. Revision of the generic classification of the hawkfishes (Cirrhitidae), with descriptions of three new genera. *Zootaxa* 12:1–12.

———. 2001b. *Surgeonfishes of Hawai'i and the world.* Mutual Publishing and Bishop Museum Press, Hawai'i. 123 p.

———. 2005. A review of mimicry in marine fishes. *Zool. Stud.* 44:299–328.

Randall, J. E., K. Aida, T. Hibiya, N. Mitsuura, H. Kamiya, and Y. Hashimoto. 1971. Grammistin, the skin toxin of soapfishes and its significance in the classification of the Grammistidae. *Publ. Seto Mar. Biol. Lab.* 19:157–90.

Randall, J. E., and D. W. Greenfield. 1996. Revision of the Indo-Pacific holocentrid fishes of the genus *Myripristis*, with descriptions of three new species. *Indo-Pacific Fishes* 25:1–61.

Rasmussen, T. H., Å. Jespersen, and B. Korsgaard. 2006. Gonadal morphogenesis and sex differentiation in intraovarian embryos of the viviparous fish, *Zoarces viviparus* (Teleostei, Perciformes, Zoarcidae): A histological and ultrastructural study. *J. Morph.* 267:1032–47.

Rauchenberger, M. 1989. Systematics and biogeography of the genus *Gambusia* (Cyprinodontiformes: Poeciliidae). *Amer. Mus. Nov.* 2951:1–74.

Reed, D. L., K. E. Carpenter, and M. J. de Gravelle. 2002. Molecular systematics of the jacks (Perciformes: Carangidae) based on mitochondrial cytochrome b sequences using parsimony, likelihood, and Bayesian approaches. *Mol. Phylogen. Evol.* 23:513–24.

Renaud, C. B. 1997. Conservation status of Northern Hemisphere lampreys (Petromyzontidae). *J. Appl. Ichthyol.* 13:143–48.

———. 2011. *Lampreys of the world: An annotated and illustrated catalogue of lamprey species known to date.* FAO Species Catalogue for Fishery Purposes, No. 5. FAO, Rome. 109 p.

Resetarits, W. J., Jr., and C. A. Binckley. 2013. Is the pirate really a ghost? Evidence for generalized chemical camouflage in an aquatic predator, Pirate Perch *Aphredoderus sayanus*. *Amer. Nat.* 181:690–99.

Richards, W. J., and D. L. Jones. 2002. Preliminary classification of the gurnards (Triglidae: Scorpaeniformes). *Mar. Fresh. Res.* 53:275–82.

Roa-Varón, A., and G. Ortí. 2009. Phylogenetic relationships among families of Gadiformes (Teleostei, Paracanthopterygii) based on nuclear and mitochondrial data. *Mol. Phylogen. Evol.* 52:688–704.

Roberts, T. R. 1969. Osteology and relationships of characoid fishes, particularly the genera *Hepsetus, Salminius, Hoplias, Ctenolucius*, and *Acestrorhynchus*. *Proc. Cal. Acad. Sci.* 4:391–500.

———. 2012. Systematics, biology, and distribution of the species of the oceanic oarfish genus *Regalecus* (Teleostei, Lampridiformes, Regalecidae). *Mém. Mus. Nat. d'Hist. Nat., Paris (N. S.) (Série A) Zoologie* 202:1–268.

Robertson, D. R., and G. R. Allen. 2008. *Shorefishes of the tropical eastern Pacific online information system.* Version 1.0. Smithsonian Tropical Research Institute, Balboa, Panama. http://biogeodb.stri.si.edu/sftep/.

Robins, C. R. 1989. The phylogenetic relationships of the anguilliform fishes, pp. 9–23. In *Fishes of the Western North Atlantic.* E. Böhlke (ed.). Part 9. Vol. 1. Sears Foundation for Marine Research, Memoir (Yale University), New Haven, Connecticut.

Robins, C. R., and G. C. Ray. 1986. *A field guide to Atlantic coast fishes of North America.* Houghton Mifflin, Boston, Massachusetts. 354 p.

Rocha, L. A., K. C. Lindeman, C. R. Rocha, and H. A. Lessios. 2008. Historical biogeography and speciation in the reef fish genus *Haemulon* (Teleostei: Haemulidae). *Mol. Phylogen. Evol.* 48:918–28.

Roe, K. J., P. M. Harris, and R. L. Mayden. 2002. Phylogenetic relationshps of the genera of North American sunfishes and basses (Percoidei: Centrarchidae) as evidenced by the mitochondrial cytochrome b gene. *Copeia* 2002:897–905.

Rosen, D. E. 1964. The relationships and taxonomic position of the halfbeaks, killifishes, silversides, and their relatives. *Bull. Amer. Mus. Nat. Hist.* 127:217–68.

———. 1973. Interrelationships of higher euteleostean fishes, pp. 397–513. In *Interrelationships of Fishes.* P. H. Greenwood, R. S. Miles, and C. Patterson (eds.). *Zool. J. Linn. Soc.* 53. Suppl. 1.

———. 1974. Phylogeny and zoogeography of salmoniform fishes and relationships of *Lepidogalaxius salamandroides*. *Bull. Amer. Mus. Nat. Hist.* 153:265–326.

———. 1984. Zeiforms as primitive plectognath fishes. *Amer. Mus. Novit.* 2782:1–45.

————. 1985. An essay on euteleostean classification. *Amer. Mus. Novit.* 2827:1–57.

Rosen, D. E., and R. M. Bailey. 1963. The poeciliid fishes (Cyprinodontiformes) their structure, zoogeography and systematics. *Bull. Amer. Mus. Nat. Hist.* 126:1–176.

Rosen, D. E., and P. H. Greenwood. 1970. Origin of the Weberian apparatus and the relationships of ostariophysan and gonorynchiform fishes. *Amer. Mus. Novit.* 2428:1–25.

Rosen, D. E., and L. R. Parenti. 1981. Relationships of *Oryzias,* and the groups of atherinomorph fishes. *Am. Mus. Novit.* 2719:1–25.

Rosen, D. E., and C. Patterson. 1969. The structure and relationships of the paracanthopterygian fishes. *Bull. Amer. Mus. Nat. Hist.* 141:357–474.

Rosenberger, L. J. 2001. Phylogenetic relationships within the stingray genus *Dasyatis* (Chondrichthyes: Dasyatidae). *Copeia* 2001:615–27.

Ross, S. 2013. *Ecology of North American freshwater fishes.* University of California Press, Berkeley, California. 460 p.

Ruber, L., R. Britz, and R. Zardoya. 2006. Molecular phylogenetics and evolutionary diversification of labyrinth fishes (Perciformes: Anabantoidei). *Syst. Biol.* 55:374–97.

Runcie, R. M., H. Dewar, D. R. Hawn, L. R. Frank, and K. A. Dickson. 2009. Evidence for cranial endothermy in the opah (*Lampris guttatus*). *J. Exp. Biol.* 212:461–70.

Rundle, H. D., L. Nagel, J. W. Boughman, and D. Schluter. 2000. Natural selection and parallel speciation in sympatric sticklebacks. *Science* 287:306–8.

Russo, T., D. Pulcini, et al. 2012. "Right" or "wrong"? Insights into the ecology of sidedness in European flounder, *Platichthys flesus. J. Morphol.* 273:337–46.

Sadovy, Y., and T. J. Donaldson. 1995. Sexual pattern of *Neocirrhites armatus* (Cirrhitidae) with notes on other hawkfish species. *Env. Biol. Fishes* 42:143–50.

Sadovy de Mitcheson, Y., A. Cornish, M. Domeier, P. L. Colin, M. Russell, and K. C. Lindeman. 2008. A global baseline for spawning aggregations of reef fishes. *Cons. Biol.* 22:1233–44.

Sakamoto, K. 1984. Interrelationships of the family Pleuronectidae (Pisces: Pleuronectiformes). *Mem. Fac. Fish. Hokkaido Univ.* 31:95–215.

Sanciango, M. D., L. A. Rocha, and K. E. Carpenter. 2011. A molecular phylogeny of the grunts (Perciformes: Haemulidae) inferred using mitochondrial and nuclear genes. *Zootaxa* 2966:37–50.

Sanford, C. J. 1990. The phylogenetic relationships of salmonoid fishes. *Bull. Brit. Mus. (Natur. Hist.), Zool.* 56:145–53.

Sanford, C. P. J. 2000. *Salmonoid fish osteology and phylogeny (Teleostei: Salmonoidei).* Theses Zoologicae. Koeltz Scientific and A.R.G. Gantner, Ruggell, Liechtenstein. 264 p.

Santini, F., L. J. Harmon, G. Carnevale, and M. E. Alfaro. 2009. Did genome duplication drive the origin of teleosts? A comparative study of diversification in ray-finned fishes. *BMC Evol. Biol.* 9:164.

Santini, F., X. Kong, L. Sorenson, G. Carnevale, R. S. Mehta, and M. E. Alfaro. 2013. A multi-locus molecular timescale for the origin and diversification of eels (Order: Anguilliformes). *Mol. Phylogen. Evol.* 69:884–94.

Santini, F., M. T. T. Nguyen, L. Sorenson, T. B. Waltzec, J. W. Lynch, J. M. Eastman, and M. E. Alfaro. 2013. Do habitat shifts drive diversification in teleost fishes? An example from the pufferfishes (Tetraodontidae). *J. Evol. Biol.* 26:1003–18.

Santini F., and J. C. Tyler. 2002. Phylogeny of the ocean sunfishes (Molidae, Tetraodontiformes), a highly derived group of teleost fishes. *Ital. J. Zool.* 69:37–43.

————. 2003. A phylogeny of the families of fossil and extant tetraodontiform fishes (Acanthomorpha, Tetraodontiformes), Upper Cretaceous to Recent. *Zool. J. Linn. Soc.* 139:565–617.

———. 2004. The Importance of even highly incomplete fossil taxa in reconstructing the phylogenetic relationships of the Tetraodontiformes (Acanthomorpha: Pisces). *Integr. Comp. Biol.* 44:349–357.

Sasaki, K. 1989. Phylogeny of the family Sciaenidae, with notes on its zoogeography (Teleostei, Perciformes). *Mem. Fac. Fish. Hokkaido Univ.* 36:1–137.

Satoh, T. P., M. Miya, E. Hiromitsu, and N. Mutsumi. 2006. Round and pointed-head grenadier fishes (Actinopterygii: Gadiformes) represent a single sister group: Evidence from the complete mitochondrial genome sequences. *Mol. Phylogen. Evol.* 40: 129–38.

Schlupp, I. 2005. The evolutionary ecology of gynogenesis. *Annu. Rev. Ecol. Evol. Syst.* 36:399–417.

Schluter, D., K. B. Marchinko, R. D. H. Barrett, and S. M. Rogers. 2010. Natural selection and the genetics of adaptation in threespine stickleback. *Phil. Trans. Royal Soc. London,* Series B 365:2479–86.

Schnell, N. K., R. Britz, and G. D. Johnson. 2010. New insights into the complex structure and ontogeny of the occipito-vertebral gap in barbeled dragonfishes (Stomiidae, Teleostei). *J. Morph.* 271:1006–22.

Schönhuth, S., D. K. Shiozawa, T. E. Dowling, and R. L. Mayden. 2012. Molecular systematics of western North American cyprinids (Cypriniformes: Cyprinidae). *Zootaxa* 3586:281–303.

Schultz, L. P. 1958. Review of the parrotfishes family Scaridae. *Bull. U.S. Nat. Mus.* 214:1–143.

———. 1969. The taxonomic status of the controversial genera and species of parrotfishes with a descriptive list (Family Scaridae). *Smithson. Contr. Zool.* 17:1–49.

Schwarzhans, W. 1993. A comparative morphological treatise of recent and fossil otoliths of the family Sciaenidae (Perciformes). *Piscium Catalogus: Part Otolithi Piscium.* Vol. I. Verlag Dr. Freidrich Pfeil, Munchen. 245 p.

———. 1994. Sexual and ontogenetic dimorphism in otoliths of the family Ophidiidae. *Cybium* 18:71–98.

Schwarzhans, W., P. R. Møller, and J. G. Nielsen. 2005. Review of the Dinematichthyini (Teleostei, Bythitidae) of the Indo-West Pacific. Part I: *Diancistrus* and two new genera with 26 new species. *The Beugle, Rec. Mus. Art Gall. N. Terr.* 21:73–163.

Scott, W. B., and E. J. Crossman. 1973. *Freshwater fishes of Canada.* Bull. Fish. Res. Bd. Can., 184. 966 p.

Scott, W. B., and M. G. Scott. 1988. *Atlantic fishes of Canada.* Can. Bull. Fish. Fish. Aquat. Sci., 219. 731 p.

Sepulveda, C. A., K. A. Dickson, L. R. Frank, and J. B. Graham. 2007. Cranial endothermy and a putative brain heater in the most basal tuna species, *Allothunnus fallai. J. Fish Biol.* 70:1720–33.

Sepulveda, C. A., N. C. Wegner, D. Bernal, and J. B. Graham. 2005. The red muscle morphology of the thresher sharks (family Alopiidae). *J. Exp. Biol.* 208:4255–61.

Setiamarga, D. H., M. Miya, et al. 2008. Interrelationships of Atherinomorpha (medakas, flyingfishes, killifishes, silversides, and their relatives): The first evidence based on whole mitogenome sequences. *Mol. Phylogen. Evol.* 49:598–605.

Setzler, E. M., W. R. Boynton, K. V. Wood, H. H. Zion, L. Lubers, N. K. Mountford, P. Frere, L. Tucker, and J. A. Mihursky. 1980. Synopsis of biological data on striped bass, *Morone saxatilis* (Walbaum). *NOAA Technical Report NMFS Circular* 43:1–69. FAO Fish. Synop. 121.

Seymour, R. S., N. C. Wegner, and J. B. Graham. 2008. Body size and the air-breathing organ of the Atlantic tarpon, *Megalops atlanticus. Comp. Biochem. Physiol.,* Part A 150:282–87.

Shedlock, A. M., T. W. Pietsch, M. G. Haygood, P. Bentzen, and M. Hasegawa. 2004. Molecular systematics and life history evolution of anglerfishes (Teleostei: Lophiiformes): Evidence from mitochondrial DNA. *Steenstrupia* 28:129–44.

Shimada, K. 2005. Phylogeny of lamniform sharks (Chondrichthyes: Elasmobranchii) and the contribution of dental characters to lamniform systematics. *Paleontol. Res.* 9:55–72.

Shimada, K., C. K. Rigsby, and S. H. Kim. 2009. Labial cartilages in the smalltooth sandtiger shark, *Odontaspis ferox* (Lamniformes: Odontaspididae), and their signficance to the phylogeny of lamniform sharks. *Anat. Rec.* 292:813–17.

Shimeld, S. M., and P. W. H. Holland. 2000. Vertebrate innovations. *Proc. Nat. Acad. Sci.* 97:4449–52.

Shinohara, G. 1994. Comparative morphology and phylogeny of the suborder Hexagrammoidei and related taxa (Pisces: Scorpaeniformes). *Mem. Fac. Fish., Hokkaido Univ.* 41:1–97.

Shinohara, G., and H. Imamura. 2007. Revisiting recent phylogenetic studies of "Scorpaeniformes." *Ichthyol. Res.* 54:92–99.

Shipp, R. L. 1974. The pufferfishes (Tetraodontidae) of the Atlantic Ocean. *Publ. Gulf Coast Res. Lab. Mus.* 41:1–162.

Shirai, S. 1992a. *Squalean phylogeny, a new framework of "squaloid" sharks and related taxa.* Hokkaido University Press, Sapporo. 151 p.

———. 1992b. Phylogenetic relationships of the angel sharks, with comments on elasmobranch phylogeny (Chondrichthyes: Squatinidae). *Copeia* 1992:505–18.

———. 1996. Phylogenetic relationships of neoselachians (Chondrichthyes: Euselachii), pp. 9–34. In *Interrelationships of fishes.* M. L. J. Stiassny, L. R. Parenti, and G. D. Johnson (eds.). Academic Press, San Diego, California.

Sloss, B. L., N. Billington, and B. M. Burr. 2004. A molecular phylogeny of the Percidae (Teleostei, Perciformes) based on mitochondrial DNA sequence. *Mol. Phylogen. Evol.* 32:545–62.

Smith, C. L. 1971. A revision of the American groupers: *Epinephelus* and allied genera. *Bull. Amer. Mus. Nat. Hist.* 146:1–241.

Smith, C. L., C. S. Rand, B. Schaeffer, and J. W. Atz. 1975. *Latimeria,* the living coelacanth, is ovoviparous. *Science* 190:1105–6.

Smith, D. G. 1989a. Family Anguillidae, pp. 24–47. In *Fishes of the western North Atlantic.* E. B. Böhlke (ed.). Part 9. Vol. 1. Orders Anguilliformes and Saccopharyngiformes. Sears Foundation for Marine Research, Memoir (Yale University), New Haven, Connecticut.

———. 1989b. Family Congridae, pp. 460–567. In *Fishes of the western North Atlantic.* E. B. Böhlke (ed.). Part 9. Vol. 1. Orders Anguilliformes and Saccopharyngiformes. Sears Foundation for Marine Research, Memoir (Yale University), New Haven, Connecticut.

———. 2012. A checklist of the moray eels of the world (Teleostei: Anguilliformes: Muraenidae). *Zootaxa* 3474:1–64.

Smith, G. R. 1992. Phylogeny and biogeography of the Catostomidae, freshwater fishes of North America and Asia, pp. 778–826. In *Systematics, historical ecology, and North American freshwater fishes.* R. L. Mayden (ed.). Stanford University Press, Stanford, California.

Smith, G. R., and R. F. Stearley. 1989. The classification and scientific names of rainbow and cutthroat trouts. *Fisheries* 14:4–10.

Smith, J. L. B. 1940. A living coelacanth from South Africa. *Trans. Royal Soc. S. Africa* 28:1–106.

———. 1956. *Old fourlegs: The story of the coelacanth.* Longmans, Green, London. 260 p.

Smith, M. M., and P. C. Heemstra (eds.). 1986. *Smiths' Sea Fishes.* Macmillan, Johannesburg, South Africa. 1,047 p.

Smith, W. D., J. J. Bizzarro, V. P. Richards, J. Nielsen, F. Márquea-Flarías, and M. S. Shivli. 2009. Morphometric convergence and molecular divergence: The taxonomic status and evolutionary history of *Gymnura crebripunctata* and *Gymnura marmorata* in the eastern Pacific Ocean. *J. Fish Biol.* 75:761–83.

Smith, W. L., P. Chakrabarty and J. S. Sparks. 2008. Phylogeny, taxonomy, and evolution of Neotropical cichlids (Teleostei: Cichlidae: Cichlinae). *Cladistics* 24:626–41.

Smith, W. L., and M. T. Craig. 2007. Casting the percomorph net widely: The importance of broad taxonomic sampling in the search for the placement of serranid and percid fishes. *Copeia* 2007:35–55.

Smith, W. L., K. R. Smith, and W. C. Wheeler. 2009. Mitochondrial intergenic spacer in fairy basslets (Serranidae: Anthiinae) and the simultaneous analysis of nucleotide and rearrangement data. *Amer. Mus. Novit.* 3652:1–12.

Smith W. L., J. F. Webb, and S. D. Blum. 2003. The evolution of the laterophysic connection with a revised phylogeny and taxonomy of butterflyfishes (Teleostei: Chaetodontidae). *Cladistics* 19:287–306.

Smith, W. L., and W. C. Wheeler. 2004. Polyphyly of the mail-cheeked fishes (Teleostei: Scorpaeniformes): Evidence from mitochondrial and nuclear sequence data. *Mol. Phylogen. Evol.* 32:627–46.

———. 2006. Venom evolution widespread in fishes: A phylogenetic road map for the bioprospecting of piscine venoms. *J. Heredity* 97:206–17.

Smith-Vaniz, W. F. 1976. The saber-toothed blennies, tribe Nemophini (Pisces: Blenniidae). *Monogr. Acad. Nat. Sci. Philadelphia* 19:1–196.

———. 1984. Carangidae: Relationships, pp. 522–30. In *Ontogeny and Systematics of Fishes.* H. G. Moser, W. J. Richards, D. M. Cohen, M. P. Fahay, A. W. Kendall, Jr., and S. L. Richardson (eds.). American Society of Ichthyologists and Herpetologists, Spec. Publ. No. 1. Allen Press, Lawrence, Kansas.

———. 1989. Revision of the jawfish genus *Stalix* (Pisces: Opistognathidae), with descriptions of four new species. *Proc. Acad. Nat. Sci. Philadelphia* 141:375–407.

Song, C. B., T. J. Near, and L. M. Page. 1998. Phylogenetic relations among percid fishes as inferred from mitochondrial Cytochrome b DNA sequence data. *Mol. Phylogen. Evol.* 10:349–53.

Sparks, J. S., and W. L. Smith. 2004. Phylogeny and biogeography of cichlid fishes (Teleostei: Perciformes: Cichlidae). *Cladistics* 20:501–17.

Springer, V. G. 1968. Osteology and classification of the fishes of the family Blenniidae. *Bull. U.S. Nat. Mus.* 284:1–85.

———. 1993. Definition of the suborder Blennioidei and its included families (Pisces: Perciformes). *Bull. Mar. Sci.* 52:472–95.

Springer, V. G., and T. H. Fraser. 1976. Synonymy of the fish families Cheilobranchidae (= Alabetidae) and Gobiesocidae, with descriptions of two new species of *Alabes. Smithson. Contr. Zool.* 234:1–23.

Springer, V. G., and W. C. Freihofer. 1976. Study of the monotypic fish family Pholidichthyidae (Perciformes). *Smithson. Contr. Zool.* 216:1–43.

Springer, V. G., and G. D. Johnson. 2004. Study of the dorsal gill-arch musculature of teleostome fishes, with special reference to the Actinopterygii. *Bull. Biol. Soc. Wash.* 11:1–260.

Springer, V. G., and H. K. Larson. 1996. *Pholidichthys anguis,* a new species of pholidichthyid fish from Northern Territory and Western Australia. *Proc. Biol. Soc. Wash.* 109:353–65.

Springer, V. G., and W. F. Smith-Vaniz. 2008. Supraneural and pterygiophore insertion patterns in carangid fishes, with description of a new Eocene carangid tribe, Paratrachinotini, and a survey of anterior and anal-fin pterygiophore insertion patterns in Acanthomorpha. *Bull. Biol. Soc. Wash.* 16:1–73.

Starnes, W. C. 1988. Revision, phylogeny and biogeographical comments on the circumtropical marine percoid family Priacanthidae. *Bull. Mar. Sci.* 43:117–203.

Stearley, R. F., and G. R. Smith. 1993. Phylogeny of the Pacific trouts and salmons *(Oncorhynchus)* and the genera of the family Salmonidae. *Trans. Amer. Fish. Soc.* 122:1–33.

Stein, D. L. 2012. Snailfishes (Family Liparidae) of the Ross Sea, Antarctica, and closely adjacent waters. *Zootaxa* 3285:1–120.

Stein, D. L., N. V. Chernova, and A. P. Andriashev. 2001. Snailfishes (Pisces: Liparidae) of Australia, including descriptions of thirty new species. *Rec. Aust. Mus.* 53:341–406.

Stephens, J. S. 1963. A revised classification of the blennioid fishes of the American family Chaenopsidae. *Univ. California Publ. Zool.* 68:1–165.

Stiassny, M. L. J. 1986. The limits and relationships of the acanthomorph teleosts. *J. Zool., London* B 1986:411–60.

———. 1991. Phylogenetic intrarelationships of the family Cichlidae: An overview, pp. 1–35. In *Cichlid fishes: Behaviour, ecology, and evolution*. M. H. A. Keenleyside (ed.). Chapman and Hall, London.

———. 1993. What are grey mullets? *Bull. Mar. Sci.* 52:197–219.

———. 1996. Basal ctenosquamate relationships and the interrelationships of the myctophiform (scopelomorph) fishes, pp. 405–26. In *Interrelationships of fishes*. M. L. J. Stiassny, L. R. Parenti, and G. D. Johnson (eds.). Academic Press, San Diego, California.

Stiassny, M. L. J., and J. S. Jensen. 1987. Labroid intrarelationships revisited: Morphological complexity, key innovations, and the study of comparative diversity. *Bull. Mus. Comp. Zool.* 151:269–319.

Stiassny, M. L. J., L. R. Parenti, and G. D. Johnson (eds.). 1996. *Interrelationships of fishes*. Academic Press, San Diego, California. 496 p.

Stiassny, M. L. J., E. O. Wiley, G. D. Johnson, and M. R. de Caravalho. 2004. Gnathostome fishes, pp. 410–29. In *Assembling the tree of life*. J. Cracraft and M. Donoghue (eds.). Oxford University Press, New York.

Strauss, R. E., and C. E. Bond. 1990. Taxonomic methods: Morphology, pp. 109–40. In *Methods for fish biology*. C. B. Schreck and P. B. Moyle (eds.). American Fisheries Society, Bethesda, Maryland.

Strauss, S. E. 1993. Relationships among the cottid genera *Artedius, Clinocottus,* and *Oligocottus* (Teleostei: Scorpaeniformes). *Copeia* 1993:518–22.

Sulak, K. J., R. E. Crabtree, and J.-C. Hureau. 1984. Provisional review of the genus *Polyacanthonotus* (Pisces, Notacanthidae) with description of a new Atlantic species, *Polyacanthonotus merretti. Cybium* 8:57–68.

Sullivan, J., J. Lundberg, and M. Hardman. 2006. A phylogenetic analysis of the major groups of catfishes (Teleostei: Siluriformes) using rag1 and rag2 nuclear gene sequences. *Mol. Phylogen. Evol.* 41:636–62.

Tan, Y. Y., R. Kodzius, B. Tay, A. Tay, S. Brenner, and B. Venkatesh. 2012. Sequencing and analysis of full-length cDNAs, 5'-ESTs and 3'-ESTs from a cartilaginous fish, the elephant shark (*Callorhinchus milii*). *PLoS ONE* 7:e47174.

Tang, K. L. 2001. Phylogenetic relationships among damselfishes (Teleostei: Pomacentridae) as determined by mitochondrial DNA data. *Copeia* 2001:591–601.

Tang, K. L., P. B. Berendzen, E. O. Wiley, J. F. Morrissey, R. Winterbottom, and G. D. Johnson. 1999. The phylogenetic relationships of the suborder Acanthuroidei (Teleostei: Perciformes) based on molecular and morphological evidence. *Mol. Phylogen. Evol.* 11:415–25.

Tang, K. L., and C. Fielitz. 2013. Phylogeny of moray eels (Anguilliformes: Muraenidae), with a revised classification of true eels (Teleostei: Elopomorpha: Anguilliformes). *Mitochond. DNA* 24:55–66.

Tarp, F. H. 1952. A revision of the family Embiotocidae (The Surfperches). *Cal. Dept. Fish and Game, Fish Bull.* 88:1–99.

Taylor, G. R., J. A. Wittington, and H. J. Grier. 2000. Age, growth, maturation and protandric sex reversal in the common snook, *Centropomus undecimalis,* from the east and west coasts of south Florida. *Fish. Bull.* 98:612–24.

Taylor, W. R. 1969. A revision of the catfish genus *Noturus* Rafinesque with an analysis of higher groups in the Ictaluridae. *Bull. U.S. Nat. Mus.* 282:1–315.

Tee-Van, J. et al. (eds.). 1948–1989. *Fishes of the western North Atlantic.* Parts 1–9. Sears Foundation for Marine Research, Memoir 1 (Yale University), New Haven, Connecticut.

Teletchea, F., V. Laudet, and C. Hänni. 2006. Phylogeny of the Gadidae (sensu Svetovidov, 1948) based on their morphology and two mitochondrial genes. *Mol. Phylogen. Evol.* 38:189–99.

Tesch, F.-W. (transl.). 1977. *The eel.* Chapman and Hall, London. 434 p.

Teske, P. R., and L. B. Beheregaray. 2009. Evolution of seahorses' upright posture was linked to Oligocene expansion of seagrass habitats. *Biol. Lett.* 5:521–23.

Thacker, C. E. 2009. Phylogeny of Gobioidei and placement within Acanthomorpha, with a new classification and investigation of diversification and character evolution. *Copeia* 2009:93–104.

———. 2011. Systematics of Butidae and Eleotridae, pp. 79–85. In *The biology of gobies.* R. A. Patzner, J. L. Van Tassel, M. Kovačić, and B. G. Kapoor (eds.). Science Publishers, Enfield, New Hampshire.

Thacker, C. E., and D. M. Roje. 2009. Phylogeny of cardinalfishes (Teleostei: Gobiiformes: Apogonidae) and the evolution of visceral bioluminescence. *Mol. Phylogen. Evol.* 52:735–45.

———. 2011. Phylogeny of Gobiidae and identification of gobiid lineages. Syst. Biodivers. 9:329–347.

Thompson, B. 1998. Redescription of *Aulopus bajacali* Parin and Kotlar, 1984, comments on its relationships and new distribution records. *Ichthyol. Res.* 45:43–51.

Thompson, D. 1945. *On growth and form.* Cambridge University Press, New York. 1,116 p.

Thomson, D. A. 1969. Toxic stress secretions of the boxfish, *Ostracion meleagris* Shaw. *Copeia* 1969:335–52.

Thomson, D. A., L. T. Findley, and A. N. Kerstitch. 2000. *Reef fishes of the Sea of Cortez: The rocky-shore fishes of the Gulf of California.* Rev. ed. University of Texas Press, Austin, Texas. 353 p.

Thomson, J. M. 1997. The Mugilidae of the world. *Mem. Queensland Mus.* 41:457–562.

Thomson, K. S. 1991. *Living fossil, the story of the coelacanth.* W. W. Norton and Co., New York. 252 p.

Thorson, T. B. 1976. Observations on the reproduction of the sawfish, *Pristis perotteti,* in Lake Nicaragua, with recommendations for its conservation, pp. 641–50. In *Investigations of the ichthyofauna of Nicaraguan Lakes.* T. B. Thorson (ed.). University of Nebraska, Lincoln, Nebraska.

Tibbetts, I. R., and L. Carseldine, 2003. Anatomy of a hemiramphid pharyngeal mill with reference to *Arrhamphus sclerolepis krefftii* (Steindachner) (Teleostei: Hemiramphidae). *J. Morph.* 25:228–43.

Tighe, K. A., and J. G. Nielsen. 2000. *Saccopharynx berteli,* a new gulper eel from the Pacific Ocean (Teleostei, Saccopharyngidae). *Ichthyol. Res.* 47:39–41.

Travers, R. A. 1984a. A review of the Mastacembeloidei, a suborder of synbranchiform teleost fishes. Part I: Anatomical descriptions. *Bull. Brit. Mus. (Nat. Hist.), Zool.* 46:1–133.

———. 1984b. A review of the Mastacembeloidei, a suborder of synbranchiform teleost fishes. Part II: Phylogenetic analysis. *Bull. Brit. Mus. (Nat. Hist.), Zool.* 47:83–150.

Trewavas, E. 1977. The sciaenid fishes (croakers or drums) of the Indo-West-Pacific. *Trans. Zool. Soc. London* 33:253–541.

Tsukamoto, K., S. Chow, T. Otake, H. Kurogi, N. Mochioka, M. J. Miller, J. Aoyama, et al. 2011. Oceanic spawning ecology of freshwater eels in the western North Pacific. *Nat. Commun.* 2:179.

Turner, G. F. 2007. Adaptive radiation of cichlid fish. *Cur. Biol.* 17:R827-R831.

Turner, G. F., O. Seehausen, M. E. Knight, C. J. Allender, and R. L. Robinson. 2001. How many species of cichlid fishes are there in African lakes? *Molec. Ecol.* 10:793–806.

Tyler, J. C. 1980. Osteology, phylogeny, and higher classification of the fishes of the order Plectognathi (Tetraodontiformes). *NOAA Tech. Rept. NMFS Circ.* 434:1–422.

Tyler, J. C., G. D. Johnson, E. B. Brothers, D. M. Tyler, and C. L. Smith. 1993. Comparative early life histories of western Atlantic squirrelfishes (Holocentridae): Age and settlement of rhynchichthys, meeki, and juvenile stages. *Bull. Mar. Sci.* 53:1126–50.

Tyler, J. C., G. D. Johnson, I. Nakamura, and B. B. Collette. 1989. Morphology of *Luvarus imperialis* (Luvaridae), with a phylogenetic analysis of the Acanthuroidei. *Smithson. Contr. Zool.* 485:1–78.

Tyler, J. C., B. O. O'Toole, and R. Winterbottom. 2003. Phylogeny of the genera and families of zeiform fishes, with comments on their relationships with tetraodontiforms and caproids. *Smithson. Contr. Zool.* 618:1–110.

Tyler, J. C., and F. Santini. 2005. A phylogeny of the fossil and extant zeiform-like fishes, Upper Cretaceous to Recent, with comments on the putative zeomorph clade (Acanthomorpha). *Zool. Scripta* 34:157–75.

Uiblein, F. 2007. Goatfishes (Mullidae) as indicators in tropical and temperate coastal habitat monitoring and management. *Mar. Biol. Res.* 3:275–88.

Vari, R. P. 1979. Anatomy, relationships and classification of the families Citharinidae and Distichodontidae (Pisces: Characoidei). *Bull. Brit. Mus (Nat. Hist.), Zool.* 36:261–344.

———. 1989. A phylogenetic study of the neotropical characiform family Curimatidae (Pisces: Ostariophysi). *Smithson. Contr. Zool.* 471:1–71.

Vélez-Zuazo, X., and I. Agnarsson. 2011. Shark tales: A molecular species-level phylogeny of sharks (Selachimorpha, Chondrichthyes). *Mol. Phylogen. Evol.* 58:207–17.

Venkatesh, B., E. F. Kirkness, Y. H. Loh, A. L. Halpern, A. P. Lee, J. Johnson, N. Dandona, et al. 2007. Survey sequencing and comparative analysis of the elephant shark (*Callorhinchus milii*) genome. *PLoS Biol* 5:e101.

Venkatesh, B., and W. H. Yap. 2004. Comparative genomics using fugu: A tool for the identification of conserved vertebrate cis-regulatory elements. *Bioessays* 27:100–7.

Vergara-Chen, C., W. E. Aguirre, M. González-Wangu, and E. Bermingham. 2009. A mitochondrial DNA based phylogeny of weakfish species of the *Cynoscion* group (Pisces: Sciaenidae). *Mol. Phylogen. Evol.* 53:602–7.

Vidal, O., E. Garcia-Berthou, P. A. Tedesco, and J.-L. Garcia-Martin. 2010. Origin and genetic diversity of mosquitofish (*Gambusia holbrooki*) introduced to Europe. *Biol. Invas.* 12:841–51.

von der Heyden, S., and C. A. Matthee. 2008. Towards resolving familial relationships within the Gadiformes, and the resurrection of the Lyconidae. *Mol. Phylogen. Evol.* 48:764–69.

Wade, R. A. 1962. The biology of the tarpon, *Megalops atlanticus,* and the ox-eye, *Megalops cyprinoides,* with emphasis on larval development. *Bull. Mar. Sci.* 12:545–99.

Wainwright, P. C., and R. G. Turingan. 1997. Evolution of pufferfish inflation behavior. *Evolution* 51:506–18.

Wainwright, P. C., W. L. Smith, S. A. Price, K. L. Tang, J. S. Sparks, L. A. Ferry, K. L. Kuhn, R. I. Eytan, and T. J. Near. 2012. The evolution of pharyngognathy: A phylogenetic and functional appraisal of the pharyngeal jaw key innovation in labroid fishes and beyond. *Syst. Biol.* 61:1001–27.

Walker, H. J., Jr., and J. Bollinger. 2001. A new species of *Trinectes* (Pleuronectiformes: Achiridae), with comments on the other eastern Pacific species of the genus. *Rev. Biol. Trop.* 49 (Suppl. 1):177–85.

Walker, H. J., Jr., and R. H. Rosenblatt. 1988. Pacific toadfishes of the genus *Porichthys* (Batrachoididae) with descriptions of three new species. *Copeia* 1988:887–904.

Warner, R. R., and R. K. Harlan. 1982. Source sperm competition and sperm storage as determinants of sexual dimorphism in the dwarf surfperch, *Micrometrus minimus. Evolution* 36:44–55.

Watson, W., and H. J. Walker, Jr. 2004. The world's smallest vertebrate, *Schindleria brevipinguis,* a new paedomorphic species in the family Schindleriidae (Perciformes: Gobioidei). *Rec. Aust. Mus.* 56:139–42.

Webb, J. F. 1989. Gross morphology and evolution of the mechanosensory lateral line system in teleost fishes. *Brain Beh. Evol.* 33:34–53.

———. 2013. Morphological diversity, development, and evolution of the mechanosensory lateral line system, pp. 17–72. In *The Lateral Line.* S. Coombs, H. Bleckmann, R. Fay, and A. N. Popper (eds.). Springer Handbook of Auditory Research. Vol. 48. Springer, New York.

Webb, J. F., and W. L. Smith. 2000. The laterophysic connection in chaetodontid butterflyfish: Morphological variation and speculations on sensory function. *Phil. Trans. Roy. Soc. London* B 355:1125–29.

Webb, J. F., W. L. Smith, D. R. Ketten. 2006. The laterophysic connection and swim bladder in butterflyfishes in the genus *Chaetodon* (Perciformes: Chaetodontidae). *J. Morph.* 267:1338–55.

Weitzman, S. H. 1962. The osteology of *Brycon meeki,* a generalized characoid fish, with a definition of the family. *Stanford Ichthyol. Bull.* 8:1–77.

———. 1974. Osteology and evolutionary relationships of the Sternoptychidae, with a new classification of stomiatoid families. *Bull. Amer. Mus. Nat. Hist.* 153:327–478.

Westneat, M. W., and M. E. Alfaro. 2005. Phylogenetic relationships and evolutionary history of the reef fish family Labridae. *Mol. Phylogen. Evol.* 36:370–90.

Westneat, M. W., M. E. Alfaro, P. C. Wainwright, D. R. Bellwood, J. R. Grubich, J. L. Fessler, K. D. Clements, et al. 2005. Local phylogenetic divergence and global evolutionary convergence of skull function in reef fishes of the family Labridae. *Proc. Royal Soc.* B 272:993–1000.

Westphal M. F., S. R. Morey, J. C. Uyeda, and T. J. Morgan. 2011. Molecular phylogeny of the subfamily Amphistichinae (Teleostei: Embiotocidae) reveals parallel divergent evolution of red pigmentation in two rapidly evolving lineages of sand-dwelling surfperch. *J. Fish Biol.* 79:313–30.

Whitehead, A. 2010. The evolutionary radiation of diverse osmotolerant physiologies in killifish (*Fundulus* spp.). *Evolution* 64:2070–85.

Whitehead, P. J. P. 1962. The species of *Elops* (Pisces: Elopidae). *Ann. Mag. Nat. Hist.* 13:323–29.

———. 1963. A contribution to the classification of clupeoid fishes. *Ann. Mag. Nat. Hist.,* Ser. 13:737–50.

———. 1985. *Clupeoid fishes of the world (suborder Clupeoidei). Part 1: Chirocentridae, Clupeidae and Pristigasteridae.* FAO species catalogue. *FAO Fish. Synop.* 125, Vol. 7, Part 1:1–303. FAO, Rome.

Whitehead, P. J. P., M.-L. Bauchot, J.-C. Hureau, J. G. Nielsen and E. Tortonese (eds.). 1986. *Fishes of the North-eastern Atlantic and the Mediterranean.* Vols. 1–3. UNESCO, Paris. 1,473 p.

Whitehead, P. J., G. J. Nelson, and T. Wongratana. 1988. *Clupeoid fishes of the world (suborder Clupeoidei). Part 2. Engraulididae.* FAO Species Catalogue. *FAO Fish. Synop.* 125, Vol. 7, Part 2:305–579. FAO, Rome.

Whiteman, E. A., and I. M. Côté. 2004. Monogamy in marine fishes. *Biol. Rev.* 79:351–75.

Wiley, E.O. 1976. The systematics and biogeography of fossil and recent gars (Actinopterygii: Lepisosteidae). *Misc. Publ. Mus. Nat. Hist. Univ. Kansas* 64:1–111.

———. 1986. A study of evolutionary relationships of *Fundulus* topminnows (Teleostei: Fundulidae). *Am. Zool.* 26:121–30.

———. 1992. Phylogenetic relationships of the Percidae (Teleostei: Perciformes): A preliminary examination, pp. 247–67. In *Systematics, historical ecology, and North American freshwater fishes.* R.L. Mayden (ed.). Stanford University Press, Stanford, California.

Wiley, E.O., and G.D. Johnson, 2010. A teleost classification based on monophyletic groups, pp. 123–82. In *Origin and phylogenetic interrelationships of teleosts.* J.S. Nelson, H.-P. Schultze, and M.V.H. Wilson (eds.). Verlag Dr. Friedrich Pfeil, Munchen.

Wiley, E.O., G.D. Johnson, and W.W. Dimmick. 1998. The phylogenetic relationships of lampridiform fishes (Teleostei: Acanthomorpha), based on a total evidence analysis of morphological and molecular data. *Mol. Phylogen. Evol.* 10:417–25.

———. 2000. The interrelationships of acanthomorph fishes: A total evidence approach using molecular and morphological data. *Biochem. Syst. Ecol.* 28:319–50.

Wiley, M.L., and B.B. Collette. 1970. Breeding tubercles and contact organs in fishes: Their occurrence, structure, and significance. *Bull. Amer. Mus. Nat. Hist.* 143:145—216.

Wilkens, L., M. Hoffman, and W. Wojtenek. 2002. The electric sense of the paddlefish: A passive system for the detection and capture of zooplankton prey. *J. Physiol.* 96:363–77.

Williams, J.T., and J.C. Tyler. 2003. Revision of the western Atlantic clingfishes of the genus *Tomicodon* (Gobiesocidae), with descriptions of five new species. *Smithson. Contr. Zool.* 621:1–26.

Wilson, A.B., and J.W. Orr. 2011. The evolutionary origins of Syngnathidae: Pipefishes and seahorses. *J. Fish Biol.* 78:1603–23.

Wilson, M.V.D., and A.M. Murray. 2008. Osteoglossomorpha: Phylogeny, biogeography, and fossil record and the significance of key African and Chinese fossil taxa. *Geol. Soc. London, Spec. Publ.* 295:185–219.

Wilson, M.V.H., and P. Veilleux. 1982. Comparative osteology and relationships of the Umbridae (Pisces: Salmoniformes). *Zool. J. Linn. Soc.* 76:321–52.

Wilson, N.G., and G.W. Rouse. 2010. Convergent camouflage and the non-monophyly of "seadragons" (Syngnathidae: Teleostei): Suggestions for a revised taxonomy of syngnathids. *Zool. Scripta* 39:551–58.

Wilson, R.R. 1994. Interrelationships of the subgenera of *Coryphaenoides* (Gadiformes: Macrouridae): Comparison of protein electrophoresis and peptide mapping. *Copeia* 1994:42–50.

Wilson, R.R., and P. Attia. 2003. Interrelationships of the subgenera of *Coryphaenoides* (Teleostei: Gadiformes: Macrouridae): Synthesis of allozyme, peptide mapping, and DNA sequence data. *Mol. Phylogen. Evol.* 27:343–47.

Wilson, R.R., J.F. Siebenaller, and B.J. Davis. 1991. Phylogenetic analysis of species of three subgenera of *Coryphaenoides* (Teleostei: Macrouridae) by peptid mapping of homologs of LDH-A4. *Biochem. Syst. Ecol.* 18:565–72.

Wilson, S.K. 2009. Diversity in the diet and feeding habits of blennies. pp. 139–162. In *The biology of blennies,* R.A. Patzner, E.J. Gonçalves, P.A. Hastings, and B.G. Kapoor (eds.). Science Publishers, Enfield, New Hampshire.

Winfield, I.J., and J.S. Nelson (eds.). 1991. *Cyprinid fishes: Systematics, biology and exploitation.* Chapman and Hall, London. 667 p.

Winterbottom, R. 1974a. A descriptive synonymy of the striated muscles of the Teleostei. *Proc. Acad. Nat. Sci. Philadelphia* 125:225–317.

————. 1974b. The familial phylogeny of the Tetraodontiformes (Acanthopterygii: Pisces) as evidenced by their comparative myology. *Smithson. Contr. Zool.* 155:1–201.

————. 1993a. Search for the gobioid sistergroup (Actinopterygii: Percomorpha). *Bull. Mar. Sci.* 52:395–414.

————. 1993b. Myological evidence for the phylogeny of the recent genera of the surgeonfishes (Percomorpha, Acanthuridae), with comments on the Acanthuroidei. *Copeia* 1993:21–39.

Winterbottom, R., and J. C. Tyler. 1983. Phylogenetic relationships of aracanin genera of boxfishes (Ostraciidae: Tetraodontiformes). *Copeia* 1983:902–17.

Wisner, R. L. 1974. The taxonomy and distribution of lanternfishes (Family Myctophidae) of the eastern Pacific Ocean. *NORDA Rep. 3*. Navy Ocean Res. Dev. Activity, Bay St. Louis, Mississippi. 229 p.

Wisner, R. L., and C. B. McMillan. 1995. Review of new world hagfishes of the genus *Myxine* (Agnatha, Myxinidae) with descriptions of nine new species. *Fish. Bull.* 93:530–50.

Wootton, R. J. 1984. *The biology of the sticklebacks* (2nd ed.). Academic Press, London. 387 p.

Wright, J. J. 2009. Diversity, phylogenetic distribution, and origins of venomous catfishes. *BMC Evol. Biol.* 9:282.

Yabe, M. 1985. Comparative osteology and mycology of the superfamily Cottoidea (Pisces: Scorpaeniformes), and its phylogenetic classification. *Mem. Fac. Fish. Hokkaido Univ.* 32:1–130.

Yagishita, N., T. Kobayashi, and T. Nakabo. 2002. Review of monophyly of the Kyphosidae (sensu Nelson, 1994), inferred from the mitochondrial ND2 gene. *Ichthyol. Res.* 49:103–8.

Yamanoue, Y., M. Miya, K. Matsuura, M. Katoh, H. Sakai, and M. Nishida. 2004. Mitochondrial genome and phylogeny of the ocean sunfishes (Tetraodontiformes: Molidae). *Ichthyol. Res.* 51:269–73.

————. 2008. A new perspective on phylogeny and evolution of tetraodontiform fishes (Pisces: Acanthopterygii) based on whole mitochondrial genome sequences: Basal ecological diversification? *BMC Evol. Biol.* 8:212.

Yamanoue, Y., M. Miya, K. Matsuura, N. Yagishita, K. A. Mabuchi, H. Sakai, M. Katoh, and M. Nishida. 2007. Phylogenetic position of tetraodontiform fishes within the higher teleosts: Bayesian inferences based on 44 whole mitochondrial genome sequences. *Mol. Phylogen. Evol.* 45:89–101.

Yokoyama, R., and A. Goto. 2005. Evolutionary history of freshwater sculpins, genus *Cottus* (Teleostei: Cottidae) and related taxa, as inferred from mitochondrial DNA phylogeny. *Mol. Phylogen. Evol.* 36:654–68.

Yopak, K. E., and L. R. Frank. 2009. Brain size and brain organization of the whale shark, *Rhincodon typus*, using magnetic resonance imaging. *Brain Behav. Evol.* 74:121–42.

Zander, C. D. 2011. Gobies as predator and prey, pp. 291–344. In *The biology of gobies*. R. A. Patzner, J. L. Van Tassel, M. Kovačić, and B. G. Kapoor (eds.). Science Publishers, Enfield, New Hampshire.

Zehren, S. J. 1979. *The comparative osteology and phylogeny of the Beryciformes (Pisces: Teleostei)*. Evol. Monogr., Vol. 1. University of Chicago Press, Chicago, Illinois. 389 p.

————. 1987. Osteology and evolutionary relationships of the boarfish genus *Antigonia* (Teleostei: Caproidae). *Copeia* 1987:564–92.

Zhang, H., Q. Wei, H. Du, L. Shen, Y. Li, and Y. Zhao. 2009. Is there evidence that the Chinese paddlefish (*Psephurus gladius*) still survives in the upper Yangtze River? Concerns inferred from hydroacoustic and capture surveys, 2006–2008. *J. Appl. Ichthyol.* 25 (Suppl. 2):95–99.

Zhang, J.-Y. 2006. Phylogeny of Osteoglossomorpha. *Vert. Pal. Asiatica.* 44:43–59.

INDEX

Bold indicates primary discussion where there are multiple discussions.

emperors, 170
Enchelycore octaviana, 73
Engraulidae, 77–8
Engraulis, 77
Engraulis mordax, 78
Engraulis ringens, 77
Engyophrys, 225
Enneacanthus, 150
Enneanectes, 187
Enneanectes glendae, 188
Enneapterygius, 187
Eopsetta, 224
Ephippidae, 171
Epibulus insidiator, 177
Epigonidae, 170
Epinephelidae, 138, **141–2**
Epinephelini, 141
Epinephelus, 141
Epinephelus labriformis, 142
Epiplatys sexfasciatus, 131
Eptatretus, 15
Eptatretus stoutii, 16
Erimyzon, 84
Erpetoichthys, 57
Erpetoichthys calabaricus, 57
Erythrinidae, 86
Esocidae, 94, **96–7**
Esociformes, 94, **96**
Esox, 96
Esox americanus, 97
Esox masquinongy, 96
Etheostoma, 151
Etheostoma blennioides, 151
Etheostomatinae, 151
Etmopteridae, 36
Etropilinae, 179
Etropus, 223
Eucinostomus, 156
Eucinostomus entomelas, 156
Eucyclogobius newberryi, 194
Eugerres, 156
Eugerres lineatus, 156
Euleptorhamphus, 125
Eurypharyngidae, 76
Eurypharynx, 76
Eurypharynx pelecanoides, 76
Eusphyra, 33
Euteleostei, 62, 92

Euthynnus, 201
Euthynnus affinis, 201
Eutrigla gurnardus, 141
Evermannellidae, 104
Eviota, 193
Exocoetidae, 124, **126–7**
Exocoetus, 126

false brotulas, 214
fangtooths, xix, **117**
featherfin knifefishes, 64
filefishes, 230
Fistularia corneta, 135
Fistulariidae, 133, **135**
flagfins, 102
flashflightfishes, 116, **119**
flatfishes, 119, 145, 172, 200, **222**, 227
flatheads, 138, 145
flattails, 163
Floridichthys, 129
flying gurnards, 137
flyingfishes, 124, **126**
Foa, 153
Fodiator, 126
Forcipiger, 164
Forcipiger flavissimus, 164
Freshwater Butterflyfish, 63
Freshwater Drum, 160
freshwater eels, 71–2
freshwater halfbeaks, 124
freshwater hatchetfishes, 87
freshwater smelts, 93
frill sharks, 34
frogfishes, 218–20
frogmouth catfishes, 90
Fugu, 234
Fundulidae, 127–8
Fundulus, 128
Fundulus notatus, 128
Fundulus parvipinnis, 128

Gadidae, 113
Gadiformes, 111–4
Gadus, 113
Gadus morhua, 113
Galaxias, 93
Galaxias brevipinnis, 94
Galaxiidae, 93–4

marblefishes, 171
marine hatchetfishes, 99
marine smelts, 93
marlins, 172
Mastacembelidae, 136
Mastacembelus, 136
Masturus, 236
medusafishes, 203
Megachasma pelagios, 30
Megachasmidae, 28, **30**
Megalaspis cordyla, 174
Megalomycteridae, 115
Megalopidae, 68
Megalops, 68
Megalops atlantica, 68
Megamouth Shark, 28
megamouth sharks, 28, **30**
Melamphaes, 115
Melamphaes acanthomus, 116
Melamphaidae, 115–6
Melanocetidae, 221
Melanocetus, 220
Melanocetus johnsonii, 221
Melanochromis, 178
Melanostomiidae, 100
Melanostomiinae, 100
Melanotaenia splendida, 124
Melanotaeniidae, 122, **124**
Melichthys, 228
Melichthys niger, 229
Menidia, 122
Menidiinae, 122–3
Mesobius berryi, 112
Mexican Tetra, 87
Microcanthinae, 163
Microdesmidae, 193, **195–6**
Microgadus, 113
Micrometrus minimus, 183
Micropterus, 149–50
Micropterus dolomieu, 149
Microstoma, 93
Microstoma microstoma, 92
Microstomatidae, 92–3
Microstomus, 224
midshipmen, 217–8
Milkfish, 81
minnows, 81, 83
Mirapinnidae, 115

Misgurnus anguillicaudatus, 85
Mitsukurina owstoni, 28, **30**
Mitsukurinidae, 28, **30**
Mnierpini, 189
Mobula, 49
Mobula tarapacana, 50
Mobulinae, 49
Mochokidae, 90
mojarras, 156–7
Mola, 236
Mola mola, 55, **236**
molas, 236–7
Molidae, 228, **236–7**
mollies, 131
Monacanthidae, 228, **230–1**
Monacanthus, 230
Monacanthus ciliatus, 231
Monocentridae, 116, **119**
Monocirrhus polyacanthus, 170
Monodactylidae, 170
Monognathus, 76
Monolene, 225
mooneyes, 63, **65**
moonfishes, 170
Moorish Idol, 197
Moorish idols, 198
moray eels, 72
Mordaciidae, 16
Moridae, 114
Mormyridae, 64
Mormyrus, 63
Morone, 147
Morone saxatilis, 147
Moronidae, 147
morwongs, 170
Mosquitofish, 131
Mote Sculpin, 213
Moxostoma, 84
mudminnows, 94, **96**
Mugil, 120
Mugilidae, 120–1
Mugiliformes, 119–20
mullets, 119–21
Mullidae, 162
Mulloidichthys, 162
Mullus, 162
Mullus surmulatus, 162
Muraena, 72

Plesiobatidae, 46
Plesiopidae, 171
Plesiops nigricans, 171
Pleurogramma, 183
Pleurogrammus, 207
Pleurogrammus monopterygius, 208
Pleuronectes, 224
Pleuronectidae, 223–4
Pleuronectiformes, 119, **222–7**
Pleuronichthys coenosus, 224
Pliotrema warreni, 40
plownose chimaeras, 22
poachers, 213
Poecilia, 130
Poecilia formosa, 130
Poecilia latipinna, 130
Poeciliidae, 127, **130**
Poeciliopsis, 130
Polycentridae, 170
Polydactylus, 159
Polydactylus approximans, 159
Polyipnus, 99
Polymixia, 109
Polymixia nobilis, 109
Polymixiinidae, 109
Polymixiiniformes, 109
Polynemidae, 159–60
Polynemus, 159
Polyodon spathula, 60
Polyodontidae, 59–60
Polyptera palmas, 57
Polypteridae, 57
Polypteriformes, 56–7
Polypterus, 57
Pomacanthidae, 166
Pomacanthus, 166
Pomacentridae, 176, 180–1
Pomacentrinae, 181
Pomacentrus, 180
Pomadasys, 157
Pomoxis, 149–50
pompanos, 174
ponyfishes, 170
porcupinefishes, 234
porgies, 169
Porichthyinae, 218
Porichthys, 217
Porichthys myriaster, 217

Potamotrygonidae, 46
Premnas, 180
Priacanthidae, 152
Priacanthus, 152
priapiumfishes, 122, **124**
pricklebacks, 213
pricklefishes, 115
Prionace, 32
Prionotus, 140
Prionurus, 197
Pristidae, 43
Pristiformes, 40, **43**
Pristigenys, 152
Pristigenys serrula, 152
Pristiophoridae, 40
Pristiophoriformes, 40
Pristiophorus japonicus, 40
Pristipomoides, 154
Pristipomoides zonatus, 155
Pristis, 43
Pristis pectinata, 43
Prognathodes, 164
Prognathodes falcifer, 164
Promethichthys prometheus, 202
Pronotogrammus multifasciatus, 143
Proscylliidae, 31
Protoanguilla palau, 71
Protoanguillidae, 71
Protopteridae, 53–4
Protopterus, 53
Protopterus aethiopicus, 54
prowfishes, 213
Psammoperca, 145
Psectrogaster rutiloides, 86
Psenes, 202
Psenes cyanophrys, 203
Psenopsis anomala, 203
Psettina, 225
Pseudalutarius nasicornis, 231
Pseudamia, 153
Pseudamia zonata, 154
Pseudanthias, 143
Pseudanthias squamipinnis, 143
Pseudocarcharias kamoharai, 30
Pseudocarchariidae, 28, **30**
Pseudochromidae, 171
Pseudocrenilabrinae, 179
Pseudoginglymostoma, 26

Siganidae, 197–8
Siganus canaliculatus, 198
Siluriformes, 80–1, **87–9**
silversides, 119, **122**
Sinobdella sinensis, 136
six-gill sharks, 34–5
sixgill stingrays, 46
skates, **41**, 44–6
sleeper sharks, 36
sleepers, 195–6
slickheads, 92
slimeheads, 116, **119**
smallscale pike characins, 86
smelts, 93–4
snaggletooths, 100
snailfishes, 211
snake eels, 73
snake mackerels, 202
snakeheads, 206
snappers, **154–5**, 157
snipe eels, 75
snipefishes, 133
snooks, 145
soapfishes, 141–2
Solenostomidae, 133
Somniosidae, 36
South Amerian Lungfish, 53–4
spadefishes, 171
Sparidae, 169–70
Sparus, 169
Sphoeroides, 233
Sphoeroides lispus, 233
Sphyraena, 168
Sphyraena argentea, 168
Sphyraena barracuda, 168
Sphyraenidae, **168**, 200
Sphyrna, 33
Sphyrna zygaena, 33
Sphyrnidae, 33
spikefishes, 237
Spinachia, 132
spiny eels, **70**, 136
splitfins, 131
Squalidae, 36, **37**
Squaliformes, 34, **36–8**
Squaliolus, 38
Squaliolus aliae, 36
Squalomorphii, 24

Squalus, 37
Squalus suckleyi, 37
squaretails, 203
Squatina, 39
Squatina californica, 39
Squatinidae, 39
Squatiniformes, 39
squeakers, 90
squirrelfishes, 118
Stalix, 148
stardrums, 161
stargazers, 184–5
Starksia, 189
Starksiini, 189–90
Steatogenys, 91
Stegastes, 180
Stegastes nigricans, 181
Stegastinae, 181
Stegostomatidae, 26
Stellerina xyosterna, 213
Stelliferinae, 161
Stephanoberyciformes, 115–6
Sternoptychidae, 99
Sternoptyx, 99
Stethaprion crenatum, 86
Stichaeidae, 213
sticklebacks, 119, **131–2**
stingrays, 46–50
Stomias, 100
Stomias atriventer, 100
Stomiidae, 100–1
Stomiiformes, **97**, 99–100
Stomiinae, 100
stonefishes, 145
Striped Bass, 147
Striped Mullet, 162
Stromateidae, 203
Stromateiformes, 202–4
Stromateus, 202
Stromateus stellatus, 203
Strongylura, 124
Strongylura exilis, 125
sturgeons, 58
Stylephoridae, 108–9
Stylephorus, 107
Stylephorus chordatus, 108
suckers, 84
Sufflamen, 228

Xiphophorus, 130

Yellowhead Jawfish, 148

Zalembius, 182
Zanclidae, 197
Zanclus cornutus, 198
Zaniolepis, 207
Zaniolepis frenata, 208
Zaprora silenus, 213
Zaproridae, 213
Zapteryx, 44
Zebrafish, 84

Zebrasoma, 197
Zebrasoma veliferum, 197
Zeidae, 115
Zeiformes, **114–5**, 206
Zenarchopteridae, 124
Zeus, 114
Zeus faber, 114–5
Zoarces, 212–3
Zoarces viviparus, 213
Zoarcidae, 212–3
Zoarcoidei, 138, **207**
Zu, 107
Zu cristatus, 108